Macquarie Monographs in Cognitive Science

A Special Issue of
Cognitive Neuropsychology

Cognitive Neuropsychology twenty years on

Edited by

Max Coltheart
Macquarie University, Sydney, Australia

and

Alfonso Caramazza
Harvard University, Cambridge, MA, USA

Routledge
Taylor & Francis Group
LONDON AND NEW YORK

First published 2006 by Psychology Press

Published 2018 by Routledge
2 Park Square, Milton Park, Abingdon, Oxon OX14 4RN
52 Vanderbilt Avenue, New York, NY 10017

Routledge is an imprint of the Taylor & Francis Group, an informa business

British Library Cataloguing in Publication Data
A catalogue record for this book is available from the British Library

This book is the seventh volume in the series *Macquarie Monographs in Cognitive Science* and is also a special issue of the journal *Cognitive Neuropsychology* being Volume 23 (2006) Issue 1.

Cover by Anú Design, Tara, Co. Meath, Ireland
Typeset by Techset Composition Limited, Salisbury, Wiltshire, UK

ISSN 0264–3294
ISBN 13: 978-1-84169-972-1 (hbk)
ISBN 13: 978-1-138-88333-8 (pbk)

COGNITIVE NEUROPSYCHOLOGY TWENTY YEARS ON

Macquarie Monographs in Cognitive Science

Series Editor: MAX COLTHEART

The Macquarie Monographs in Cognitive Science series will publish original monographs dealing with any aspect of cognitive science.

Each volume in the series will cover a circumscribed topic and will provide readers with a summary of the current state-of-the-art in that field.

A primary aim of the volumes is also to advance research and knowledge in the field through discussion of new theoretical and experimental advances.

Titles in the series:

Routes to reading success and failure: Toward an integrated cognitive psychology of atypical reading
Nancy Ewald Jackson and Max Coltheart

Cognitive neuropsychological approaches to spoken word production
Lyndsey Nickels (Ed.)

Rehabilitation of spoken word production in aphasia
Lyndsey Nickels (Ed.)

Masked priming: The state of the art
Sachiko Kinoshita and Stephen J. Lupker (Eds.)

Individual differences in theory of mind: Implications for typical and atypical development
Betty Repacholi and Virginia Slaughter (Eds.)

From mating to mentality: Evaluating evolutionary psychology
Julie Fitness and Kim Sterelny (Eds.)

***Cognitive Neuropsychology* twenty years on**
Max Coltheart and Alfonso Caramazza (Eds.)

COGNITIVE NEUROPSYCHOLOGY

Volume 23 Issue 1 February 2006

Special issue: *Cognitive Neuropsychology* twenty years on

Guest editors: Max Coltheart and Alfonso Caramazza

Contents

COGNITIVE NEUROPSYCHOLOGY, 2006, 23 (1), 3–12

Cognitive Neuropsychology twenty years on

Alfonso Caramazza
Cognitive Neuropsychology Laboratory, Harvard University, Cambridge, MA, USA

Max Coltheart
Macquarie Centre for Cognitive Science, Macquarie University, Sydney, NSW, Australia

Cognitive neuropsychology began in the second half of the nineteenth century when neurologists such as Lichtheim, Wernicke, Bastian, and others began to make inferences about the cognitive architecture of the intact language-processing system from studying the different ways in which spoken or written language abilities broke down after brain damage. They even began to express their proposals about this architecture by means of explicit box-and-arrow diagrams: hence the term "the diagram-makers" that was applied to them.

These cognitive neuropsychologists were also cognitive neuroscientists: They were interested not only in the functional architecture of cognition, but also in how the components of such an architecture were localized in the brain. Their cognitive neuropsychology was successful (their diagrams of the language-processing system are simplified versions of diagrams that enjoy contemporary support; see Coltheart, Rastle, Perry, Langdon, & Ziegler, 2001, for examples) but their cognitive neuroscience was not. They did not succeed in localizing in the brain any of the hypothesized components of a functional architecture of cognition, and this failure exposed the whole enterprise to damning criticisms from noncognitively-oriented neurologists such as Head (1926). This, plus the demise of cognitive psychology itself consequent upon of the rise of behaviourism at the beginning of the twentieth century, saw cognitive neuropsychology practically vanish from the scientific scene for the first half of the twentieth century.

However, after the advent of the so-called "Cognitive Revolution" in the middle of the twentieth century (Broadbent, 1956; Chomsky, 1959; Miller, Galanter, & Pribram, 1960), cognitive neuropsychology awoke from its slumbers, aroused by seminal papers from Marshall and Newcombe (1966, 1973) on the cognitive neuropsychology of reading and from Shallice and Warrington (1970) on the cognitive neuropsychology of memory. Also important were developments in the area of sentence processing in aphasia where linguistic and psycholinguistic theory played a crucial role in guiding the analysis of aphasic symptoms (Caramazza & Zurif, 1976; Marin, Saffran, & Schwartz, 1976). The first conference solely devoted to cognitive neuropsychology was held at Oxford in 1979 (the conference was on deep dyslexia, one of the three forms of acquired dyslexia defined by Marshall & Newcombe, 1973), and its proceedings were published as a book in the following year (Coltheart, Patterson, & Marshall, 1980). The field was burgeoning rapidly; it needed its own journal, and *Cognitive Neuropsychology* began publication in 1984. The field also needed an undergraduate text, and *Human Cognitive Neuropsychology* (Ellis &

Correspondence should be addressed to Alfonso Caramazza, Cognitive Neuropsychology Laboratory, Harvard University, Cambridge, MA 02138, USA (Email: caramazz@fas.harvard.edu).

DOI:10.1080/02643290500443250

Young, 1986) was published shortly after the journal first appeared. Shortly after that, an advanced text (Shallice, 1988) appeared.

The 20th birthday of the journal was celebrated by a symposium at the 22nd European Workshop on Cognitive Neuropsychology at Bressanone, Italy, in February 2004, and from that symposium came the papers collected together here.

Cognitive neuropsychology and cognitive neuroscience

It is useful, if one wants to make some remarks about the past twenty years of cognitive neuropsychology, to begin by discussing the distinction, already alluded to above, between cognitive neuropsychology and cognitive neuroscience. Cognitive neuroscience is the study of those neural systems of the brain that subserve cognition, and so it is a branch of neuroscience. Cognitive neuropsychology is the use of investigations of people with impairments of cognition (acquired or developmental) to learn more about normal cognitive processes, and so it is a branch of cognitive psychology, just as Rapp and Goldrick (2006) say. This view of cognitive neuropsychology as a branch of cognitive psychology and as distinct from cognitive neuroscience is widely accepted: "The term cognitive neuropsychology often connotes a purely functional approach to patients with cognitive deficits that does not make use of, or encourage interest in, evidence and ideas about brain systems and processes" (Schacter, 1992, p. 560); or for a more nuanced position: "Cognitive Neuropsychology's domain of inquiry concerns the structure of normal perceptual, motor, and cognitive processes. As such, it constitutes a branch of cognitive science. What distinguishes cognitive neuropsychology from other branches of cognitive science is the type of observations that it uses in developing and evaluating theories of normal cognition. The data used in cognitive neuropsychology are the patterns of performance produced by brain-damaged subjects. Because the basic data used in cognitive neuropsychology are the result of a biological manipulation—a brain lesion—these data will be directly relevant to claims about the functional organization of the brain. Hence cognitive neuropsychology may also be considered to be a branch of cognitive neuroscience. However, depending on the particular proclivities of individual cognitive neuropsychologists, there is considerable variation in the specific weight given by any one investigator to the cognitive or the neural part of the brain/cognition equation" (Caramazza, 1992, pp. 80–81).

The distinction between cognitive neuroscience and cognitive neuropsychology is an important one, but it is sometimes overlooked or neglected. There seem to be several reasons for this.

First, cognitive neuropsychologists are usually studying people with brain disorders (even though not studying those disorders).

Second, some cognitive neuropsychologists are also cognitive neuroscientists (just as some mathematicians are also physicists, and some psychologists are also economists) whereas others are not. For example, the papers in this collection by Rapp and Goldrick, Martin, Coltheart, and Miceli and Capasso are solely cognitive-neuropsychological; they have nothing at all to say about the brain system subserving cognition. In contrast the papers by Caramazza and Mahon, Humphreys and Riddoch, Buxbaum, Vallar, and Schwartz are not only cognitive-neuropsychological (seeking to reach conclusions about the nature of normal cognitive mechanisms from studying people in whom these mechanisms have broken down) but also cognitive-neuroscientific (seeking to reach conclusions about the neural organization of those mechanisms).

Third, there often seem to be (generally unstated) assumptions that conclusions about the functional architecture of cognition have implications for what the brain must be like, and conclusions about what the brain is like have implications for theories of the functional architecture of cognition (see Caramazza, 1992, for discussion of this issue). If these assumptions were correct, then anyone doing cognitive neuropsychology would also ipso facto be doing cognitive neuroscience. But are these assumptions correct? Do facts about the mind constrain possible theories about what the brain could be like? One of the present authors has expressed doubts about that (Coltheart, 1982, 2004, in press), and so did one of the papers in the very first issue of *Cognitive Neuropsychology* (Mehler, Morton, & Jusczyk, 1984). Consider, for example, such claims about cognitive architecture as those made in Figure 5 of Rapp and Goldrick (2006, which depicts a detailed model of how speech production is achieved), Figure 3 of Coltheart (2006, which depicts a detailed model of how visual word recognition and reading aloud are achieved), or Figure 5 of Miceli and Capasso (2006, which depicts a detailed model of the structure of the orthographic representations used in spelling). Here we have strong, explicit, and detailed claims about certain cognitive architectures. Suppose we were convinced that the claims were true in all three cases; would anything about the brain follow, and, if so, what? And what do we currently know about the brain that importantly constrains theories about cognitive architecture? It is clear that many people do believe that there are such constraints in both directions—for example, "because I believe that neural constraints can be important for cognitive theorizing, I use the term cognitive neuroscience rather than cognitive neuropsychology" (Schacter, 1992, p. 560). However, whether there are actually such constraints is currently still a matter of controversy: See, for example, the symposium on "What has cognitive neuroimaging told us about the mind (so far)?" in the journal *Cortex* this year (see Coltheart, in press, and the commentaries upon that paper). So there is no general agreement as to how cognitive neuropsychology is related to cognitive neuroscience.

Nonetheless, as already noted, it is equally clear that neuropsychological data are by their very nature function–brain pairings, and therefore they offer a potential window into the neural organization of cognitive systems (Caramazza, 1992). Neuropsychological data have been used to formulate hypotheses about the relationship between particular brain areas and processing components in cognitive theories. For example, the elegant and detailed studies of patient DF (Goodale, Milner, Jakobson, & Carey, 1991) have been used to distinguish between the perception of form within the object recognition process and the "perception" of form used to guide motor behaviour. However, this work has also led to the proposal that a specific part of lateral occipital cortex is crucial for a perceptual process tied to the conscious recognition of form but that these processes are not needed for guiding motor behaviour (see Humphreys & Riddoch, 2006, for further discussion of these issues).

Syndromes, symptoms, and single case studies: The patient as a snowflake

It is deeply characteristic of cognitive neuropsychology that it studies symptoms rather than syndromes and carries out single case studies rather than group studies. These issues surface in many of the papers collected here; For example, it is precisely these issues that underlie the comment, made in relation to hemispatial neglect by Buxbaum (2006), who observes that "nearly every possible fractionation of the disorder has been reported, raising the possibility that each patient may be as unique as a snowflake".

This is not true just of neglect: It is true of every disorder that has been studied in any detail by cognitive neuropsychologists. Broca's aphasia is a classic example, as is very clearly documented by Martin (2006). Early work (Caramazza & Zurif, 1976) began with a hypothesis that attributed this disorder to a single cause: a defect of a syntactic processing system that is used both for understanding sentences and for constructing them. Soon, however, it became clear that agrammatic comprehension and agrammatic production do not always co-occur; indeed, they doubly dissociate (see, e.g., Caramazza & Berndt, 1985). Given this double dissociation, the disorder of language seen in Broca's aphasia cannot always be ascribed to damage to a single syntactic system used for both sentence comprehension and sentence construction. Distinct explanations are thus needed for agrammatic sentence comprehension and agrammatic sentence construction, and so it cannot be right to seek *the* explanation of the syndrome of Broca's aphasia.

But perhaps a more restricted aphasic syndrome—agrammatic sentence construction, say—might be a suitable subject of scientific study? This soon turned out not to be so either, because the various symptoms of even this more restricted syndrome doubly dissociate too, to a remarkably refined degree. For example, some patients with agrammatic sentence constructions are impaired at the use of function words but not at the use of affixes, whereas others show the opposite dissociation (Berndt, 1987; Miceli, Silveri, Romani, & Caramazza, 1989; Parisi, 1987).

This is one reason why cognitive neuropsychologists study symptoms, not syndromes. What, they might say here, can we infer about the architecture of the sentence construction system from the fact that it can be damaged in such a way that function words suffer but affixes do not, and also can be damaged in such a way that function words do not suffer but affixes do? Which theories about this architecture are ruled out by this double dissociation, and which are compatible with it? Questions like this can be asked if the data from which inferences are to be drawn are data about single symptoms. Such questions cannot be asked if the data are data about syndromes such as Broca's aphasia (Caramazza, 1984).

The focus on studying single symptoms rather than syndromes (groups of symptoms) goes hand-in-hand with the strategy of carrying out single case studies rather than group studies. It is easy to collect together a group of people all diagnosed as exhibiting Broca's aphasia; it is very unlikely that one could collect together a group of people with intact sentence comprehension and impaired ability to generate affixes correctly but otherwise intact sentence construction (including intact use of function words).

To pursue Buxbaum's (2006) analogy, that would be like trying to collect together a group of snowflakes that all had exactly the same morphology. The number of possible different shapes for snowflakes is so large that the likelihood of obtaining two snowflakes with the same morphology is quite small, which makes the prospect of group studies of snowflake morphology an impractical one. Marshall (1984), Coltheart (1984), and Howard and Franklin (1988) make exactly the same point about the impracticality of group studies in cognitive neuropsychology. Suppose the model of some cognitive domain that one wishes to investigate contains n processing components and m pathways of communication between them. If each component or pathway can be independently impaired by brain damage, then the number of different patterns of impairment of the system that can arise is $2^{(m+n)}$. The values of m and n do not have to be very large for $2^{(m+n)}$ to become astronomically large. Since the probability of seeing two consecutive patients with the same pattern of impairment (which is the only justification for treating them as a group) is $1/2^{(m+n)}$, this probability is infinitesimal for values of m and n that are typical of current models of cognition. That is why it is appropriate to think of patients as being as unique as snowflakes.

If, for practical purposes, every patient is unique, how can we amass knowledge that generalizes across people? This requires what Caramazza (1986) referred to as the "universality assumption": the assumption that there is no qualitative variation across neurologically intact people in the architecture of the cognitive system that these people use to perform in a certain cognitive domain. This allows us to infer that, although patient X and patient Y currently have very different systems as a consequence of their brain damage, they had the same system premorbidly, and it is about that system that we want to make inferences from studying patients X and Y. Cognitive neuropsychology cannot be practised unless this universality assumption is made, but nor can cognitive psychology. This is how, even though every patient is essentially unique, we can seek generalizable knowledge from studying them (Caramazza, 1986).

The same is true for snowflakes. There is a falsifiable theory of snowflake generation that makes the following predictions (and no doubt many others):

1. All snowflakes are snow crystals, or are composed of snow crystals.
2. Any snow crystal can have 3 or 6 or 12 sides, but none can have 4 or 5 or 8 sides.

Just as a single patient can refute some hypothesis about cognitive architecture by yielding a pattern of data that according to that hypothesis could never occur, so a single snow crystal can refute this theory of snowflake generation—a crystal with 4 or 5 or 8 sides is all that is needed (see Caramazza, 1986, for detailed discussion).

None of this is meant to be a claim that syndromes and group studies have no role to play at all in cognitive neuropsychology. Indeed, the first studies of the cognitive neuropsychology of reading in the modern era were studies of syndromes of acquired dyslexia: surface dyslexia, deep dyslexia, and visual dyslexia (Marshall & Newcombe, 1966, 1973). What this work showed us was that there are subtypes of acquired dyslexia (Castles & Coltheart, 1993, did the same thing with reference to developmental dyslexia). That allowed work on dyslexia to focus on the subtypes rather than on some undifferentiated entity "acquired dyslexia". Work on each subtype then revealed subtypes of the subtypes: different subtypes of deep dyslexia and different subtypes of surface dyslexia, for example. So the data compelled finer and finer fractionations of the syndromes, until eventually what is being studied is not a small group of symptoms (a small syndrome) but a single symptom.

The moral is clear: In any field of cognition where cognitive neuropsychology is underdeveloped, starting with small group studies of symptom collections (syndromes) might prove to be a useful ground-clearing exercise. Mature development of the cognitive neuropsychology of that domain of cognition is signalled by the replacement of this approach in favour of research in which inferences about the intact cognitive system are made on the basis of data from studies of individual symptoms. In all bar one of the papers in this volume, this is the approach that is taken. The one exception is the work on the cognitive neuropsychology of everyday action discussed by Myrna Schwartz. She discusses explicitly the use of group studies. However, the aim of the research that she discusses is not to infer something about the cognitive architecture of the intact action-planning system from studying people with acquired deficits of action planning: It is to investigate "hypotheses about brain–behaviour or deficit–behaviour correlations" (Schwartz, 2006).

Modularity

With his *The Modularity of Mind* (1983), Jerry Fodor did cognitive neuropsychology a great service by elucidating a concept that has played an important role in the development of the subject—namely, modularity. It should be no surprise, then, to find an extended review of Fodor's book in the first issue of the journal (Schwartz & Schwartz, 1984).

Some cognitive neuropsychologists (e.g., Coltheart, 1999) are completely committed to the view that the mind is modular (in the Fodorian sense). Others (e.g., Caramazza, 1992) assume only a weak form of modularity—that is, that the mind is componentially structured (in the sense used by Simon, 1969, and Marr, 1982). In all the papers in this volume, theories about the architecture of a cognitive system are postulated in which that system is considered to be composed of information-processing components each responsible for one of the information-processing jobs that need doing if cognition is to run smoothly. It is this property of cognitive systems that makes them amenable to neuropsychological investigation. In other words, it is because cognitive systems are composed of relatively autonomous processing components that "local" brain damage can result in dissociation of functions.

That we do not need the strong modularity assumption in order to make progress in cognitive neuropsychology is well illustrated by the case of belief formation. On a Fodorian account of modularity, belief formation is supported by nonmodular central cognitive processes, and Fodor claims that these nonmodular central processes are not amenable to scientific study. Yet the cognitive neuropsychology of belief formation actually seems to be progressing rather well (see, e.g., Coltheart, 2005; Coltheart & Davies, 2000).

Computational cognitive neuropsychology

A computational model of cognition is a computer program that is capable of performing the cognitive task in question and, more importantly, performs the task in exactly the same way as, according to the theory that is instantiated by the model, people perform that task. There are major virtues associated with computational modelling:

1. Attempting to implement any theory of cognition as a working computer program always identifies a host of hitherto unsuspected ways in which the theory was underspecified or incomplete—problems that have to be fixed if the theory is to claim viability.
2. Once the theory has been made "fully complete", and the program is executable, one can see immediately whether the theory does in fact offer an adequate account of this domain of cognition: Can the program actually do the task?
3. Theory testing can be done with great rigour: Is the behaviour of the programs affected by all and only those stimulus properties that affect human performance in this cognitive domain, and in the same way?
4. Even if the answer to the question above is "Yes" in relation to a particular computational model, there may be other computational models in that cognitive domain, implementations of competing theories, which are equally successful in simulating the relevant facts. So theory adjudication is needed. It is much easier to discover experimental outcomes about which competing theories make different predictions if these theories are expressed as computer programs.

The papers in this volume show that computational modelling is rapidly becoming important in cognitive neuropsychology. If the theory of which the model is an instantiation is correct, that theory ought to be able to offer an explanation of abnormal as well as of normal cognition: When the theory has been translated into computational terms, it should be possible to "lesion" the computational model so that it shows symptoms that are also shown by patients. Whenever this is achieved, further support for the underlying theory has been obtained. This is computational cognitive neuropsychology; and this kind of work is reported in a number of the papers in this volume.

A widely used model of speech productions is that of Gary Dell and his associates (see, e.g., Dell, 1986). This has been used not only to offer an account of normal speech production, but also to model speech production in aphasia (Dell, Schwartz, Martin, Saffran, & Gagnon, 1997), and data from aphasia have been essential in development of the model. Rapp and Goldrick (2006) discuss in detail the implications of data from aphasic speech production for fundamental computational properties of the speech production system such as whether there is feedback in the system and whether processing is cascaded or thresholded. Coltheart (2006) argues in his chapter that data from acquired dyslexia have played a crucial role in evaluating competing computational models of reading. Schwartz (2006) discusses in detail in her chapter acquired disorders of action and how they can be accounted for in relation to an explicit computational model of everyday action and planning, the CS model, and in the chapter by Miceli and Capasso (2006) we see data on acquired dysgraphia beginning to exert constraints on an explicit computational model of spelling.

Although it is indisputable that computational modelling provides an especially useful extra tool in the toolbox of cognitive neuropsychologists, this is not to say that the interpretation of modelling results is any less problematic than the interpretation of other experimental results. Thus, for example, there are open and difficult issues concerning how one determines whether a computational model can generate the patterns of results seen in brain-damaged subjects. Some theorists are content with a general qualitative fit of the data (e.g., Dell et al., 1997) while others consider it crucial that the fit be quantitatively appropriate (e.g., Coltheart et al., 2001; Ruml & Caramazza, 2000; Ruml, Caramazza, Capasso, & Miceli, 2005). Thus, for example, Ruml et al. (2005) have argued that the strongly interactive model of lexical access proposed by Dell and colleagues is undermined by the fact that it fails to account for the detailed distribution of naming error types in aphasic patients. Independently of how this issue is resolved, the important point here is that increasingly precise theoretical proposals are possible in the context of computational models.

From boxes and arrows to the structure of representations

Much of the early development in modern cognitive neuropsychology was concerned with the articulation of the functional architecture of specific cognitive systems (e.g., the spelling system). These theories were formulated in terms of the components of processing implicated in a task and their organization—the so-called box-and-arrow models. Although often denigrated for their relatively general nature, these models played (and continue to play) an important role in formulating hypotheses about the general architecture of cognitive systems (see Coltheart et al., 2001). In fact, such cognitive architectures are inescapable features of all cognitive theories (even of those proposed by denigrators of the so-called box-and-arrow theories) for the simple reason that any nontrivial aspect of cognition will involve a number of processing components and their associated representations. Be this as it may, the crucial question is whether cognitive neuropsychological data can be used to inform cognitive theory beyond the general level of functional architecture.

Some theorists (e.g., Shallice, 1988) have suggested that cognitive neuropsychological data are too "noisy" for use beyond the level of functional architecture. Others (McCloskey & Caramazza, 1991) have argued instead that there is no a priori restriction on the usefulness of such data for the purpose of developing cognitive theory at any arbitrary level of detail. They offered as existence proof for this position the case of spelling, where significant progress has been made in characterizing the structure of the orthographic representations computed at various levels of the spelling process. Crucially, McCloskey and Caramazza noted that the kind of data that were used for the latter purpose consisted of the detailed analyses of error distributions and not simply the patterns of dissociations across tasks (the more common type of data reported in neuropsychological investigations). Caramazza and Miceli (1990) reported that there were precise constraints on the occurrence and distribution of error types in the spelling performance of their dysgraphic subject, LB. For example, they noted that LB's letter substitution and transposition errors were strictly constrained by their consonant–vowel (CV) status: Consonants were exchanged/transposed only with consonants, and vowels were exchanged/transposed only with vowels. This constraint, together with other converging evidence, was taken as indicating that the orthographic representation used at the level of the graphemic buffer specified not only the identity and order of graphemes but also their CV structure. This conclusion has since received wide confirmation (for review, see Miceli & Capasso, 2006; Tainturier & Rapp, 2001), validating the claim that cognitive neuropsychological data can be used to constrain theories beyond the level of functional architecture to inform the types of representation used at various levels of processing. Indeed, there is a growing body of literature focusing on the implications for the structure and content of cognitive representations from the patterns of deficits in brain-damaged individuals (see, e.g., Nickels, 2001; Rapp & Goldrick, 2006).

By way of conclusion: More on cognition and the brain

As already noted, cognitive neuropsychology can be considered a branch of cognitive psychology where subjects' performance is used to inform theories of normal cognition. However, as also already noted, there is increasing interest in relating cognitive neuropsychological investigations to developments in cognitive neuroscience (e.g., papers by Buxbaum, 2006, Caramazza & Mahon, 2006, Humphreys & Riddoch, 2006; and Vallar, 2006). In fact, *Cognitive Neuropsychology* has recently published a good number of papers that focus on the interface of cognitive neuropsychology and neuroscience (see, e.g., papers in two special issues edited by Martin & Caramazza, 2003, *The organization of conceptual knowledge in the brain: Neuropsychological and neuroimaging perspectives*, and by Rumiati & Caramazza, 2005, *The multiple functions of sensory-motor representations*), and the composition of the Editorial Board increasingly reflects this slight repositioning of the journal vis-à-vis strictly cognitive versus neuroscience accounts of cognitive processes. This is a healthy development, and we think it reflects the recognition that cognitive neuropsychological data play a central role not only in developing theories of normal cognition but also in validating conclusions reached on the basis of neuroimaging and other neuropsychological data. This development in no way represents a rejection or even a dilution of the original motivation for the creation of a journal devoted to classical cognitive neuropsychology. As can be seen from the papers included in this volume, the principal objective of cognitive neuropsychology remains the formulation and evaluation of cognitive theories. The data from cognitive neuropsychology are extremely rich in terms of the constraints that they provide for cognitive theory but are rather weak as the basis for constraining theories of the functional organization of the brain. Still, it is important that a mature cognitive neuropsychology should reach out to cognitive neuroscience in their common effort to understand the mind-brain.

PrEview proof published online 8 December 2005

REFERENCES

Berndt, R. S. (1987). Symptom co-occurrence and dissociation in the interpretation of agrammatism. In M. Coltheart, G. Sartori, & R. Job (Eds.), *The cognitive neuropsychology of language*. Hove, UK: Lawrence Erlbaum Associates Ltd.

Broadbent, D. E. (1956). *Perception and communication*. London: Pergamon Press.

Buxbaum, L. J. (2006). On the right (and left) track: Twenty years of progress in studying hemispatial neglect. *Cognitive Neuropsychology, 23*, 156–173.

Caramazza, A. (1984). The logic of neuropsychological research and the problem of patient classification in aphasia. *Brain & Language, 21*, 9–20.

Caramazza, A. (1986). On drawing inferences about the structure of normal cognitive systems from the analysis of patterns of impaired performance: The case for single-patient studies. *Brain & Cognition, 5*, 41–66.

Caramazza, A. (1992). Is cognitive neuropsychology possible? *Journal of Cognitive Neuroscience, 4*, 80–95.

Caramazza, A., & Berndt, R. S. (1985). A multicomponent deficit view of agrammatic Broca's aphasia. In M.-L. Kean (Ed.), *Agrammatism*. New York: Academic Press.

Caramazza, A., & Mahon, B. Z. (2006). The organization of conceptual knowledge in the brain: The future's past and some future directions. *Cognitive Neuropsychology, 23*, 13–38.

Caramazza, A., & Miceli, G. (1990). The structure of graphemic representations. *Cognition, 37*, 243–297.

Caramazza, A., & Zurif, E. B. (1976). Dissociation of algorithmic and heuristic processes in language comprehension: Evidence from aphasia. *Brain & Language, 3*, 572–582.

Castles, A., & Coltheart, M. (1993). Varieties of developmental dyslexia. *Cognition, 47*, 149–180.

Chomsky, N. (1959). Review of B. F. Skinner's *Verbal Behavior. Language, 35*, 26–58.

Coltheart, M. (1982). The psycholinguistic analysis of acquired dyslexias: Some illustrations. *Philosophical Transactions of the Royal Society, B298*, 151–164.

Coltheart, M. (1984). Acquired dyslexias and normal reading. In R. N. Malatesha & H. A. Whitaker (Eds.), *Dyslexia: A global issue*. The Hague, The Netherlands: Martinus Nijhoff.

Coltheart, M. (1999). Modularity and cognition. *Trends in Cognitive Sciences, 3*, 115–120.

Coltheart, M. (2004). Brain imaging, connectionism, and cognitive neuropsychology. *Cognitive Neuropsychology, 21*, 21–25.

Coltheart, M. (2005). Delusional belief. *Australian Journal of Psychology, 57*, 72–76.
Coltheart, M. (2006). Acquired dyslexias and the computational modelling of reading. *Cognitive Neuropsychology, 23*, 96–109.
Coltheart, M. (in press). What has cognitive neuroimaging told us about the mind (so far)? *Cortex.*
Coltheart, M., & Davies, M. (Eds). (2000). *Pathologies of belief.* Oxford, UK: Blackwells.
Coltheart, M., Patterson, K., & Marshall, J. C. (Eds.). (1980). *Deep dyslexia.* London: Routledge and Kegan Paul.
Coltheart, M., Rastle, K., Perry, C., Langdon, R., & Ziegler, J. (2001). DRC: A dual route cascaded model of visual word recognition and reading aloud. *Psychological Review, 108*, 204–256.
Dell, G. S. (1986). A spreading-activation theory of retrieval in sentence production. *Psychological Review, 93*, 283–321.
Dell, G. S., Schwartz, M. F., Martin, N., Saffran, E. M., & Gagnon, D. A. (1997). Lexical access in aphasic and nonaphasic speakers. *Psychological Review, 104*, 801–838.
Ellis, A. W., & Young, A. W. (1986). *Human cognitive neuropsychology.* Hove, UK: Lawrence Erlbaum Associates Ltd.
Fodor, J. A. (1983). *The modularity of mind.* Cambridge, MA: Bradford Books.
Goodale, M. A., Milner, A. D., Jakobson, I. S., & Carey, D. P. (1991). A neurological dissociation between perceiving objects and grasping them. *Nature, 349*, 154–156.
Head, H. (1926). *Aphasia and kindred disorders of speech.* London: Cambridge University Press.
Howard, D., & Franklin, S. (1988). *Missing the meaning? A cognitive neuropsychological study of processing of words by an aphasic patient.* Cambridge, MA: MIT Press.
Humphreys, G. W., & Riddoch, M. J. (2006). Features, objects, action: The cognitive neuropsychology of visual object processing, 1984–2004. *Cognitive Neuropsychology, 23*, 174–201.
Marin, O. S. M., Saffran, E. M., & Schwartz, M. F. (1976). Dissociations of language in aphasia: Implications for normal function. *Annals of the New York Academy of Sciences, 280*, 868–884.
Marr, D. (1982). *Vision.* New York: W.H. Freeman and Company.
Marshall, J. C. (1984). Toward a rational taxonomy of the developmental dyslexias. In R. N. Malatesha & H. A. Whitaker (Eds.), *Dyslexia: A global issue.* The Hague, The Netherlands: Martinus Nijhoff Publishers.
Marshall, J. C., & Newcombe, F. (1966). Syntactic and semantic errors in paralexia. *Neuropsychologia, 4*, 169–176.
Marshall, J. C., & Newcombe, F. (1973). Patterns of paralexia: A psycholinguistic approach. *Journal of Psycholinguistic Research, 2*, 175–199.
Martin, R. C. (2006). The neuropsychology of sentence processing: Where do we stand? *Cognitive Neuropsychology, 23*, 74–95.
Martin, A., & Caramazza, A. (Eds.). (2003). The organisation of conceptual knowledge in the brain: Neuropsychological and neuroimaging perspectives [Special issue]. *Cognitive Neuropsychology, 20*, 195–592.
McCloskey, M., & Caramazza, A. (1991). On crude data and impoverished theory. *Behavioral and Brain Sciences, 14*, 453–454.
Mehler, J., Morton, J., & Jusczyk, P. W. (1984). On reducing language to biology. *Cognitive Neuropsychology, 1*, 83–116.
Miceli, G., & Capasso, R. (2006). Spelling and dysgraphia. *Cognitive Neuropsychology, 23*, 110–134.
Miceli, G., Silveri, M. C., Romani, C., & Caramazza, A. (1989). Variation in the pattern of omissions and substitutions of grammatical morphemes in the spontaneous speech of so-called agrammatic patients. *Brain and Language, 36*, 447–492.
Miller, G. A., Galanter, E., & Pribram, K. H. (1960). *Plans and the structure of behavior.* New York: Holt, Rinehart & Winston.
Nickels, L. (2001). Spoken word production. In B. Rapp (Ed.), *The handbook of cognitive neuropsychology: What deficits reveal about the human mind.* Philadelphia: Psychology Press.
Parisi, D. (1987). Grammatical disturbances of speech production. In M. Coltheart, G. Sartori, & R. Job (Eds.), *The cognitive neuropsychology of language.* Hove, UK: Lawrence Erlbaum Associates Ltd.
Rapp, B., & Goldrick, M. (2006). Speaking words: Contributions of cognitive neuropsychological research. *Cognitive Neuropsychology, 23*, 39–73.
Rumiati, R.I., & Caramazza, A. (Eds.). (2005). The multiple functions of sensory-motor representations [Special issue]. *Cognitive Neuropsychology, 22*, 259–496.
Ruml, W., Caramazza, A., Capasso, R., & Miceli, G. (2005). A test of models of naming using Italian fluent aphasics. *Cognitive Neuropsychology, 22*, 131–168.

Ruml, W., & Caramazza, A. (2000). An evaluation of a computational model of lexical access: Comments on Dell et al. *Psychological Review*, *107*, 609–634.

Schacter, D. L. (1992). Understanding implicit memory: A cognitive neuroscience approach. *American Psychologist*, *47*, 559–569.

Schwartz, M. F. (2006). The cognitive neuropsychology of everyday action planning. *Cognitive Neuropsychology*, *23*, 202–221.

Schwartz, M. F., & Schwartz, B. (1984). In defence of organology. *Cognitive Neuropsychology*, *1*, 25–42.

Shallice, T. (1988). *From neuropsychology to mental structure.* Cambridge, UK: Cambridge University Press.

Shallice, T., & Warrington, E. K. (1970). Independent functioning of verbal memory stores. *Quarterly Journal of Experimental Psychology*, *22*, 261–273.

Simon, H. A. (1969). *The sciences of the artificial.* Cambridge, MA: The MIT Press.

Tainturier, M.-J., & Rapp, B. (2001). The spelling process. In B. Rapp (Ed.), *The handbook of cognitive neuropsychology: What deficits reveal about the human mind.* Philadelphia: Psychology Press.

Vallar, G. (2006). Memory systems: The case of phonological short-term memory. A festschrift for *Cognitive Neuropsychology*. *Cognitive Neuropsychology*, *23*, 135–155.

COGNITIVE NEUROPSYCHOLOGY, 2006, 23 (1), 13–38

The organisation of conceptual knowledge in the brain: The future's past and some future directions

Alfonso Caramazza and Bradford Z. Mahon
Harvard University, Cambridge, MA, USA

We review the development and current status of theories of the organisation and representation of conceptual knowledge in the human brain. The currently known facts from optic aphasia, category-specific semantic deficits, and functional neuroimaging are consistent with a framework in which the first-order constraint on the organisation of conceptual knowledge is domain. Data from functional neuroimaging suggests additionally a framework characterised by both domain- and modality-specific constraints. Work in congenital disorders and in apraxia indicate that the content of conceptual knowledge is not exhausted by modality-specific input/output processes. It is concluded that future empirical and theoretical work on the organisation and representation of conceptual knowledge will profit from a reorientation of the problem from the organisation of distinct processing systems to the content of information represented internal to such systems.

INTRODUCTION

Modern theories of the organisation of conceptual knowledge in the brain can be divided into two groups, depending on their underlying principles. One group of theories, based on the *neural structure principle*, assumes that the organisation of conceptual knowledge is governed by representational constraints imposed by the brain itself. A second group of theories, based on the *correlated structure principle*, assumes that the organisation of conceptual knowledge in the brain is a reflection of the statistical co-occurrence of object properties in the world. Two types of hypotheses have appealed to the neural structure principle: the Modality-Specific Semantic Hypothesis and the Domain-Specific Hypothesis. Theories based on the correlated structure principle can be distinguished from one another by the types of feature properties (e.g., correlation, distinctiveness) to which they appeal, and how such properties are (assumed to be) distributed in the world.

The modern study of the organisation of conceptual knowledge in the brain began with the work of Warrington, McCarthy, and Shallice in category-specific semantic deficits (Warrington & McCarthy, 1983, 1987;

Correspondence should be addressed to Alfonso Caramazza, Department of Psychology, William James Hall, Harvard University, 33 Kirkland Street, Cambridge, MA 02138, USA (Email: Caram@wjh.harvard.edu).

AC was supported in part by NIH grant DC04542. BZM was supported in part by a Fulbright Grant from the United States and Spanish governments. We would like to thank Erminio Capitani, Argye Hillis, Marcella Laiacona, Gabriele Miceli, Brenda Rapp, Jennifer Shelton, and Naomi Zack for many clarifying discussions. We would also like to thank Jorge Almeida, Max Coltheart, and Alex Martin for comments on an earlier version of this manuscript. We are grateful to Alex Martin for making available the graphics for Figure 3.

DOI:10.1080/02643290542000021

Warrington & Shallice, 1984). However, an important precursor is the work on optic aphasia (Lhermitte & Beauvois, 1973), which provided the initial impetus for modality-based theories of conceptual representation. Following the lead of these first reports, the case study approach in cognitive neuropsychology has been central to the development and evaluation of extant hypotheses about the organisation of conceptual knowledge. More recently, researchers have sought convergent evidence in functional neuroimaging. In this article we review the development and current evidential state of extant theories and outline some future theoretical directions.

OPTIC APHASIA

Optic aphasia is a modality-specific naming impairment specific to visually presented objects that cannot be reduced to a general visual agnosia or a general anomia. Lhermitte and Beauvois' (1973) patient was 73% correct at naming objects to visual presentation but 91% correct to tactile naming, 96% correct to definition, and 100% correct at gesturing the correct use associated with visually presented objects (see also, e.g., Campbell & Manning, 1996; Coslett & Saffran, 1989; Hillis & Caramazza, 1995; Riddoch & Humphreys, 1987; for reviews, see Plaut, 2002; Riddoch, 1999). Beauvois (1982) explained this pattern of behaviour by assuming that the conceptual system is organised into visual and verbal semantics: Optic aphasia would result from a disconnection between the two semantic systems. The typical lesion profile of optic aphasics is consistent with this hypothesis: Optic aphasic patients typically present with left occipital lesions extending to the splenium, effectively cutting off the left-hemisphere language centres from visual input.

There were early critiques of the modality-specific semantics proposal (Caramazza, Hillis, Rapp, & Romani, 1990; Riddoch, Humphreys, Coltheart, & Funnell, 1988). It was argued that the phenomenon of optic aphasia did not compel a division between visual and verbal semantics, for two independent reasons. First, the phenomenon of optic aphasia could be explained by assuming that (intact) pre-semantic representations of the visual structure of objects were disconnected from the semantic system but still connected to motor programmes supporting correct object use (Riddoch et al., 1988). A second critique of the Modality-Specific Semantic Hypothesis was that the notion of modality-specificity at the conceptual level had not been sufficiently fleshed out (Caramazza et al., 1990). It was not specified what it was about the information stored in modality-specific semantic subsystems that made them "modality-specific." Is it the format in which information is represented? Is it the content of the information (i.e., what that information is *about*) but not the format in which it is stored? Is it the modality through which the information was acquired?

Caramazza and colleagues (1990) outlined an alternative interpretation of the performance of optic aphasics (the Organized Unitary Content Hypothesis: OUCH), which assumes that certain types of input/output modalities have a privileged relationship with, or privileged access to, certain types of semantic information. To the degree that the notion of privileged access was sufficient to account for the dissociation between naming and gesturing to visually presented objects, there was no need to make the additional assumption of a functional division between visual and verbal modality-specific semantic subsystems.[1]

Further proposals were made, including the view that optic aphasia reflected a disconnection

[1] We use the terms "conceptual" and "semantic" interchangeably: i.e., category-specific semantic deficits (see below) could equally be called category-specific conceptual deficits. We will use the term "modality-specific semantic subsystem" to refer to information (or process) that mediates between pre- and post-semantic representations (e.g., between visual structural descriptions and phonological/orthographical lexical representations). We will refer to the latter type of representations as "modality-specific input/output" representations. The distinction between modality-specific input/output representations and modality-specific semantic representations will be examined more closely below.

of left-from right-hemisphere semantics (Coslett & Saffran, 1989, 1992). On this view, right-hemisphere semantics is sufficient to support relatively coarse, visually based semantic processing (e.g., gesturing the use of objects) but cannot communicate with the left-hemisphere language centres. Hillis and Caramazza (1995) argued that if the right hemisphere were disconnected from the left, then the naming errors of optic aphasics should bear no resemblance to the target. However, it is normally observed that the majority of naming errors made by optic aphasics are semantically related to the target. In other words, the nature of the errors made by optic aphasics would compel the assumption that the right hemisphere is not *completely* disconnected from the left hemisphere. However, at this point, the proposal becomes indistinguishable from both the multiple semantics view and the OUCH.

Some data germane but not decisive in regard to this debate were reported by Hillis and Caramazza (1995) in their study of an optic aphasic patient, DHY. This patient presented with the same pattern of performance that was originally used to motivate the assumption of separate (modality- or hemisphere-specific) semantic systems. DHY was impaired for naming to visual presentation but not for naming to definition or in response to tactilely presented stimuli, and showed normal performance in tasks that measured the ability to construct (visual) structural representations of objects. And, as had been reported for other optic aphasics, DHY did not demonstrate any marked semantic impairment on relatively "easy" semantic tasks with visually presented stimuli, such as word–picture matching. However, further testing demonstrated that when visually presented semantic tasks were made more difficult, such that fine-grained discriminations were required between similar concepts, DHY was impaired. The implication of these data is that semantic processing of visually presented stimuli in optic aphasics is not "intact." This follows both from the dominant type of naming errors (semantically related to the target, or to a previous response) and the presence of a mild semantic impairment in processing visual stimuli. A plausible interpretation of these results is that DHY's impairment reflects the failure to normally access a modality independent semantic representation from an intact structural description of visually presented objects.

Recently, Plaut (2002) proposed a reorientation of the debate with an implemented account of optic aphasia in which the semantic system is claimed to exhibit a "graded degree of modality-specificity." The implemented model's representation of the semantic system can be schematically captured in terms of a two-dimensional grid, with two input modalities (vision and touch) and two output modalities (oral naming and gesture). (Vision and touch were equidistant from phonology, and vision was equidistant from action and phonology.) The model was trained using a topographic learning bias that favoured short over long connections between a given input modality (i.e., vision or touch) and semantic representations (i.e., patterns of activation over units on the two-dimensional grid). Subsequent to training, the major quantitative patterns of dissociation observed in optic aphasia could be simulated by lesioning the shorter connections between the visual input modality and the semantic system.

The question is: which theoretical account, the Modality-Specific Semantics Hypothesis or OUCH, is supported by the simulations reported by Plaut (2002)? The answer to this question is clear: When the overall levels of activation of semantic units were inspected subsequent to either visual or tactile input "...there was no reliable effect of the horizontal position of the unit [closer to vision or closer to touch], the modality of presentation, or the interaction of these factors ($ps > .29$). Thus, there is no difference in the extent to which the two modalities [vision and touch] generate greater activation over closer semantic units compared to more distant units; the entire[2] semantic system is

[2] Note that amodal theories of conceptual representation (e.g., OUCH) are not committed to this (unnecessarily) strong claim that the *entire* semantic system is involved in representing *all* types of information (e.g., see discussion of OUCH below).

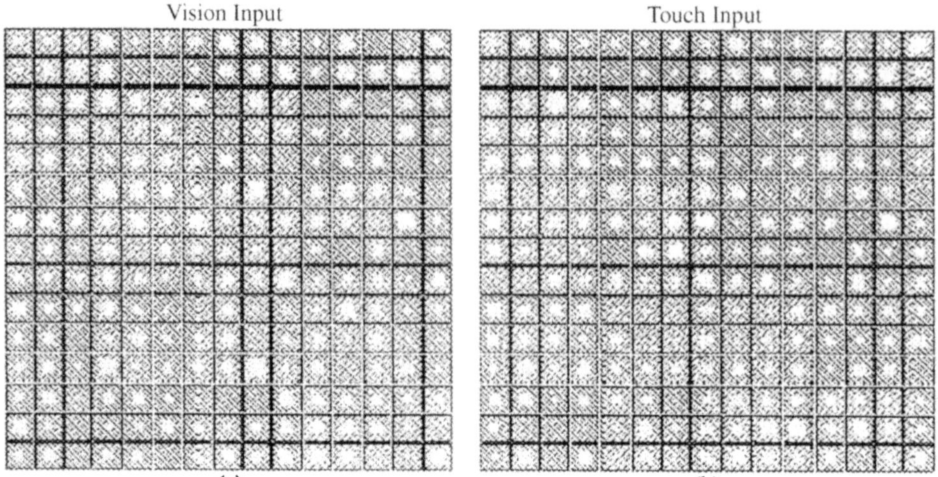

Figure 1. *Mean semantic activation for each unit as a function of modality of presentation. (From Figure 4, Plaut, 2002, p. 614; with permission of the author and Psychology Press http://www.psypress.co.uk/journals.asp)*

involved in representing both visual and tactile input" (p. 615). Since there is no modality specificity (graded or otherwise) in the semantic representations implemented in the model, the reported simulations constitute an existence proof of OUCH. Or stated differently, the topographic learning bias modelled by Plaut (2002) is one way to implement the notion of privileged accessibility, which is the basic assumption made by OUCH in order to account for the performance of optic aphasic patients.

The early critiques of the multiple semantics approach (Caramazza et al., 1990; Riddoch et al., 1988) did not go unanswered (Shallice, 1993) and the answers were further engaged (Hillis, Rapp, & Caramazza, 1995). However, as is evident from this brief overview, the phenomenon of optic aphasia has been a fertile area for developing hypotheses, but has not been nearly as decisive in their evaluation. Much of the empirical and theoretical work in cognitive neuropsychology about the organisation of conceptual knowledge in the brain shifted its focus to the phenomenon of category-specific semantic deficits and to functional neuroimaging studies in normal subjects.

CATEGORY-SPECIFIC SEMANTIC DEFICITS

The phenomenon of category-specific semantic deficits frames what has proven to be a rich question: How could the conceptual system be organised such that various conditions of damage can give rise to conceptual impairments that disproportionately affect specific semantic categories? There is emerging consensus that any viable answer to this question must be able to account for the following three facts (e.g., Caramazza & Shelton, 1998; Cree & McRae, 2003; Moss & Tyler, 2003; Samson & Pillon, 2003; for recent reviews, see Capitani, Laiacona, Mahon, & Caramazza, 2003; Humphreys & Forde, 2001; Tyler & Moss, 2001; see Figure 2 for an example of a patient with a selective deficit to living animate things):

1. The *grain* of the phenomenon: Patients can be disproportionately impaired for either living animate things (animals) compared to living inanimate things (fruits and vegetables) (KR: Hart & Gordon, 1992; EW: Caramazza & Shelton, 1998) or living inanimate things

Picture Naming: EW's naming deficit was restricted to the category "animals" and did not extend to the other living things such as "fruit/vegetables," for which performance was at ceiling. On subsets of the Snodgrass & Vanderwart (1980) picture set matched jointly for familiarity and frequency, and visual complexity and familiarity, EW was disproportionately impaired at naming animals compared to non-animals.

Matched: Familiarity and Frequency

	Animals	Non-Animals
EW	12/22 (55%)	18/22 (82%)
Controls	11/11 (100%)	10.8 (98%)
Range	11	10-11

Matched: Visual Complexity and Familiarity

	Animals	Non-Animals
EW	7/17 (41%)	16/17 (94%)
Controls	16.6/17 (98%)	16/17 (94%)
Range	16-17	16-17

Object Decision: EW was asked to decide (yes or no) whether a depicted object was real (see below for examples of stimuli). Performance on this task is interpreted as reflecting the integrity of the visual/structural description system (i.e., the modality-specific input system that stores representations corresponding to the form or shape of objects, and which is used to access conceptual information). EW performed significantly below the normal range for differentiating real from unreal animals (36/60; 60% correct; control mean: 54/60; 90%) but within the normal range for differentiating real from unreal non-animals (55/60; 92% correct; control mean: 50.5/60; 84% correct). However, EW performed within the normal range on complex visual processing tasks, such as visual matching and face recognition. These data indicate that EW does not have a general deficit for processing visually complex stimuli, and suggest that the patient's impairment for object reality decision for animals is categorically base

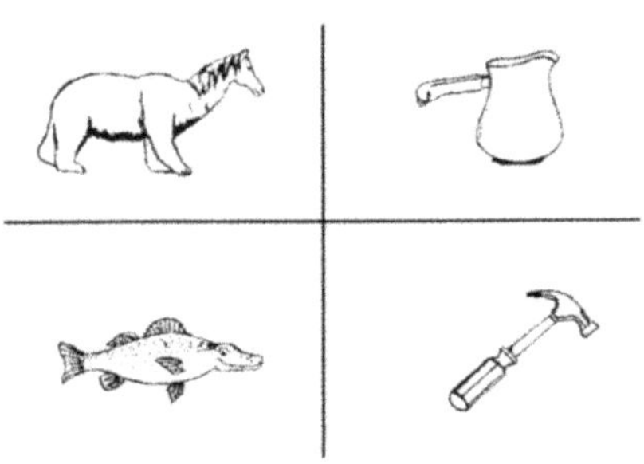

Sound Identification: EW was also impaired at naming animals compared to non-animals based on their characteristic sounds (8/32; 25% correct vs. 20/32; 63% correct: $z = 3.06$, $p < .05$) indicating that the patient's naming impairment is not restricted to visual input.

Parts Decision: EW was asked to decide which of two heads (or parts) went with a headless body (or object missing a part) (see below for examples of stimuli). EW was severely impaired on this task for animals (60% correct; normal mean: 100%) but performed within the normal range for artifacts (97% correct; normal mean: 97%).

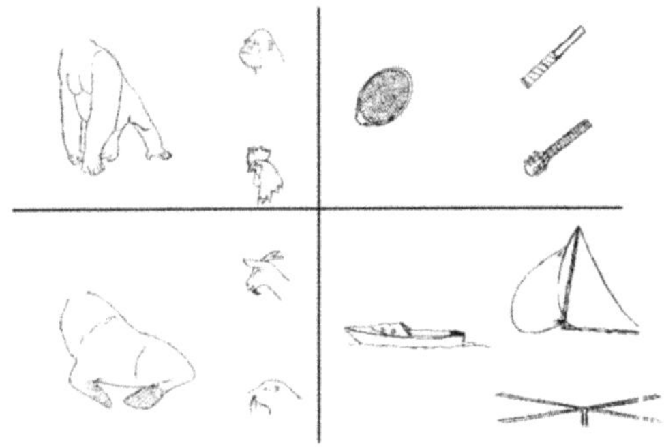

Central Attribute Judgments: EW was asked to decide whether a given attribute was true of a given item (see below for examples of stimuli). EW was severely impaired for attributes pertaining to animals (65% correct; control range: 85 – 100%) but within the normal range for non-animals (95% correct; control range: 86-100%). EW was equivalently impaired for both visual/perceptual and functional/associative knowledge of living things (65% correct for both types of knowledge) but within the normal range for both types of knowledge for non-animals (visual/perceptual: 93.5% correct; control range: 86-100%; functional/ associative: 98% correct; normal range: 92-100%).

Examples of Central Attribute Questions

Visual/perceptual	Functional/associative
Does a cow have a mane?	Does a whale fly?
Does a whale have a large tail fin?	Does an eagle lay eggs?
Does a whale have eight legs?	Is a cow a farm animal?

Figure 2. *An illustrative case of category-specific semantic deficit: Patient EW (Caramazza & Shelton, 1998). Reprinted from Trends in Cognitive Sciences, 7, Caramazza & Mahon, "The organization of conceptual knowledge: The evidence from category-specific semantic deficits", pp. 354–361, © 2003, with permission from Elsevier.*

compared to living animate things (e.g., MD: Hart, Berndt, & Caramazza, 1985; JJ: Hillis & Caramazza, 1991; TU: Farah & Wallace, 1992; FAV: Crutch & Warrington, 2003; RS: Samson & Pillon, 2003).

2. The *profile* of the phenomenon: Category-specific semantic deficits are not associated with disproportionate impairments for modalities or types of information (e.g., FM: Laiacona, Barbarotto, & Capitani, 1993; SB: Sheridan & Humphreys, 1993; EA: Barbarotto, Capitani, & Laiacona, 1996; Laiacona, Capitani, & Barbarotto, 1997; EW: Caramazza & Shelton, 1998; CN98: Gaillard, Auzou, Miret, Ozsancak, & Hannequin, 1998; Jennifer: Samson, Pillon, & De Wilde, 1998; PL: Laiacona & Capitani, 2001). Conversely, disproportionate impairments for modalities or types of information are not necessarily associated with category-specific semantic deficits (e.g., AC: Coltheart, Inglis, Cupples, Michie, Bates, & Budd, 1998; IW: Lambon Ralph, Howard, Nightingale, & Ellis, 1998).
3. The *severity* of overall impairment: The direction of category-specific semantic deficits (i.e., living things worse than nonliving things, or vice versa) is not related to the overall severity of knowledge impairment (Garrard, Patterson, Watson, & Hodges, 1998; Zannino, Perri, Carlesimo, Pasqualettin, & Caltagirone, 2002).

Most of the empirical and theoretical work in category-specific semantic deficits has been driven by an attempt to evaluate a theoretical proposal first advanced by Warrington, Shallice, and McCarthy: the Sensory/Functional Theory. The Sensory/Functional Theory is an extension of the Modality-Specific Semantic Hypothesis proposed by Beauvois (1982). In addition to assuming that the semantic system is functionally organised by modality or type of information, the Sensory/Functional Theory assumes that the recognition/identification of items from different semantic categories (e.g., living things compared to nonliving things) differentially depends on different modality-specific semantic subsystems (i.e., visual/perceptual information compared to functional/associative information) (for data and/or discussion pertaining to this latter assumption, see Caramazza & Shelton, 1998; Cree & McRae, 2003; Farah & McClelland, 1991; Garrard, Lambon Ralph, Hodges, & Patterson, 2001; Mahon & Caramazza, 2006; Tyler & Moss, 2001; Vinson, Vigliocco, Cappa, & Siri, 2003). Category-specific semantic deficits were thus explained by assuming damage to the modality or type of information upon which recognition/identification of items from the impaired category differentially depends.

The original formulation of the Sensory/Functional Theory is inconsistent with facts (1) and (2). The fact that living animate things and living inanimate things doubly dissociate is at variance with the assumption that both categories are processed/represented by the same semantic system. The fact that category-specific semantic deficits are not associated with deficits to a modality or type of knowledge indicates that the phenomenon is not caused by impairments to modalities or types of knowledge. There have been a number of theoretical responses to this state of the field.[3]

[3] Some early discussions of category-specific semantic deficits raised concerns that the phenomenon may be the result of uncontrolled stimulus variables: e.g., visual, complexity, familiarity, frequency (e.g., Funnell & Sheridan, 1992; Gaffan & Heywood, 1993; Stewart, Parkin, & Hunkin, 1992). Subsequent case reports in which the materials were carefully controlled for the relevant variables, as well as the observation of double dissociations over the same materials (e.g., Hillis & Caramazza, 1991) rule out this possibility (for review, see Capitani et al., 2003). More recently, Sartori and Lombardi (2004) have argued that category-specific semantic deficits for living things compared to nonliving things in a naming to definition task may arise due to a failure to "control" for the variable feature relevance, a measure of how predictable a concept is from a single feature. However, category-specific semantic deficits can manifest in a number of tasks besides naming to definition (e.g., picture naming, picture–word matching; picture matching; part–whole matching; category fluency; semantic attribute questions; see Figure 2). (For discussion and analysis of other stimulus variables, see Cree & McRae, 2003; Howard, Best, Bruce, & Gatehouse, 1995; for further discussion of the relevance framework, see Mahon & Caramazza, 2006).

Reformulating the sensory/functional theory

One proposal has been that specific types of visual/perceptual information are differentially important for living animate things or living inanimate things. For instance, it has been proposed that colour information is more important for fruits than animals (e.g., Cree & McRae, 2003; Crutch & Warrington, 2003; Humphreys & Forde, 2001) while biological motion information is more important for living animate things than living inanimate things (e.g., Cree & McRae, 2003). While it is certainly true that biological motion is crucial for animals while colour may be quite important in distinguishing among fruits, it does not follow that damage to mechanisms dedicated to processing these perceptual attributes would necessarily result in differential impairments for the categories of animals and fruits. Thus, for example, there is ample evidence that selective damage of colour knowledge does not result in disproportionate difficulty for the category of living inanimate objects. For instance, Miceli, Fouch, Capasso, Shelton, Tamaiuolo, and Caramazza (2001) reported a patient with a selective deficit for knowledge of object colour, but no associated disproportionate deficit for fruit/vegetables (see also Luzzatti & Davidoff, 1994). Samson and Pillon (2003) reported a patient with a disproportionate impairment for fruit/vegetables who was unimpaired at attributing the correct colour to fruit/vegetable stimuli (see also Crutch & Warrington, 2003). Similarly, motion-impaired patients have been reported who do not present with difficulties in naming pictured animals (Vaina, Makris, Kennedy, & Cowey, 1998), while patients who were not motion impaired have been reported to be selectively impaired for living animate things (e.g., Caramazza & Shelton, 1998).

Humphreys and Forde (2001) proposed a further modification of the Sensory/Functional Theory, in which it was assumed that there is greater perceptual crowding (due to greater perceptual overlap) between the visual structural descriptions of living things than nonliving things. Thus, damage to the visual structural system will disproportionately affect living things compared to nonliving things (see also Gale, Done, & Frank, 2001; Laws, Gale, Frank, & Davey, 2002; Tranel, Logan, Frank, & Damasio, 1997). As an account of the existence of category-specific semantic impairments for living things, this proposal is inconsistent with observations of patients with conceptual impairments for living things in the absence of damage to the visual structural description system (FM: Laiacona et al., 1993; EA: Barbarotto et al., 1996; Laiacona et al., 1997; SB: Sheridan & Humphreys, 1993; Jennifer: Samson et al., 1998; for discussion, see Mahon & Caramazza, 2001).

The Correlated Structure Principle

A second line of research has sought an account of category-specific semantic deficits based on the Correlated Structure Principle. For instance, the OUCH model (Caramazza et al., 1990) makes two principal assumptions. First, conceptual features corresponding to object properties that often co-occur will be stored close together in semantic space; and second, focal brain damage can give rise to category-specific semantic deficits either because the conceptual knowledge corresponding to objects with similar properties is stored in adjacent neural areas, or because damage to a given property will propagate damage to highly correlated properties. While the original OUCH model is not inconsistent with the currently available data from category-specific semantic deficits, it is too unconstrained to provide a principled answer to the question of *why* the various facts are as they are.

The most developed extension of OUCH is the Conceptual Structure Account of Tyler, Moss, and colleagues (Tyler & Moss, 2001; Tyler, Moss, Durrant-Peatfield, & Levy, 2000; for similar proposals, see Devlin, Gonnerman, Anderson, & Seidenberg, 1998; Garrard et al., 2001; Gonnerman, Andersen, Devlin, Kempler, & Seidenberg, 1997; McRae & Cree, 2002; Vinson et al., 2004). The Conceptual Structure

Account makes three assumptions. (1) Living things have more shared features than nonliving things, or put differently, nonliving things have more distinctive/informative features than living things. (2) For living things, biological function information is highly correlated with shared perceptual properties (e.g., can see/has eyes). For artifacts, function information is highly correlated with distinctive perceptual properties (e.g., used for spearing/has tines). (3) Features that are highly correlated with other features will be more resistant to damage than features that are not highly correlated. This hypothesis thus explains the cause of category-specific semantic deficits by assuming random (or diffuse) damage to a conceptual system that is not organised by modality or object domain.

The Correlated Structure Account (i.e., the conjunction of the above three assumptions) predicts that a disproportionate deficit for living things will be observed when damage is relatively mild, while a disproportionate deficit for nonliving things will only arise when damage is so severe that all that is left in the system are the highly correlated shared perceptual and function features of living things. This prediction is not consistent with the observation of a severe deficit for nonliving things in the context of relatively spared knowledge of animals (Hillis & Caramazza, 1991). Perhaps even more problematic is that the central prediction of the theory is not confirmed by cross-sectional analyses of patients at varying stages of Alzheimer's disease: there is no interaction between the severity of overall impairment and the direction of category-specific semantic deficits (Garrard et al., 1998; Zannino et al., 2002).[4]

The Domain-Specific Hypothesis

The third route that has been pursued is the Domain-Specific Hypothesis (Caramazza & Shelton, 1998). The Domain-Specific Hypothesis as developed in the context of category-specific semantic deficits must be distinguished from other possible ways in which a domain-specific approach might be articulated. Common to all approaches is the assumption that a given cognitive process is domain specific if the scope of its extension is delimited by the semantic (i.e., content-defined) class membership of the objects that it processes. This assumption by itself, however, leaves unaddressed three (at least logically) orthogonal extant issues.

First is the issue of whether conceptual processing of objects is domain specific because of *innately* determined neural constraints. We assume that domain-specific constraints on the organisation of conceptual knowledge are innately determined. This assumption makes a strong prediction: The categories of category-specific semantic deficits will be restricted to only those categories that, if identified could have had survival or reproductive value. An example of such a category is that of "conspecifics," which we discuss below.

A second issue is whether modality-specific *input* processes are organised by object domain. We assume that object domain is an innately determined parameter of neural organisation at both the conceptual and perceptual level. In other words, the claim here is that stimuli are categorised (e.g., as animate) at a stage of processing prior to the conceptual level. Below we review functional imaging and neuropsychological data consistent with this assumption.

A third issue is whether the content of object concepts is exhausted by information/processes internal to the sensory-motor systems of the brain. We show that the content of object concepts cannot be reduced to sensory-motor processing; to do this we argue that various patterns of neuropsychological impairments *could not exist* if it were the case that conceptual knowledge was exhausted by information internal to modality-specific input/output representations.

[4] One study (Gonnerman et al., 1997) reported an association between the severity of conceptual impairment and the direction of category-specific deficit, but the reported interaction has subsequently been shown to be an artifact of ranking the patients according to performance on only one object category (see Zannino et al., 2002, for discussion).

Innately determined domain-specific processes

The proposal that innate structure in the central nervous system can lead to domain-specific organisation is by no means new; there is a range of evidence for innate neural structure in non-human species. For instance, Emlen (1967, 1969) investigated the navigational strategies of a migratory species of bird (Indigo Buntings, *Passerina Cyanea)* and found that a crucial factor is the relative position of the stars (see also work in the navigational abilities of desert ants, e.g., Collett, Collett, Bisch, & Wehner, 1998). The claim that a migratory species of bird such as the Indigo Bunting might be born with innately determined "hardware" in order to solve complex navigational processes based on the relative positions of stars would seem to be relatively uncontroversial. But if we are willing to grant innate constraints on the structure of the bird mind, why not on that of the human mind?

Another example from animal studies comes from a series of studies by Mineka and colleagues (e.g., Cook & Mineka, 1989; Mineka, Davidson, Cook, & Keir, 1984). Rhesus monkeys captured in the wild, but not naïve, laboratory-raised Rhesus monkeys, show a strong fear response when presented with a snake (even a plastic model). Mineka and colleagues found that when naïve, laboratory-raised Rhesus monkeys observe a video of another monkey displaying a fear response to a snake, they will subsequently display the same fear response when confronted with a snake. However, naïve, laboratory-raised monkeys do not show a transfer of fear reaction if they view videos of other monkeys displaying fear responses to flowers or to rabbits. The interesting implication of these data is that the transfer of fear response is mediated, and actually constrained, by the Rhesus monkeys' categorisation of the stimulus.

Knowledge of conspecifics in humans: Does this category behave like a domain?. It follows from the assumption that domain-specific processing of objects is innately constrained that there will be domain-specific processes only for those categories that could have conferred a survival and/or reproductive value. The category of conspecifics is arguably one of the most salient categories of things in the world that satisfies this condition (Shelton, Fouch, & Caramazza, 1998). Paradoxically, data from processing of conspecifics has until now figured only marginally in discussions of the organisation of conceptual knowledge (but see Haslam, Kay, & Hanley, 2002, for review and discussion). Here we use the category "conspecifics" as a test case and ask whether similar profiles of impairment are observed for this category as are observed for living animate and living inanimate things. A further expectation is that knowledge of conspecifics will include a system for attributing intentional content to other minds: that is, Theory of Mind.

Patient APA (Miceli, Capasso, Daniele, Esposito, Magarelli, & Tomaiuolo, 2000) was impaired for knowledge of people (e.g., 10/32 famous face naming) but did not present with significant (or differential) difficulties naming objects/animals (e.g., 74/80) (see also Kay & Hanley, 1999). Furthermore, APA was normal on the Benton Face Recognition Test, indicating that the patient was not impaired in processing the structural characteristics of faces. A contrasting semantic impairment was reported by Kay and Hanley (2002): Patient ML did not have an impairment in recognising faces as familiar or not, was normal for a number of stringent tests requiring identification of famous people, but was impaired (equivalently) for objects and animals. This contrasting pattern reflected in patients APA and ML has recently been reported over the same battery of materials (Thompson, Graham, Williams, Patterson, Kapur, & Hodges, 2004; patients MA and JP).[5] Thus, the domain of conspecifics can be spared or impaired

[5] JP was impaired for knowledge of people but relatively spared for knowledge of objects and animals, while MA showed the reverse profile. Both patients were normal on a range of visuospatial tests, including object decision and perception of unfamiliar faces, and neither patient was more impaired for living or nonliving things.

independently of both objects and other living things (e.g., living animate); importantly, an impairment for people is not associated with a general impairment for living things compared to nonliving things. This observation is significant in the context of the proposal discussed above, that the primary deficit in patients with impairments for living animate things is to knowledge of the typical ways in which living animate things move (e.g., Cree & McRae, 2003). If this were the case, then the category "living animate things" should not fractionate into the domains of animals and conspecifics. Thus, there is a tripartite distinction within the category "living things": animals, fruit/vegetables, and conspecifics.

It is important that at least some impairments for person-specific knowledge are not reducible to a general anomia for proper names or a modality-specific visual input impairment (i.e., prosopagnosia). As noted above, patient APA was not prosopagnosic, and, although impaired in famous face naming, was normal for naming monuments and geographical places (Miceli et al., 2000).[6] Patient JP's impairment (Thompson et al., 2004) was observed across several matching and sorting tasks, using pictures as well as spoken and written names, suggesting that the semantic impairment for person knowledge observed in this patient cannot be reduced to a problem with face recognition (see also Ellis, Young, & Critchley, 1989; Verstichel, Cohen, & Crochet, 1996).

Is the domain of conspecifics a special domain? The domain of conspecifics is unique in that the object of processing is of the same type as the agent of processing. The ability to attribute intentional states to other people has been termed "Theory of Mind" (Premack & Woodruff, 1978). If the domain of conspecifics constitutes an evolutionarily defined domain, then Theory of Mind abilities is one use to which such processing might be uniquely applied. What is the evidence that Theory of Mind abilities are functionally isolable from other processes?

Recent neuropsychological work indicates that theory of mind abilities do not depend on intact general executive functioning abilities (Fine, Lumsden, & Blair, 2001; Gregory et al., 2002; Lough, Gregory, & Hodges, 2001; Varley, Siegal, & Want, 2001; for review, see Siegal & Varley, 2002). For instance, Fine and colleagues (2001) reported a patient with either congenital or early amygdala damage, impaired Theory of Mind abilities, but spared executive functioning. The reverse functional dissociation has also been reported: impaired executive functioning but intact Theory of Mind abilities (Varley et al., 2001).

An important area for investigating the functional and neural bases of Theory of Mind comes from the study of individuals with congenitally impaired Theory of Mind abilities (Asperger syndrome and autism; for review, see U. Frith, 2001; Gallagher & Frith, 2002). For instance, Castelli, Frith, Happé, and Frith (2002) compared the activation observed for 10 normal adults and 10 individuals with Asperger syndrome when watching animated geometric shapes moving as "biological stimuli" compared to shapes moving randomly. Replicating previous findings (Castelli, Happé, Frith, & Frith, 2000) normals showed increased activation for the animated condition in the (all primarily right-sided) basal temporal region, superior temporal sulcus, medial prefrontal cortex, and extra-striate cortex (V3). Compared to normals, individuals with Asperger syndrome showed reduced activation in basal and superior temporal regions and in medial prefrontal cortex. The same level of activation was observed in extra-striate cortex.

Other functional neuroimaging work with healthy subjects supports the role of the right ventromedial prefrontal cortex in mental state

[6] This performance profile can be contrasted with that observed for case GR (Lucchelli, Muggia, & Spinnler, 1997), who presented with an anomia for people's names, but spared name recognition and face–name matching. The patient could give detailed semantic information about people he could not name, and was in the normal range for visual and verbal naming of other proper name categories (Italian cities, monuments, European cities, rivers, mountains, currencies, commercial brands).

attribution. Mitchell and colleagues (2002) had participants make yes/no judgments on noun/adjective word pairs presented visually (Could the adjective be true of the noun? e.g., Emily–beautiful; orange–seedless). Nouns were taken from the categories clothing, fruit, and proper names of people. It was observed that the person versus object contrast resulted in areas that were less deactivated compared to baseline in (among other areas) dorsal and ventral aspects of medial prefrontal cortex. Consistent findings were obtained in the study by Gallagher, Jack, Roepstorff, and Frith (2002), in which participants played the game "paper, rock, scissors." The crucial contrast was between a condition in which participants thought they were playing against a computer (computer stance) and one in which they thought they were playing against a person (human stance). Actually, in both conditions they were playing against a computer operating randomly. The only area that was more activated in the "human stance" condition vs. the "computer stance" condition was the anterior aspect of the paracingulate cortex (see also McCabe, Houser, Ryan, Smith, & Trouard, 2001).

The relation between biological motion processing, affect, and Theory of Mind has emerged as a central area of study in attempts to understand the functional and neuroanatomical organisation of Theory of Mind abilities (C. D. Frith & Frith, 1999). This brief overview highlights a primarily right-sided network specialised for processing information about conspecifics.

Are modality-specific input processes organised by object domain?

The Domain-Specific Hypothesis assumes that perceptual (i.e., pre-conceptual) stages of object recognition may be functionally organised by domain-specific constraints. With respect to the visual modality, this assumption generates the prediction that patients may present with category-specific visual agnosia (a deficit in recognising visually presented objects despite intact elementary visual processing). Tentative evidence for this prediction is provided by the observation of patients with equivalent impairments to visual/perceptual and functional/associative knowledge of living things, but a visual agnosia for living things compared to nonliving things (Barbarotto et al., 1996; Barbarotto, Capitani, Spinnler, & Trivelli, 1995; Capitani, Laiacona, & Barbarotto, 1993; Caramazza & Shelton, 1998; Laiacona et al., 1993; Lambon Ralph et al., 1998; but see Capitani et al., 2003, for critical review). Also consistent with such an organisation is the observation that patients can present with impairments for recognizing faces but not visually presented objects (e.g., Newcombe, Mehta, & De Haan, 1994) as well as the reverse: spared face recognition but impaired object recognition (Moscovitch, Winocur, & Behrmann, 1997).

There is a large body of evidence from functional neuroimaging that demonstrates differentiation by semantic domain within modality-specific systems specialised for processing object form (ventral temporal cortex) and object associated motion (lateral temporal cortex). Items from living animate categories (animals, human faces) differentially activate the superior temporal sulcus (right > left) (e.g., Chao, Haxby, & Martin, 1999a; Chao, Martin, & Haxby, 1999b; Haxby, Ungerleider, Clark, Schouten, Hoffman, & Martin, 1999; Hoffman & Haxby, 2000; Kanwisher, McDermott, & Chun, 1997) and the lateral aspect of the fusiform gyrus (e.g., Chao et al., 1999a; Chao, Weisberg, & Martin, 2002; Kanwisher et al., 1997; Kanwisher, Stanley, & Harris, 1999; McCarthy, Puce, Gore, & Allison, 1997), while items corresponding to nonliving things differentially activate the middle temporal gyrus (left > right) (e.g., Chao et al., 1999a; Devlin et al., 2002a; Martin, Wiggs, Ungerleider, & Haxby, 1996) and the medial aspect of the fusiform gyrus (Chao et al. 1999a, 2002; but see Devlin et al., 2002b). For a domain-specific interpretation of these findings, see Kanwisher (2000); for an alternative interpretation see Tarr and Gauthier (2000).

In a recent series of studies, Beauchamp and colleagues (Beauchamp, Lee, Haxby, & Martin, 2002, 2003) demonstrated that, in addition to the category-specific foci of activation just reviewed, lateral temporal cortex prefers object-associated motion whereas ventral temporal cortex is more sensitive to object-associated form and texture. In their 2002 report, it was found that lateral temporal cortex responded more to moving images than to static images, while ventral temporal regions responded equally to both. In their 2003 report, it was found that ventral temporal cortex preferred videos of moving images to point light displays of the same moving images, while lateral temporal cortex responded either more to point light displays than videos (superior temporal sulcus: humans) or equivalently to videos and point light displays (middle temporal gyrus: tools). Furthermore, within lateral temporal cortex, the superior temporal sulcus preferred human stimuli moving in an articulated manner (e.g., jumping jacks) than an unarticulated manner (e.g., rotating about the centre of mass) (Beauchamp et al., 2002) as well as point light displays of moving humans to point light displays of moving tools (Beauchamp et al., 2003). In contrast, the middle temporal gyrus preferred point light displays of moving tools to point light displays of moving humans (Beauchamp et al., 2003; see also Grossman & Blake, 2002; Grossman et al., 2000; Kourtzi & Kanwisher, 2000; Senior et al., 2000).

A crucial issue concerning these differential patterns of activation in ventral and lateral temporal areas is whether they reflect conceptual or modality-specific input processing (see Whatmough, Chertkow, Murtha, & Hanratty, 2002, for some empirical work on this issue in ventral temporal cortex). At minimum,[7] these functional neuroimaging data are consistent with the proposal that modality-specific input systems are organised by object domain. Consensus on this interpretation has been, at best, reluctant; a number of researchers have argued that differential effects of object category (in, e.g., ventral temporal areas) are driven by object-specific features and not object domain (e.g., Bookheimer, 2002; Gerlach, Law, Gade, & Paulson, 2000; Ishai, Ungerleider, Martin, Schouten, & Haxby, 1999; Kraut, Moo, Segal, & Hart, 2002; Martin & Chao, 2001; Moore & Price, 1999; Mummery, Patterson, Hodges, & Price, 1998; Perani et al., 1995; Thompson-Schill, 2003; but see, e.g., Kanwisher, 2000). For instance, Martin and colleagues (e.g., Martin & Chao, 2001) have argued that it is not the case that (e.g.) the right superior temporal sulcus processes "biological motion" per se, but rather that it processes "articulated motion." And it just so happens that biological motion is articulated while mechanical motion is not. The crucial point of difference, then, between this proposal and the Domain-Specific Hypothesis is not necessarily the content of what is processed in a given area, but how a given area comes to process the information that it does. The sensory/motor account of Martin and colleagues is one implementation of OUCH: Objects that share properties are represented close together in the brain. In this regard developmental findings will play an important role in adjudicating between theories (e.g., Farah & Rabinowitz, 2003).[8]

The primary empirical motivation for the Sensory/Motor Theory is the observation that the areas of activation observed for living and non-living things are differential and not selective. Thus, the strongest evidence for a domain-specific interpretation of these findings would come from a demonstration that the same patterns of activation can be observed when object-associated features are removed from the stimuli.

In a recent study by Martin and Weisberg (2003), participants viewed three types of computer-animated displays, all consisting of the

[7] "At minimum" because all extant theoretical interpretations assume that modality-specific input representations exist, while there is not consensus that concepts are represented independently of modality-specific input/output representations.

[8] The Domain-Specific Hypothesis must assume that there is some innate content that allows a given domain-specific system to become "locked" to the right category of objects (i.e., a triggering mechanism). The claim is not that individual object concepts are given innately; rather, the claim is more along the lines of the type of content assumed to be localised in the theory of Martin and colleagues.

same geometric shapes, but differing in the type of motion in which the shapes were engaged (for discussion of such stimuli, see Heider & Simmel, 1944; Scholl & Tremoulet, 2000). The geometric shapes could be (1) moving as biological entities (e.g., chasing, playing a game; "biological motion" condition); (2) moving as mechanical entities (e.g., cue balls, bowling balls; "mechanical motion" condition); or (3) moving randomly; "random motion" condition. In lateral temporal cortex, it was observed that the superior temporal sulcus responded more to the "biological motion" condition, while the middle temporal gyrus responded more to the "mechanical motion" condition, indicating that the corresponding neural regions for processing motion were engaged by these stimuli. More striking were the findings in ventral temporal cortex: Lateral regions of ventral temporal cortex responded more to the "biological motion" condition while medial regions responded more to the "mechanical motion" condition. Given that ventral temporal regions are most responsive to object form and texture (e.g., Beauchamp et al., 2003), and given that object form and texture were exactly the same between the "biological" and "mechanical" motion conditions, these data indicate that the activation observed in ventral temporal areas can be driven by higher-order "interpretations" of the semantic domain to which the geometric shapes belong, and not by object-specific features. Perhaps relevant to this inference is the observation that activation associated with the "biological motion" compared to the "mechanical motion" condition was also observed in the right ventromedial prefrontal cortex and the amygdala (for related findings in right superior temporal sulcus using acoustic stimuli, see Kriegstein, Eger, Kleinschmidt, & Giraud, 2003). As discussed above, both of these areas have been implicated in the ability to attribute intentions to (at least) conspecifics.

The activation observed in lateral and ventral temporal areas by Martin and Weisberg (2003) was not only differential for one type of stimulus (e.g., biological) compared to the other (i.e., mechanical): When an area in ventral temporal cortex responded to one type of stimulus (biological or mechanical motion) it did not respond to the other type of stimulus more than to the random motion baseline.[9] The observation that different areas of cortex within modality-specific input systems (e.g., form, motion) respond differentially to different semantic categories is contrary to the assumption of modality-specific input systems not internally organised by object domain.

It might be argued that a combination of the Sensory/Functional Theory and a theory based on the Correlated Structure Principle (e.g., OUCH; see also Levy, Hasson, Avidan, Hendler, & Malach, 2001) could accommodate such patterns of activation (e.g., Cree & McRae, 2003; Vinson et al., 2003). However, the observation that such patterns of activation can be demonstrated to not only be "differential" but also "selective" (Martin & Weisberg, 2003) is contrary to an interpretation that assumes that object-specific features are driving the observed effects. The possibility of observing "selective" patterns of category-specific activation is uniquely afforded by the Domain-Specific Hypothesis.

Conceptual content cannot be reduced to modality-specific input/output content

Our discussion of the Domain-Specific Framework has assumed that the content of object concepts is not reducible to information/processes internal to modality-specific input/output systems of the brain. However, and as noted above, this view on the nature of conceptual content is not entailed in any way by the assumption of domain-specific neural circuits for representing/processing conceptual knowledge of some types of objects. Here we distinguish two possible architectures that might be articulated within a Domain-Specific Framework, and we argue for the second.

[9] This was the case bilaterally in ventral temporal cortex, in the right superior temporal sulcus (social > mechanical), and in the left middle temporal gyrus (mechanical > social).

1. Modality-specific input/output systems are both format and content[10] specific for the modality of information they process, and they are organised by object domain. All conceptual content is grounded in such modality-specific input/output systems. In other words, domain-specific simulation.
2. Modality-specific input/output systems are both format and content specific for the modality of information they process, and they are organised by object domain. There is a level of conceptual content represented independently of modality-specific input/output systems that is organised by object domain.

The issue of whether conceptual information is exhausted by modality-format-specific input/output systems has been a topic of recent research and debate, in large part due to theories developed within the "Simulationist Framework." The central assumption of the strong form of the Simulationist Framework is that in order to go from a physical stimulus to "understanding" one must internally "run" or "simulate" the production processes (i.e., modality-format-specific representations) that would mediate production of an event that is the same as that which is understood (see also the motor theory of speech perception, e.g., Liberman, Cooper, Shankweiler, & Studdert-Kennedy, 1967, for a similar proposal at the level of recognition).[11] Consider the observation of someone drinking a glass of water. The visual input of this event would result in a simulation of the observed motor action. In the course of simulating the action of drinking a glass of water, a number of conceptual states might be invoked. For instance, one might read off the intentions that could plausibly have formed the basis of the simulation (were the simulation an actual production event). One such state might be, for example, ⟨being thirsty⟩ (for recent reviews, see Blakemore & Decety, 2001; Gallese & Goldman, 1998; see also Adolphs, 2003; Allison, Puce, & McCarthy, 2000). Such intentional states could then be attributed to the observed agent.[12] Another conceptual state that might depend on the simulation of drinking a glass of water would be the concept GLASS (e.g., Allport, 1985; Barsalou, Simmons, Barbey, & Wilson, 2003; Martin, Ungerleider, & Haxby, 2000).

One issue that arises is whether such a framework would provide the means for individuating distinct mental states that are coextensive with the same motor programme (for discussion, see Jacob & Jeannerod, 2005). So for instance, if one observed a person who was about to give a presentation taking a drink of water, the resulting simulation would presumably involve the same motor programme(s); however, it might be inferred that the observed drinker was nervous (instead of thirsty). Or, if one observed the same person taking a drink of water *during* a presentation, again, the motor simulation would be identical, but it might be inferred that that person's throat was parched, but not that they were thirsty or nervous, etc. This would seem to indicate that attributions of mental

[10] Such systems might be modality-context-specific as well; i.e., information is represented in a format congruent with the modality through which the information was acquired, and the only modality in which information can be stored is that through which it was acquired; see Caramazza and colleagues (1990) for discussion. The present point can be made independently of the stronger (i.e., modality-context-specific) assumption.

[11] How the notion of "same" is fleshed out is an important aspect of simulation theory. In particular, is the simulation driven by first acknowledging the "type" of event that is occurring, and then a "type" identical simulation is run? If so, the claim would be that modality-format-specific representations are type identified. If not, then the simulation must be in terms of some particular past experience. If the former, then the question becomes how "abstract" is the information internal to modality-format-specific representations assumed to be? If the latter, then the question becomes: How does one learn anything to begin with? See the discussion of developmental data below.

[12] For example: "Inferring intentions from observed actions might depend on the same mechanism that labels the consequences of one's own actions as being produced by one's own intentions" (Blakemore & Decety, 2001, p. 563). Or similarly, Gallese and Goldman (1998) write: "In the present article we will propose that humans' mind-reading abilities rely on the capacity to adopt a simulation routine" (p. 493).

states depend on background knowledge. But how is "background knowledge" embedded in modality-specific input/output systems?

However, the strong form of the Simulationist Framework is not only about the order in which information is processed; it is also about the representation of intentional content (including both object concepts and the representations underlying theory of mind abilities). The Simulationist Framework makes the stronger claim that not only is understanding necessarily *mediated* by production processes; understanding is the running of modality-specific input/output information (e.g., Barsalou et al., 2003; Gallese & Goldman, 1998).

There is a very appealing aspect to the simulation theory, the notion that understanding is a form of action. However, as a hypothesis about the representation of intentional content, the Simulationist Framework depends on at least two things: empirical demonstrations that (1) production programs are run in the course of recognition, and (2) such production programs are sufficient to ground conceptual content. We focus the discussion to follow on studies that have looked at biological motion processing and the representation of object concepts. There are two questions to be addressed: First, does recognition of biological motion involve the processes required to produce such motion? Second, what happens to recognition and access to conceptual knowledge when modality-specific output processes are damaged?

One line of evidence marshalled in support of the Simulationist Framework comes from developmental work indicating that the capacity/proclivity to imitate is "innate." For instance, babies imitate facial gestures from a very early age and they also will imitate people but not robots (e.g., trying to pick up a dumbbell) (Meltzoff, 1995; Meltzoff & Moore, 1977). But there are also developmental findings cited in support of the Simulationist Framework that would seem to create more problems than they could engender support (see also discussion in Jacob & Jeannerod, 2005). For instance, 8 to 16-week-old infants prefer point-light walking figures to dynamic noise or the same figure rotated 180 degrees (Fox & McDaniel, 1982) while 3- to 5-month-old infants can discriminate between a point light walker and similar figures with scrambled spatial relationships between the moving dots (Bertenthal, Proffitt, & Kramer, 1987). Given that 3- to 5-month-old babies do not have experience with walking, what are they simulating in order that they can recognise walking?

It is likely that there is much about walking that is innate, and so it might be argued that 3- to 5-month-olds are simulating whatever is given innately in respect of the ability to walk. If this is the direction of argument, then the substrate of simulation is not modality-format-specific information learned through experience, but something more abstract. Alternatively, it might be pointed out that 3- to 5-month-olds have plenty of experience watching people walk, and so they are simulating their past visual (and not motor) experiences. If this is the direction of argument, then why not assume that adults recognise biological movement the way babies do?

Experimental work with adults converges with the developmental findings. For instance, when adult subjects view a specific facial expression, the corresponding muscles in the observer's face are "activated" (recorded via EMG reactions) (Lundqvist & Dimberg, 1995). A simulationist interpretation of such findings assumes that the capacity to infer emotion from the observation of others' faces depends on the ability to simulate the observed expressions. What happens to the ability to attribute emotions to faces when the capacity to move one's face is not present?

Calder, Keane, Cole, Campbell, and Young (2000) report the performance of an individual, LP, who had bilateral paralysis of the face from infancy (Mobius syndrome). LP was normal on an unfamiliar face-matching test (Benton Test), impaired on Warrington's Recognition Test for unfamiliar faces, and borderline impaired on a test requiring recognition of famous people. On a test of facial affect recognition (apply one of the six basic emotions to a face) LP was not impaired, but was slightly impaired on a more difficult version (constructed from morphs). The patient was normal in applying one of the six affect labels to typical corresponding sounds (e.g., laughter = happiness). From the performance of patient LP there would seem to be something right about the Simulationist Framework: LP was not unimpaired on all administered tests. However, at the same time, the fact that the patient *could succeed at all* on tasks of emotion attribution based on facial affect indicates that the ability to attribute intentional states is not exhausted by simulation of the observed behaviour.

Another line of evidence that has been marshalled in support of the Simulation Framework is based on apparent motion effects: the (seemingly) perceptual experience of movement induced by static images presented rapidly and alternatingly in different places. Apparent motion effects induced with geometric shapes follow the shortest possible path. However, when normal subjects observe apparent motion effects that involve parts of the human body, the geometrically longer but biomechanically plausible path is perceived over the geometrically shorter but biomechanically impossible path (e.g., Shiffrar & Freyd, 1993). Furthermore, motor execution areas are activated only during the biomechanically possible conditions (Stevens, Fonlupt, Shiffrar, & Decety, 2000).[13] An interpretation of these data in terms of the Simulation Framework assumes that recognition of biological motion of conspecifics involves covert production of the same movements.

Servos, Osu, Santi, and Kawato (2002) asked whether the neural areas mediating biological motion perception overlap with areas activated during motor imagery. The authors compared observation of Johansson point-light displays with a motor imagery task (e.g., scratch back with right arm). There was no overlap observed between biological motion perception and motor imagery, even at a liberal alpha level.

Pavlova, Staudt, Sokolov, Birbaumer, and Krageloh-Mann (2003) asked whether the degree of motor impairment in 13- to 16-year-old children with congenital motor disorders was inversely related to sensitivity to Johansson point-light displays. There was no significant relationship between visual sensitivity to the point-light displays and severity of motor impairments, while there was a relation between degree of motor impairment and the volume of periventricular lesions in parietal-occipital areas. As in the study of Calder and colleagues (2000) a trend is reported by Pavlova and colleagues that is in the direction predicted by the Simulationist Framework. Nevertheless, and as in the study of Calder and colleagues (2000), the data from Pavlova and colleagues indicate that recognition of biological motion does not depend on *ever* having produced such motion in one's life. In other words, there is no past experience to simulate.[14] This implies that modality-specific output representations are not sufficient to ground conceptual knowledge.

Stronger support for this conclusion comes from the study of apraxia and the proposal that the ability to recognise/identify manipulable objects depends on information that is active during the use of such objects: Specifically, that conceptual knowledge of manipulable objects is represented in terms of modality-specific output representations that code the motor movements

[13] Why were premotor areas activated only during biomechanically *possible* action? In other words, if activation in premotor cortex is the criterion (here) for simulation, then given that there was no such activation for biomechanically impossible actions, wouldn't this suggest that the possible/impossible classification happened somewhere else? In fact, there was some activation in orbito-prefrontal cortex that was greater for biomechanically implausible motion over plausible motion. But if the decision (or "filter") happens outside the simulator, then the simulator is outside the system it is supposed to replace (see Blakemore & Decety, 2001, for discussion).

[14] It might be argued that the simulation in these cases is more abstract and does not occur over representations corresponding to past experiences. At this point, however, the basic assumption of the Simulationist Framework would have been abandoned, since the claim was that information encoded during past experiences with similar events is "re-activated" in order to understand the present event.

associated with the use of such objects (Barsalou et al., 2003; Gallese & Goldman, 1998). It is important to distinguish the Simulationist Framework from a closely related theory, the Sensory/Motor Theory of Martin and colleagues (2000; see also Allport, 1985). The difference between the Sensory/Motor Theory and the Simulationist Framework is that the Sensory/Motor Theory is not committed to the claim that the same representation underlies production and recognition. Thus, the observation of a dissociation between the ability to use objects and the ability to recognise the correct gestures associated with objects (for reviews, see, e.g., Cubelli, Marchetti, Boscolo, & Della Sala, 2000; Johnson-Frey, 2004; Rothi, Ochipa, & Heilman, 1991) is at variance with the Simulationist Framework, but not the Sensory/Motor Theory.

The claim that information required to use manipulable objects grounds conceptual knowledge of such objects has been motivated primarily by results from functional neuroimaging. A well-documented finding is that left premotor cortex is differentially activated when subjects perform various tasks over tool stimuli compared to nonmanipulable stimuli (e.g., animals, houses) (e.g., Chao & Martin, 2000; Chao et al., 2002; Gerlach et al., 2000; Gerlach, Law, Gade, & Paulson, 2002; Grabowski, Damasio, & Damasio, 1998; Martin, Wiggs, Ungerleider, & Haxby, 1996; for review, see Grèzes, & Decety, 2001; Martin & Chao, 2001; see Gallese & Goldman, 1998, and Rizzolatti, Fogassi, & Gallese, 2001, for review of work in the monkey model). The area activated in the left premotor cortex is activated when subjects are asked to imagine grasping objects, but not to actually do so (Decety et al., 1994) (see Figure 3).

However, there are also functional neuroimaging data that are inconsistent with an interpretation of such premotor activation as a necessary step in object recognition. Johnson-Frey, Maloof, Newman-Norlund, Farrer, Inati, and Grafton (2003) found greater activation in inferior frontal regions (precentral and inferior frontal gyri, bilaterally) for photographs of a

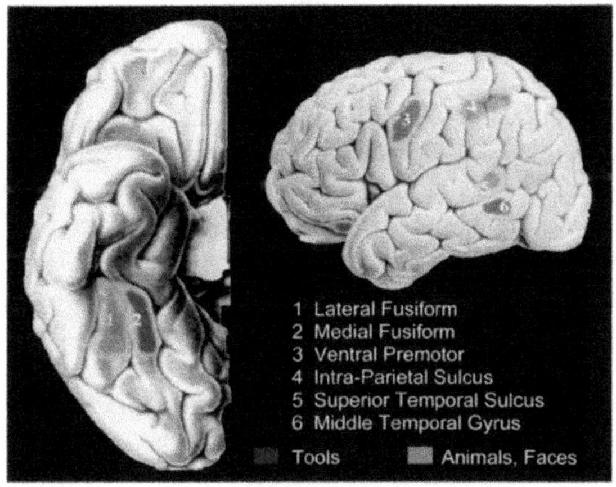

Figure 3. *This schematic of activation patterns by semantic category is based on a recent review and discussion of the functional neuroimaging literature by Martin and Chao (2001). Figure provided by Alex Martin.*

hand grasping an object compared to photographs of a hand touching the same objects. This activation remained when the objects were nontools (i.e., novel shapes) and when the hand was grasping the object in a way that would not serve the function of the object (see also recent work by Kellenbach, Brett, & Patterson, 2003; Phillips, Noppeney, Humphreys, & Price, 2002).

The decisive issue is what happens to the ability to recognise tools when the ability to use them is impaired. There are now a number of reports of patients with impairments for using objects, but spared recognition/identification (Buxbaum & Saffran, 2002; Buxbaum, Sirigu, Schwartz, & Klatzky, 2003; Buxbaum, Veramonti, & Schwartz, 2000; Cubelli et al., 2000; Hodges, Spatt, & Patterson, 1999; Montomura & Yamadori, 1994; Moreaud, Charnallet, & Pellat, 1998; Ochipa, Rothi, & Heilman, 1989; Rosci, Valentina, Laiacona, & Capitani, 2003; Rumiati, Zanini, Vorano, & Shallice, 2001; for review, see Johnson-Frey, in press; for discussion, see Dumont, Ska, & Joanette, 2000; Hodges, Bozeat, Lambon Ralph, Patterson, & Spatt, 2000; Mahon & Caramazza, 2003). For instance, the patient reported by Ochipa and colleagues (1989) was 17/20 for naming real objects, but could use only 2 of the 20 objects correctly. The performance of patients such as that reported by Ochipa and colleagues falsifies the claim that conceptual knowledge of manipulable objects is grounded in modality-specific output representations required to use them.

We are compelled to assume an architecture in which modality-specific input/output systems are functionally isolable from conceptual knowledge. One implication of the conclusion that modality-specific input/output representations do not exhaust the content of concepts is that simulations over such representations will also not be sufficient to ground conceptual content. If simulations are not sufficient to ground the conceptual content of object concepts held in the first person, then they are not going to be sufficient to attribute intentional content to other individuals. The argument has not been, nor does it imply, that "simulations" do not exist.

What, then, is the role of simulation in a theory of the organisation and representation of conceptual knowledge? One possibility is that the observed activation of (supposed) modality-specific input/output representations, and which is argued to support the Simulationist Framework, is completely epiphenomenal to conceptual processing. For instance, it could be that biological motion recognition and production (e.g., motor movements associated with language production, facial affect, bodily movements, manipulation of objects) are related to Theory of Mind in a way analogous to how phonological/orthographical processes are related to lexical meaning in language: They are the usual way to get from a physical stimulus to meaning, and from meaning to a physical stimulus, but they do not *ground* meaning or *contain* meaning. This is not to say that the development of Theory of Mind abilities (cf. lexical meaning) does not depend on modality-specific input; rather its acquisition doesn't depend on a *specific* modality, and its subsequent representation, once acquired, is not exhausted by the modality through which it was acquired. On this account, the activation of modality-specific output information (e.g., activation in left premotor cortex while naming manipulable objects) would be analogous to observations of phonological activation of unproduced words (e.g., Costa, Caramazza, & Sabastián-Galles, 2000; Peterson & Savoy, 1998).

A second possibility is that while simulations over modality-specific input/output representations are not sufficient to ground conceptual content, such "simulations" may contribute in important ways to the "full" meaning of object concepts. In other words, while one's concept of HAMMER is not represented in terms of information required to use hammers, it might be that information required to use hammers nevertheless adds in important ways to our understanding of hammers.

CONCLUSIONS

Beginning with the seminal work of Lhermitte and Beauvois in optic aphasia and Warrington, McCarthy, and Shallice in category-specific semantic deficits, issues of the organisation and representation of conceptual knowledge in the human brain have become central foci of research and discussion in cognitive science. The development of theoretical positions over the last several decades has been greatly influenced by the hypothesis that the semantic system is functionally organised by modality or type of information. There are, at present, many extant models of the organisation and representation of conceptual knowledge in the brain. Different proposals appeal to different principles of organisation, and one way to organise the extant space of hypotheses is to acknowledge where the various proposals fit within a common hierarchy of questions (for discussion, see Caramazza & Mahon, 2003). It has been the structure of this article to outline a framework that can be characterised at three levels of analysis. At the broadest level is the issue of whether conceptual and perceptual processes are organised by object domain. We have argued that the first-order constraint on the organisation of such processes are the domains living animate, living inanimate, conspecifics, and possibly tools. At the second level is the issue of whether information is organised within domains by modality or type of information. We have argued that there is some evidence for assuming that modality-specific input systems are organised by object domain. It remains an open (and independent) question whether a secondary organisational principle of conceptual knowledge of objects is the modality of content about which the information is concerned. At the most fine-grained level in this hierarchy of questions is the issue of the organisation of information within a given domain-(modality-) specific system. Hypotheses developed on the basis of the correlated structure principle should prove useful for proposing answers to this issue.

Issues of conceptual content can also be examined by looking at the tiers of processing internal to a given "vertical channel." For instance, internal to the domain "conspecifics" there will be, by hypothesis, distinct systems dedicated to analysis of visual form, visual motion, and conceptual knowledge, as well the attribution of intentional content.

The big questions about the organisation and representation of conceptual knowledge in the brain will have to be approached, simultaneously, from a number of different methodological perspectives. The goal of such work is to seek convergence, both empirically and theoretically, across different perspectives. Such convergence will be attained when extant theories take into account the relative strengths and weaknesses of the various methodological approaches.

PrEview proof published online 2 September 2005

REFERENCES

Adolphs, R. (2003). Cognitive neuroscience of human social behaviour. *Nature Reviews Neuroscience, 4*, 165–178.

Allison, T., Puce, A., & McCarthy, G. (2000). Social perception from visual cues: Role of the STS region. *Trends in Cognitive Science, 4*, 267–278.

Allport, D. A. (1985). Distributed memory, modular subsystems and dysphasia. In S. K. Newman & R. Epstein (Eds.), *Current perspectives in dysphasia*. New York: Churchill Livingstone.

Barbarotto, R., Capitani, E., & Laiacona, M. (1996). Naming deficit in herpes simplex encephalitis. *Acta Neurologica Scandinavica, 93*, 272–280.

Barbarotto, R., Capitani, E., Spinnler, H., & Trivelli, C. (1995). Slowly progressive semantic impairment with category specificity. *Neurocase, 1*, 107–119.

Barsalou, L. W., Simmons, W. K., Barbey, A. K., & Wilson, C. D. (2003). Grounding conceptual knowledge in the modality-specific systems. *Trends in Cognitive Sciences, 7*, 84–91.

Beauchamp, M. S., Lee, K. E., Haxby, J. V., & Martin, A. (2002). Parallel visual motion processing streams for manipulable objects and human movements. *Neuron, 34*, 149–159.

Beauchamp, M. S., Lee, K. E., Haxby, J. V., & Martin, A. (2003). FMRI responses to video and point-light displays of moving humans and manipulable objects. *Journal of Cognitive Neuroscience, 15*, 991–1001.

Beauvois, M.-F. (1982). Optic aphasia: A process of interaction between vision and language. *Proceedings of the Royal Society (London), B298*, 35–47.

Bertenthal, B. I., Proffitt, D. R., & Kramer, S. J. (1987). Perception of biomechanical motions by infants: Implementation of various processing constraints. *Journal of Experimental Psychology: Human Perception and Performance, 13*, 577–585.

Blakemore, S. J., & Decety, J. (2001). From the perception of action to the understanding of intention. *Nature Reviews Neuroscience, 2*, 561–567.

Bookheimer, S. (2002). Functional MRI of language: New approaches to understanding the cortical organization of semantic processing. *Annual Review of Neuroscience, 25*, 151–188.

Buxbaum, L. J., & Saffran, E. M. (2002). Knowledge of object manipulation and object function: Dissociations in apraxic and non-apraxic subjects. *Brain and Language, 82*, 179–199.

Buxbaum, L. J., Sirigu, A., Schwartz, M. F., & Klatzky, R. (2003). Cognitive representations of hand posture in ideomotor apraxia. *Neuropsychologia, 41*, 1091–1113.

Buxbaum, L. J., Veramonti, T., & Schwartz, M. F. (2000). Function and manipulation tool knowledge in apraxia: Knowing "what for" but not "how." *Neurocase, 6*, 83–97.

Brugger, P., Kollias, S. S., Muri, R. M., Crelier, G., Hepp-Reymond, M., & Regard, M. (2000). Beyond re-membering: Phantom sensations of congenitally absent limbs. *Proceedings of the National Academy of Sciences, 97*, 6167–6172.

Calder, A. J., Keane, J., Cole, J., Campbell, R., & Young, A. W. (2000). Facial expression recognition in people with Mobius syndrome. *Cognitive Neuropsychology, 17*, 73–87.

Campbell, R., & Manning, L. (1996). Optic aphasia: A case with spared action naming and associated disorder. *Brain and Language, 53*, 183–221.

Capitani, E., Laiacona, M., & Barbarotto, R. (1993). Dissociazioni semantiche intercategoriali. Parte II: procedura automatica di analisi di una batteria standardizzata. *Archivio di Psicologia, Neurologia e Psichiatria, 54*, 457–476.

Capitani, E., Laiacona, M., Mahon, B., & Caramazza, A. (2003). What are the facts of category-specific deficits? A critical review of the clinical evidence. *Cognitive Neuropsychology, 20*, 213–262.

Caramazza, A., Hillis, A. E., Rapp, B. C., & Romani, C. (1990). The multiple semantics hypothesis: Multiple confusions? *Cognitive Neuropsychology, 7*, 161–189.

Caramazza, A., & Mahon, B. Z. (2003). The organization of conceptual knowledge: The evidence from category-specific semantic deficits. *Trends in Cognitive Sciences, 7*, 325–374.

Caramazza, A., & Shelton, J. R. (1998). Domain specific knowledge systems in the brain: The animate–inanimate distinction. *Journal of Cognitive Neuroscience, 10*, 1–34.

Castelli, F., Frith, C., Happé, F., & Frith, U. (2002). Autism, Asperger syndrome and brain mechanisms for the attribution of mental states to animated shapes. *Brain, 125*, 1839–1849.

Castelli, F., Happé, F., Frith, U., & Frith., C. (2000). Movement and mind: A functional imaging study of perception and interpretation of complex intentional movement patterns. *NeuroImage, 12*, 314–325.

Chao, L. L., Haxby, J. V., & Martin, A. (1999a). Attribute-based neural substrates in posterior temporal cortex for perceiving and knowing about objects. *Nature Neuroscience, 2*, 913–919.

Chao, L. L., & Martin, A. (2000). Representation of manipulable man-made objects in the dorsal stream. *Neuroimage, 12*, 478–484.

Chao, L. L., Martin, A., & Haxby, J. V. (1999b). Are face-responsive regions selective only for faces? *Neuroreport, 10*, 2945–2950.

Chao, L. L., Weisberg, J., & Martin, A. (2002). Experience-dependent modulation of category-related cortical activity. *Cerebral Cortex, 12*, 1047–3211.

Collett, M., Collett, T. S., Bisch, S., & Wehner, R. (1998). Local and global vectors in desert ant navigation. *Nature, 394*, 269–272.

Coltheart, M., Inglis, L., Cupples, L., Michie, P., Bates, A., & Budd, B. (1998). A semantic subsystem of visual attributes. *Neurocase, 4*, 353–370.

Cook, M., & Mineka, S. (1989). Observational conditioning of fear to fear-relevant versus fear-irrelevant stimuli in rhesus monkeys. *Journal of Abnormal Psychology, 98*, 448–459.

Coslett, H. B., & Saffran, E. M. (1989). Preserved object recognition and reading comprehension in optic aphasia. *Brain, 112*, 1091–1110.

Coslett, H. B., & Saffran, E. M. (1992). Optic aphasia and the right hemisphere: A replication and extension. *Brain and Language, 43*, 143–161.

Costa, A., Caramazza, A., & Sebastián-Galles, N. (2000). The cognate facilitation effect: Implications for models of lexical access. *Journal of Experimental Psychology: Learning, Memory, and Cognition, 26*, 1283–1296.

Cree, G. S., & McRae, K. (2003). Analyzing the factors underlying the structure and computation of the meaning of chipmunk, cherry, chisel, cheese, and cello (and many other such concrete nouns). *Journal of Experimental Psychology: General, 132*(2), 163–201.

Crutch, S. J., & Warrington, E. K. (2003). The selective impairment of fruit and vegetable knowledge: A multiple processing channels account of fine-grain category specificity. *Cognitive Neuropsychology, 20*, 355–373.

Cubelli, R., Marchetti, C., Boscolo, G., & Della Sala, S. (2000). Cognition in action: Testing a model of limb apraxia. *Brain and Cognition, 44*, 144–165.

Decety, J., Perani, D., Jeannerod, M., Bettinardi, V., Tadary, B., Woods, R., Mazziotta, J. C., & Fazio, F. (1994). Mapping motor representations with positron emission tomography. *Nature, 371*, 600–602.

Devlin, J. T., Gonnerman, L. M., Anderson, E. S., & Seidenberg, M. S. (1998). Category-specific semantic deficits in focal and widespread brain damage: A computational account. *Journal Cognitive Neuroscience, 10*, 77–94.

Devlin, J. T., Moore, C. J., Mummery, C. J., Gorno-Tempini, M. L., Phillips, J. A., Noppeney, U., Frackowiak, R. S. J., Friston, K. J., & Price, C. J. (2002a). Anatomic constraints on cognitive theories of category-specificity. *Neuroimage, 15*, 675–685.

Devlin, J. T., Russell, R. P., Davis, M. H., Price, C. J., Moss, H. E., Fadili, M. J., & Tyler, L. K. (2002b). Is there an anatomical basis for category-specificity? Semantic memory studies in PET and fMRI. *Neuropsychologia, 40*, 54–75.

Dixon, M. J. (2000). A new paradigm for investigating category-specific agnosia in the new millennium. *Brain and Cognition, 42*, 142–145.

Dumont, C., Ska, B., & Joanette, Y. (2000). Conceptual apraxia and semantic memory deficit in Alzheimer's disease: Two sides of the same coin? *Journal of International Neuropsychological Society, 6*, 693–703.

Ellis, A. W., Young, A. W., & Critchley, E. M. R. (1989). Loss of memory for people following temporal lobe damage. *Brain, 112*, 1469–1483.

Emlen, S. T. (1967). Migratory orientation in the Indigo Bunting (Passerina cyanea). Part II: Mechanisms of celestial orientation. *Auk, 84*, 463–489.

Emlen, S. T. (1969). Bird migration: influence of physiological state upon celestial orientation. *Science, 165*, 716–718.

Farah, M. J., & McClelland, J. L. (1991). A computational model of semantic memory impairment: Modality specific and emergent category specificity. *Journal of Experimental Psychology General, 120*, 339–357.

Farah, M. J., & Rabinowitz, C. (2003). Genetic and environmental influences on the organization of semantic memory in the brain: Is "living things" an innate category? *Cognitive Neuropsychology, 20*, 401–408.

Farah, M. J., & Wallace, M. A. (1992). Semantically bounded anomia: Implication for the neural implementation of naming. *Neuropsychologia, 30*, 609–621.

Fine, C., Lumsden, J., & Blair, R. J. R. (2001). Dissociation between theory of mind and executive functions in a patient with early left amygdala damage. *Brain, 124*, 287–298.

Fox, R., & McDaniel, C. (1982). The perception of biological motion by human infants. *Science, 218*, 486–487.

Frith, C. D., & Frith, U. (1999). Interacting minds: A biological basis. *Science, 286*, 1692–1695.

Frith, U. (2001). Mind blindness and the brain in autism. *Neuron, 32*, 969–979.

Funnell, E., & Sheridan, J. (1992). Categories of knowledge? Unfamiliar aspects of living and nonliving things. *Cognitive Neuropsychology, 9*, 135–153.

Gaffan, D., & Heywood, C. A. (1993). A spurious category-specific visual agnosia for living things in normal humans and nonhuman primate. *Journal of Cognitive Neuroscience, 5*, 118–128.

Gaillard, M. J., Auzou, P., Miret, M., Ozsancak, C., & Hannequin, D. (1998). Trouble de la dénomination pour les objets manufacturés dans un cas d'encéphalite herpétique. *Révue Neurologique, 154*, 683–689.

Gale, T. M., Done, D. J., & Frank, R. J. (2001). Visual crowding and category-specific deficits for pictorial stimuli: A neural network model. *Cognitive Neuropsychology, 18*, 509–550.

Gallagher, H. L., & Frith, C. D. (2002). Functional imaging of theory of mind. *Trends in Cognitive Sciences, 7*, 77–83.

Gallagher, H. L., Jack, A. I., Roepstorff, A., & Frith, C. D. (2002). Imaging the intentional stance. *NeuroImage, 16*, 814–821.

Gallese, V., & Goldman, A. (1998). Mirror neurons and the simulation theory of mind reading. *Trends in Cognitive Sciences, 2*, 493–501.

Garrard, P., Lambon Ralph, M. A., Hodges, J. R., & Patterson, K. (2001). Prototypicality, distinctiveness and intercorrelation: Analyses of semantic attributes of living and nonliving concepts. *Cognitive Neuropsychology, 18*, 125–174.

Garrard, P., Patterson, K., Watson, P. C., & Hodges, J. R. (1998).Category specific semantic loss in dementia of Alzheimer's type. Functional-anatomical correlations from cross sectional analyses. *Brain, 121*, 633–646.

Gerlach, C., Law, I., Gade, A., & Paulson, O. B. (2000). Categorization and category effects in normal object recognition: A PET study. *Neuropsychologia, 38*, 1693–1703.

Gerlach, C., Law, I., Gade, A., & Paulson, O. B. (2002). The role of action knowledge in the comprehension of artifacts—A pet study. *NeuroImage, 15*, 143–152.

Gonnerman, L., Andersen, E. S., Devlin, J. T., Kempler, D., & Seidenberg, M. (1997). Double dissociation of semantic categories in Alzheimer's disease. *Brain and Language, 57*, 254–279.

Grabowski, T. J., Damasio, H., & Damasio, A. R. (1998). Premotor and prefrontal correlates of category-related lexical retrieval. *Neuroimage, 7*, 232–243.

Gregory, C., Lough, S., Stone, V., Erzinclioglu, S., Martin, L., Baron-Cohen, S., & Hodges, J. R. (2002). Theory of mind in patients with fontal variant frontotemporal dementia and Alzheimer's disease: theoretical and practical implications. *Brain, 125*, 752–764.

Grèzes, J., & Decety, J. (2001). Functional anatomy of execution, mental simulation, observation, and verb generation of actions: A meta-analysis. *Human Brain Mapping*, *12*, 1–19.

Grossman, E. D., & Blake, R. (2002). Brain areas active during visual motion perception of biological motion. *Neuron*, *35*, 1167–1175.

Grossman, E., Donnelly, M., Price, R., Pickens, D., Morgan, V., Neighbor, G., & Blake, R. (2000). Brain areas involved in perception of biological motion. *Journal of Cognitive Neuroscience*, *12*, 711–720.

Hanley, J. R., Young, A. W., & Pearson, N. (1989). Defective recognition of familiar people. *Cognitive Neuropsychology*, *6*, 179–210.

Harris, D. M., & Kay, J. (1995). Selective impairment of the retrieval of people's names: A case of category specificity. *Cortex*, *31*, 575–582.

Hart, J., Berndt, R. S., & Caramazza, A. (1985). Category-specific naming deficit following cerebral infarction. *Nature*, *316*, 439–440.

Hart, J., & Gordon, B. (1992). Neural subsystems for object knowledge. *Nature*, *359*, 60–64.

Haslam, C., Kay, J., & Hanley, J. R. (2002). Selective loss and preservation of biographical knowledge: Implications for representation. *Neurocase*, *8*, 169–193.

Hauser, M. D. (1997). Artifactual kinds and functional design features: What a primate understands without language. *Cognition*, *64*, 285–308.

Haxby, J. V., Ungerleider, L. G., Clark, V. P., Schouten, J. L., Hoffman, E. A., & Martin, A. (1999). The effect of face inversion on activity in human neural systems for face and object perception. *Neuron*, *22*, 189–199.

Heider, F., & Simmel, M. (1944). An experimental study of apparent behavior. *American Journal of Psychology*, *57*, 243–249.

Hillis, A. E., & Caramazza, A. (1991). Category-specific naming and comprehension impairment: A double dissociation. *Brain*, *114*, 2081–2094.

Hillis, A. E., & Caramazza, A. (1995). Cognitive and neural mechanisms underlying visual and semantic processing: Implications from "optic aphasia." *Journal of Cognitive Neuroscience*, *7*, 457–478.

Hillis, A. E., Rapp, B., & Caramazza, A. (1995). Constraining claims about theories of semantic memory: More on unitary versus multiple semantics. *Cognitive Neuropsychology*, *12*, 175–186.

Hodges, J. R., Bozeat, S., Lambon Ralph, M. A., Patterson, K., & Spatt, J. (2000). The role of conceptual knowledge in object use. Evidence from semantic dementia. *Brain*, *123*, 1913–1925.

Hodges, J. R., Spatt, J., & Patterson, K. (1999). "What" and "how": Evidence for the dissociation of object knowledge and mechanical problem-solving skills in the human brain. *Proceedings of the National Academy of Sciences, USA*, *96*, 9444–9448.

Hoffman, E. A., & Haxby, J. V. (2000). Distinct representations of eye gaze and identity in the distributed human neural system for face perception. *Nature Neuroscience*, *3*, 80–84.

Howard, D., Best, W., Bruce, C., & Gatehouse, C. (1995). Operativity and animacy effects in aphasic naming. *European Journal of Disorders of Human Communication*, *30*, 286–302.

Humphreys, G. W., & Forde, E. M. (2001). Hierarchies, similarity, and interactivity in object recognition: "Category-specific" neuropsychological deficits. *Behavioral and Brain Sciences*, *24*, 453–509.

Ishai, A., Ungerleider, L. G., Martin, A., Schouten, J. L., & Haxby, J. V. (1999). Distributed representation of objects in the human ventral visual pathway. *Proceedings of the National Academy of Sciences, USA*, *96*, 9379–9384.

Jacob, P., & Jeannerod, M. (2005). The motor theory of social cognition: A critique. *Trends in Cognitive Sciences*, *9*, 21–25.

Johnson-Frey, S. H. (2004). The neural basis of complex tool use in humans. *Trends in Cognitive Sciences*, *8*, 71–78.

Johnson-Frey, S. H., Maloof, F. R., Newman-Norlund, R., Farrer, C., Inati, S., & Grafton, S. T. (2003). Actions or hand object interactions: Human inferior frontal cortex and action observation. *Neuron*, *39*, 1053.

Kanwisher, N. (2000). Domain specificity in face perception. *Nature*, *3*, 759–763.

Kanwisher, N., McDermott, J., & Chun, M. (1997). The fusiform face area: A module in human extrastriate cortex specialized for face perception. *Journal of Neuroscience*, *17*, 4302–4311.

Kanwisher, N., Stanley, D., & Harris, A. (1999). The fusiform face area is selective for faces not animals. *NeuroReport*, *10*, 183–187.

Kay, J., & Hanley, J. R. (1999). Person-specific knowledge and knowledge of biological categories. *Cognitive Neuropsychology*, *16*, 171–180.

Kay, J., & Hanley, J. R. (2002). Preservation of memory for people in semantic memory disorder: Further category-specific semantic dissociation. *Cognitive Neuropsychology*, *19*, 113–133.

Kellenbach, M. L., Brett, M., & Patterson, K. (2003). Actions speak louder than functions: The importance of manipulability and action in tool representation. *Journal of Cognitive Neuroscience, 15*, 30–46.

Kourtzi, Z., & Kanwisher, N. (2000). Activation in human MT/MST by static images with implied motion. *Journal of Cognitive Neuroscience, 12*, 48–55.

Kraut, M. A., Moo, L. R., Segal, J. B., & Hart, J. Jr. (2002). Neural activation during an explicit categorization task: Category- or feature-specific effects? *Cognitive Brain Research, 13*, 213–220.

Kriegstein, K. V., Eger, E., Kleinschmidt, A., & Giraud, A. L. (2003). Modulation of neural responses to speech by directing attention to voices or verbal content. *Cognitive Brain Research, 17*, 48–55.

Laiacona, M., Barbarotto, R., & Capitani, E. (1993). Perceptual and associative knowledge in category specific impairment of semantic memory: A study of two cases. *Cortex, 29*, 727–740.

Laiacona, M., & Capitani, E. (2001). A case of prevailing deficit for non-living categories or a case of prevailing sparing of living categories? *Cognitive Neuropsychology, 18*, 39–70.

Laiacona, M., Capitani, E., & Barbarotto, R. (1997). Semantic category dissociations: A longitudinal study of two cases. *Cortex, 33*, 441–461.

Laine, M., Rinne, J. O., Hiltunen, J., Kaasinen, V., & Sipila, H. (2002). Different brain activation patterns during production of animals versus artifacts: A PET activation study on category-specific processing. *Cognitive Brain Research, 13*, 95–99.

Lambon Ralph, M. A., Howard, D., Nightingale, G., & Ellis, A. W. (1998). Are living and non-living category-specific deficits causally linked to impaired perceptual or associative knowledge? Evidence from a category-specific double dissociation. *Neurocase, 4*, 311–338.

Laws, K. R., Gale, T. M., Frank, R., & Davey, N. (2002). Visual similarity is greater for line drawings of nonliving than living things: The importance of musical instruments and body parts. *Brain and Cognition, 48*, 421–423.

Levy, I., Hasson, U., Avidan, G., Hendler, T., & Malach, R. (2001). Center-periphery organization of human object areas. *Nature Neuroscience, 4*, 533–539.

Lhermitte, F., & Beauvois, M.-F. (1973). A visual speech disconnection syndrome: Report of a case with optic aphasia, agnosic alexia and color agnosia. *Brain, 96*, 695–714.

Liberman, A. M., Cooper, F. S., Shankweiler, D. P., & Studdert-Kennedy, M. (1967). Perception of the speech code. *Psychological Review, 74*, 431–461.

Lough, S., Gregory, C., & Hodges, J. R. (2001). Dissociation of social cognition and executive function in frontal variant frontotemporal dementia. *Neurocase, 7*, 123–130.

Lucchelli, F., Muggia, S., & Spinnler, H. (1997). Selective proper name anomia: A case involving only contemporary celebrities. *Cognitive Neuropsychology, 14*, 881–900.

Lundqvist, L. O., & Dimberg, U. (1995). Facial expressions are contagious. *Journal of Psychophysiology, 9*, 203–211.

Luzzatti, C., & Davidoff, J. (1994). Impaired retrieval of object-color knowledge with preserved color naming. *Neuropsychologia, 32*, 1–18.

Mahon, B., & Caramazza, A. (2001). The Sensory/Functional Assumption or the data: Which do we keep? *Behavioral and Brain Sciences, 24*, 488–489.

Mahon, B. Z., & Caramazza, A. (2003). Constraining questions about the organization and representation of conceptual knowledge. *Cognitive Neuropsychology, 20*, 433–450.

Mahon, B. Z., & Caramazza, A. (2006). *Features, modalities, and domains as explanations of category-specific semantic deficits: The last shall be the first.* Manuscript submitted for publication.

Martin, A., & Chao, L. L. (2001). Semantic memory and the brain: Structure and processes. *Current Opinion in Neurobiology, 11*, 194–201.

Martin, A., Ungerleider, L. G., & Haxby, J. V. (2000). Category specificity and the brain: The Sensory/Motor Model of semantic representations of objects. In M. S. Gazzaniga (Ed.), *The new cognitive neurosciences.* Cambridge, MA: MIT Press.

Martin, A., & Weisberg, J. (2003). Neural foundations for understanding social and mechanical concepts. *Cognitive Neuropsychology, 20*, 575–587.

Martin, A., Wiggs, C. L., Ungerleider, L. G., & Haxby, J. V. (1996). Neural correlates of category-specific knowledge. *Nature, 379*, 649–652.

McCabe, K., Houser, D., Ryan, L., Smith, V., & Trouard, T. (2001). A functional imaging study of cooperation in two-person reciprocal exchange. *Proceedings of the National Academy of Science, 98*, 11832–11835.

McCarthy, C., Puce, A., Gore, J. C., & Allison, T. (1997). Face-specific processing in the human fusiform gyrus. *Journal of Cognitive Neuroscience, 9*, 605–610.

McRae, K., & Cree, G. S. (2002). Factors underlying category-specific semantic impairments. In E. M. E. Forde & G. W. Humphreys (Eds.), *Category-specificity in the brain and mind* (pp. 211–248). New York: Psychology Press.

Meltzoff, A. N. (1995). Understanding the intentions of others: Re-enactment of intended acts by 18-month-old children. *Developmental Psychology, 31*, 838–850.

Meltzoff, A. N., & Moore, M. K. (1977). Imitation of facial and manual gestures by human neonates. *Science, 198*, 75–78.

Miceli, G., Capasso, R., Daniele, A., Esposito, T., Magarelli, M., & Tomaiuolo, F. (2000). Selective deficit for people's names following left temporal damage: An impairment of domain-specific conceptual knowledge. *Cognitive Neuropsychology, 17*, 489–516.

Miceli, G., Fouch, E., Capasso, R., Shelton, J. R., Tamaiuolo, F., & Caramazza, A. (2001). The dissociation of color from form and function knowledge. *Nature Neuroscience, 4*, 662–667.

Mineka, S., Davidson, M., Cook, M., & Keir, R. (1984). Observational conditioning of snake fear in rhesus monkeys. *Journal of Abnormal Psychology, 93*, 355–372.

Mitchell, J. P., Heatherton, T. F., & Macrae, C. N. (2002). Distinct neural systems subserve person and object knowledge. *Proceedings of the National Academy of Science, 99*, 15238–15243.

Montomura, N., & Yamadori, A. (1994). A case of ideational apraxia with impairment of object use and preservation of object pantomime. *Cortex, 30*, 167–170.

Moore, C. J., & Price, C. J. (1999). A functional neuro-imaging study of the variables that generate category-specific object processing differences. *Brain, 122*, 943–962.

Moreaud, O., Charnallet, A., & Pellat, J. (1998). Identification without manipulation: A study of the relations between object use and semantic memory. *Neuropsychologia, 36*, 1295–1301.

Moscovitch, M., Winocur, G., & Behrmann, M. (1997). What is special about face recognition? Nineteen experiments on a person with visual object agnosia and dyslexia but with normal face recognition. *Journal of Cognitive Neuroscience, 9*, 555–604.

Moss, H. E., & Tyler, L. K. (2003). Weighing up the facts of category-specific semantic deficits. *Trends in Cognitive Sciences, 7*, 480–481.

Mummery, C. J., Patterson, K., Hodges, J. R., & Price, C. J. (1998). Functional neuroanatomy of the semantic system: Divisible by what? *Journal of Cognitive Neuroscience, 10*, 766–777.

Newcombe, F., Mehta, Z., & De Haan, E. H. F. (1994). Category specificity in visual recognition. In M. Farah & G. Ratcliff (Eds.), *The neuropsychology of high-level vision* (pp. 103–132). Hillsdale, NJ: Lawrence Erlbaum Associates Inc.

Ochipa, C., Rothi, L. J. G., & Heilman, K. M. (1989). Ideational apraxia: A deficit in tool selection and use. *Annals of Neurology, 25*, 190–193.

Pavlova, M., Staudt, M., Sokolov, A., Birbaumer, N., & Krageloh-Mann, I. (2003). Perception and production of biological movement in patient with early periventricular brain lesions. *Brain, 126*, 692–701.

Perani, D., Cappa, S. F., Bettinardi, V., Bressi, S., Gorno-Tempini, M. L., Matarrese, M., & Fazio, F. (1995). Different neural systems for the recognition of animals and man-made tools. *Neuroreport, 6*, 1637–1641.

Peterson, R. R., & Savoy, P. (1998). Lexical selection and phonological encoding during language production: Evidence for cascaded processing. *Journal of Experimental Psychology: Learning, Memory, and Cognition, 2*, 539–557.

Phillips, J. A., Noppeney, U., Humphreys, G. W., & Price, C. J. (2002). Can segregation within the semantic system account for category-specific effects? *Brain, 125*, 2067–2080.

Plaut, D. C. (2002). Graded modality-specific specialization in semantics: A computational account of optic aphasia. *Cognitive Neuropsychology, 19*, 603–639.

Premack, D., & Woodruff, G. (1978). Does the chimpanzee have a theory of mind? *Behavior and Brain Sciences, 4*, 515–526.

Riddoch, M. J. (1999). Optic aphasia: A review of some classic cases. In G. W. Humphreys (Ed.), *Case studies in the neuropsychology of vision* (pp. 133–160). Hove, UK: Psychology Press.

Riddoch, M. J., & Humphreys, G. W. (1987). Visual object processing in optic aphasia: A case of semantic access agnosia. *Cognitive Neuropsychology, 4*, 131–185.

Riddoch, M. J., Humphreys, G. W., Coltheart, M., & Funnell, E. (1988). Semantic systems or system? Neuropsychological evidence re-examined. *Cognitive Neuropsychology*, *5*, 3–25.

Rizzolatti, G., Fogassi, L., & Gallese, V. (2001). Neurophysiological mechanisms underlying the understanding and imitation of action. *Nature Review Neuroscience*, *2*, 661–670.

Rosci, C., Valentina, C., Laiacona, M., & Capitani, E. (2003). Apraxia is not associated to a disproportionate naming impairment for manipulable objects. *Brain and Cognition*, *53*, 412–415.

Rothi, L. J. G., Ochipa, C., & Heilman, K. M. (1991). A cognitive neuropsychological model of limb praxis. *Cognitive Neuropsychology*, *8*, 443–458.

Rumiati, R. I., Zanini, S., Vorano, L., & Shallice, T. (2001). A form of ideational apraxia as a selective deficit of contention scheduling. *Cognitive Neuropsychology*, *18*, 617–642.

Samson, D., & Pillon, A. (2003). A case of impaired knowledge for fruit and vegetables. *Cognitive Neuropsychology*, *20*, 373–401.

Samson, D., Pillon, A., & De Wilde, V. (1998). Impaired knowledge of visual and non-visual attributes in a patient with a semantic impairment for living entities: A case of a true category-specific deficit. *Neurocase*, *4*, 273–290.

Santos, L. R., & Caramazza, A. (2002). The Domain-Specific Hypothesis: A developmental and comparative perspective on category-specific deficits. In E. M. E. Forde & G. W. Humphreys (Eds.), *Category-specificity in the brain and mind.* New York: Psychology Press.

Sartori, G., & Lombardi, L. (2004). Semantic relevance and semantic disorders. *Journal of Cognitive Neuroscience*, *16*, 439–452.

Scholl, B. J., & Tremoulet, P. D. (2000). Perceptual causality and animacy. *Trends in Cognitive Sciences*, *4*, 299–308.

Senior, C., Barnes, J., Giampietro, V., Simmons, A., Bullmore, E. T., Brammer, M., & David, A. S. (2000). The functional neuroanatomy of implicit-motion perception or representational momentum. *Current Biology*, *10*, 16–22.

Servos, P., Osu, R., Santi, A., & Kawato, M. (2002). The neural substrates of biological motion perception: An fMRI study. *Cerebral Cortex*, *12*, 772–782.

Shallice, T. (1993). Multiple semantics: Whose confusions? *Cognitive Neuropsychology*, *10*, 251–261.

Shamay-Tsorry, S. G., Tomer, R., Berger, B. D., & Aharon-Peretz, J. (2003). Characterization of empathy deficits following prefrontal brain damage: The role of the right ventromedial prefrontal cortex. *Journal of Cognitive Neuroscience*, *15*, 324–337.

Shelton, J. R., Fouch, E., & Caramazza, A. (1998). The selective sparing of body part knowledge: A case study. *Neurocase*, *4*, 339–351.

Sheridan, J., & Humphreys, G. W. (1993). A verbal-semantic category-specific recognition impairment. *Cognitive Neuropsychology*, *10*, 143–184.

Shiffrar, M., & Freyd, J. J. (1993). Timing and apparent motion path choice with human body photographs. *Psychological Science*, *4*, 379–384.

Siegal, M., & Varley, R. (2002). Neural systems involved in "theory of mind." *Nature Review Neuroscience*, *3*, 463–471.

Snodgrass, J., & Vanderwart, M. (1980) A standardized set of 260 pictures: Norms for name agreement, familiarity, and visual complexity. *Journal of Experimental Psychology: Learning, Memory, and Cognition*, *6*, 174–215.

Stevens, J. A., Fonlupt, P., Shiffrar, M., & Decety, J. (2000). New aspects of motion perception: Selective neural encoding of apparent human movements. *NeuroReport*, *11*, 109–115.

Stewart, F., Parkin, A. J., & Hunkin, N. M. (1992). Naming impairments following recovery from herpes simplex encephalitis. *Quarterly Journal of Experimental Psychology*, *44A*, 261–284.

Tarr, M. J., & Gauthier, I. (2000). FFA: A flexible fusiform area for subordinate-level visual processing automatized by expertise. *Nature Neuroscience*, *3*, 764–769.

Thompson, S. A., Graham, K. S., Williams, G., Patterson, K., Kapur, N., & Hodges, J. R. (2004). Dissociating person-specific from general semantic knowledge: Roles of the left and right temporal lobes. *Neuropsychologia*, *42*, 359–370.

Thompson-Schill, S. L. (2003). Neuroimaging studies of semantic memory: Inferring "how" from "where". *Neuropsychologia*, *41*, 280–292.

Tranel, D., Logan, C. G., Frank, R. J., & Damasio, A. R. (1997). Explaining category-related effects in the retrieval of conceptual and lexical knowledge for concrete entities. *Neuropsychologia*, *35*, 1329–1339.

Tyler, L. K., & Moss, H. E. (2001). Towards a distributed account of conceptual knowledge. *Trends in Cognitive Science*, *5*, 244–252.

Tyler, L. K., Moss, H. E., Durrant-Peatfield, M. R., & Levy, J. P. (2000). Conceptual structure and the structure of concepts: A distributed account of category-specific deficits. *Brain and Language*, *75*, 195–231.

Vaina, L. M., Makris, N., Kennedy, D., & Cowey, A. (1998). The selective impairment of the perception of first-order motion by unilateral cortical brain damage. *Visual Neuroscience*, *15*, 333–348.

Varley, R., Siegal, M., & Want, S. C. (2001). Severe grammatical impairment does not preclude "theory of mind". *Neurocase*, *7*, 489–493.

Verstichel, P., Cohen, L., & Crochet, G. (1996). Associated production and comprehension deficits for people's names following left temporal lesion. *Neurocase*, *2*, 221–234.

Vinson, D. P., Vigliocco, V., Cappa, S., & Siri, S. (2003). The breakdown of semantic knowledge alone: Insights from a statistical model of meaning representation. *Brain and Language*, *86*(3), 347–441.

Warrington, E. K., & McCarthy, R. (1983). Category specific access dysphasia. *Brain*, *106*, 859–878.

Warrington, E. K., & McCarthy, R. (1987). Categories of knowledge: Further fractionations and an attempted integration. *Brain*, *110*, 1273–1296.

Warrington, E. K., & Shallice, T. (1984). Category-specific semantic impairment. *Brain*, *107*, 829–854.

Whatmough, C., Chertkow, H., Murtha, S., & Hanratty, K (2002). Dissociable brain regions process object meaning and object structure during picture naming. *Neuropsychologia*, *40*, 174–186.

Zannino, G. D., Perri, R., Carlesimo, G. A., Pasqualettin, P., & Caltagirone, C. (2002). Category-specific impairment in patients with Alzheimer's disease as a function of disease severity: A cross-sectional investigation. *Neuropsychologia*, *40*, 2268–2279.

COGNITIVE NEUROPSYCHOLOGY, 2006, 23 (1), 39–73

Speaking words: Contributions of cognitive neuropsychological research

Brenda Rapp
Johns Hopkins University, Baltimore, MD, USA
Matthew Goldrick
Northwestern University, Evanston, IL, USA

We review the significant cognitive neuropsychological contributions to our understanding of spoken word production that were made during the period of 1984 to 2004–since the founding of the journal *Cognitive Neuropsychology*. We then go on to identify and discuss a set of outstanding questions and challenges that face future cognitive neuropsychological researchers in this domain. We conclude that the last 20 years have been a testament to the vitality and productiveness of this approach in the domain of spoken word production and that it is essential that we continue to strive for the broader integration of cognitive neuropsychological evidence into cognitive science, psychology, linguistics, and neuroscience.

INTRODUCTION

The founding of *Cognitive Neuropsychology* in 1984 marked the recognition and "institutionalisation" of a set of ideas that had been crystallising for a number of years. These ideas formed the basis of the cognitive neuropsychological approach and, thus, have largely defined the journal over the past 20 years (Caramazza, 1984, 1986; Ellis, 1985, 1987; Marin, Saffran, & Schwartz, 1976; Marshall, 1986; Saffran, 1982; Schwartz, 1984; Shallice, 1979). Chief among them was an understanding of the fundamental limitations of syndromes or clinical categories as the vehicles for characterising patterns of impairment. This was complemented by the realisation that the appropriate and productive unit of analysis was the performance of the individual neurologically injured individual. Critical also was the more explicit formulation of the relationship between neuropsychology and cognitive psychology (Caramazza, 1986). The increasing application of theories of normal psychological processing to the analysis of deficits allowed neuropsychological evidence to provide significant constraints on theory development within cognitive psychology. This integration yielded the characterisation of cognitive neuropsychology as a branch of cognitive psychology.

These core ideas shaped the practice of neuropsychological research and the positive fruits of that research served, in turn, to confer greater legitimacy to and confirm these notions. One domain in which these ideas have been fruitfully applied is spoken word production. In this paper, we review the most significant cognitive

Correspondence should be addressed to Professor Brenda Rapp, Department of Cognitive Science, Johns Hopkins University, 135 Krieger Hall, Baltimore, MD 21218, USA (Email: rapp@cogsci.jhu.edu).
The first author gratefully acknowledges the support of NIH grant DC006740 for the writing of this paper.

DOI:10.1080/02643290542000049

neuropsychological findings in this domain from the period of 1984 to 2004. We then go on to discuss the research questions and challenges that we anticipate will be of interest in the next 20 years. We note that this review will be concerned solely with spoken naming of single words, and that we will exclude the related domains of sentence production and oral reading.

Spoken word production circa 1984

Ellis' (1985) review of the cognitive neuropsychological approach to spoken word production serves as an excellent snapshot of the state of cognitive neuropsychological research in spoken word production circa 1984. We will use this review as a starting point for identifying those areas in which significant progress has been made since 1984 in the cognitive neuropsychology of spoken word production.

As a backdrop to his review, Ellis used the framework depicted in Figure 1. This framework includes three major representational components: the conceptual semantic system, the speech output lexicon, and the phoneme level. This framework represents the general claim that in producing a spoken word we translate from a concept to a set of phonemes through the mediation of lexical forms. Interestingly, this general two-stage framework still underlies most current work in spoken word production. The first stage involves the selection of a lexical item to express the concept a person has in mind, and the second stage specifies the phonemes that correspond to the selected item. The objective of research on spoken word production has been to develop an increasingly more detailed understanding of the representations and processes referred to in Figure 1. We will start our review by identifying the principal issues discussed by Ellis. We have decided to group them into the following three categories: *basic architectural distinctions, the internal organisation of the speech lexicon*, and *activation dynamics* (see Table 1).

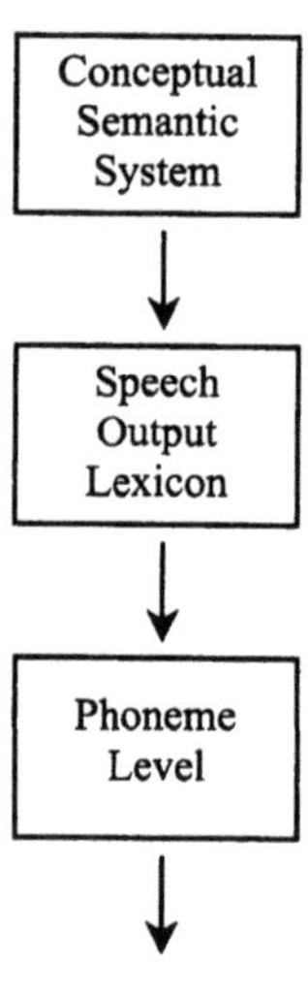

Figure 1. *Ellis (1985)'s framework for speech production.*

Questions regarding *basic architectural distinctions* concern the fundamental representational and processing distinctions that are encoded in the functional architecture. First, there is the question of whether a single store of lexical knowledge is used for word comprehension and production or if, instead, there are dual lexicons. A second question is whether the system distinguishes between representations of word meanings (lexical semantic representations) and semantic knowledge of the world, including the representation of meanings for which there may be no words (Allport, 1983; Saffran, 1982). A third question is if word meanings and word forms are represented independently, or if, instead, they are aspects of a single lexical representation. And, finally, a fourth issue concerns the content and organisation of the phonological representations and processes themselves, with particular emphasis on a possible distinction between representations/processes that are phonemic (central, abstract) versus phonetic (peripheral).

Ellis reviews two major topics in the category of *the organisation of the speech output lexicon.* First, there is the issue of whether the organisation of the speech lexicon (the long-term memory store of the sounds of familiar words) respects distinctions among grammatical categories (i.e., nouns, verbs, function words). Second, there is the

Table 1. *Open theoretical issues, circa 1984; progress on these and related issues, 1984–2004*

Major issues identified in Ellis (1985)	*Status 2004*
Basic architectural organisation:	**Basic architectural organisation:**
Are spoken comprehension and production served by a single lexicon?	Open question
Are world knowledge and lexical semantics distinct?	Open question
Are word meanings and word forms represented independently?	*The independent representation of word meaning and form*
How are phonemic and phonetic levels distinguished?	Open question
	The independent representation of word form and word syntax
Representation and processing in the speech output lexicon:	**Representation and processing in the speech output lexicon:**
Does the speech lexicon respect grammatical category distinctions?	*Grammatical category distinctions at the level of the phonological output lexicon*
Are morphologically complex words represented in a decomposed manner?	*Morphologically decomposed word forms*
	Lexical category distinctions at the level of the phonological output lexicon
Activation dynamics:	**Activation dynamics:**
What is the role of interactive activation in spoken word production?	
Specifically:	
• cascading activation	*Cascading activation from semantic-lexical-phoneme levels*
• feedback	*Feedback from phoneme to lexical levels*
• competitive inhibition	Open question

question of whether morphologically complex words are represented in a unitary (whole word) manner, or in a morphologically decomposed manner.

With regard to *activation dynamics*, Ellis (1985) discusses the possibility that various aspects of impaired word production might be understood if we make certain assumptions about the temporal characteristics of activation and information flow. In particular, in his account of form-based lexical errors and phonemic cueing, Ellis includes the notion of partial or weak activation (in contrast to all-or-none thresholded activation). He also entertains the possibility of cascading activation and feedback from the phoneme level to the speech output lexicon, as well as a mechanism of competitive inhibition among lexical representations.

While it is certainly the case that very significant cognitive neuropsychological work was carried out on all of these questions prior to 1984, the last 20 years have provided considerable advances and, in many cases, consensus regarding some of the earlier findings. Furthermore, although there are probably no findings that are uncontroversial in their interpretation, in this review we have identified findings for which there is considerable consensus regarding both their robustness and their contribution to our understanding of spoken word production. Finally there are, of course, a great number of exciting results that we will not discuss. This is in part due to space limitations, but also because our goal is not to carry out a comprehensive review of the literature but, instead, to focus on the most well-established findings from the cognitive neuropsychological literature on spoken word production.

PROGRESS: 1984–2004

Of the seven issues identified from Ellis (1985), we consider that significant progress has been made in understanding the following four: (1) the distinction between word meaning and word form, (2) grammatical category distinctions at the level of the phonological output lexicon, (3) the representation of morphologically complex words at the level of the phonological lexicon, and (4) questions of activation dynamics; the role of feedback, in particular. We consider that significant progress has also been made on two additional topics: (5) the distinction between lexical form and lexical syntax, and (6) the distinction among lexical categories at the level of phonological output lexicon (Table 1).

The basic architectural organisation

In the past 20 years, a basic focus of research interest has been to determine which of the many aspects of our word knowledge actually correspond to neurally differentiated distinctions that are respected during the course of lexical selection.

Word meaning/word form

Perhaps the most fundamental of lexical distinctions is the one between the meaning of a word and its phonological form. Psycholinguistic researchers have examined whether there are distinct lexical representations for a word's meaning and its form or whether these (and other) aspects of word knowledge are stored together under a single lexical entity (Forster, 1976; Levelt, 1989). Cognitive neuropsychological evidence has made a significant and unique contribution to answering this question.

The critical pattern of neuropsychological evidence indicating a representational and processing distinction between word meaning and word form is the following: *semantic errors in spoken naming in the face of intact word comprehension* and, additionally informative (although not obligatory) is *the absence of semantic errors in written naming.* This pattern is exemplified by the cases of RGB and HW reported by Caramazza and Hillis (1990; see also Basso, Taborelli, & Vignolo, 1978; Miceli, Benvegnú, Capasso, & Caramazza, 1997; Nickels, 1992; Rapp, Benzing, & Caramazza, 1997). For example, RGB orally named a picture of celery as "lettuce" but in written naming produced CELEY; similarly a picture of a finger was orally named as "ring" but spelled FINGER. As indicated in Table 2, RGB and HW were 100% correct in their comprehension of written and spoken words, yet they produced a large proportion (26–32%) of semantic errors in oral reading and naming. In contrast, in written naming neither of these individuals produced semantic errors.

This pattern can be understood within a functional architecture in which there is a distinction between word meaning (lexical semantics) and word form (phonological lexicon), if we assume that the neurological insult has affected the phonological lexicon or access to it. The reasoning is as follows. Errorless performance in written and spoken word comprehension tasks indicates that lexical semantics are intact. Furthermore, the fact that written spelling is free of semantic errors is additional and compelling evidence that word meaning has been adequately processed. Having established intact word comprehension, the spoken naming difficulties indicate a deficit in processing some aspect of the spoken forms. The fact that semantic errors (rather than sound-based errors) are produced allows us to reject, with some confidence, the possibility that the source of the spoken naming errors is a post-lexical impairment affecting speech production. This is because it is difficult to imagine a deficit affecting purely sound-based processing that would yield only semantic errors. In this way, the pattern clearly reveals the independence of word form and word meaning.

Additional evidence is the complementary dissociation—access to intact word forms in the face of severely impaired or absent lexical semantics. Specifically, there are cases of individuals who can read irregular words despite showing little or no evidence of understanding them (Bub, Cancelliere, & Kertesz, 1985; Cipolotti & Warrington, 1995; Coltheart, Masterson, Byng, Prior, & Riddoch, 1983; Coslett, 1991; Funnell, 1983; Hillis & Caramazza, 1991; Lambon Ralph, Ellis' & Franklin, 1995; Lambon Ralph, Ellis, & Sage, 1998; McCarthy & Warrington, 1986; Raymer & Berndt, 1996; Sartori, Masterson, & Job, 1987; Schwartz, Saffran, & Marin, 1980; Shallice, Warrington, & McCarthy, 1983; Wu, Martin, & Damian, 2002). In some cases, these individuals are also unable to correctly name the words from a picture or object stimulus (e.g., Hillis & Caramazza, 1991; Wu et al., 2002). The fact that the words are irregular makes it unlikely that they are read solely via knowledge of the systematic (or regular) relationships between graphemes and phonemes. It indicates that, instead, the word forms are recovered from the phonological lexicon either bypassing semantics or on the basis of incomplete semantic information (Hillis & Caramazza, 1995a). In either case, the striking difference observed between the paucity of lexical semantics and the integrity of lexical phonological information supports the conclusion of the independent representation of lexical semantics and lexical form.

Table 2. *RGB and HW's performance in oral and written naming*

		RGB	*HW*
Spoken	Correct	68%	65%
	Semantic[a]	32%	26%
	Omissions/Unrecognisable	0%	9%
Written	Correct	94%	91%
	Semantic[a]	0%	0%
	Omissions/Unrecognisable	6%	9%

[a]Includes definitions, morphological errors and nonwords recognisable as semantic errors; e.g., [skid](squid) for octopus.

Word form/word syntax

Another fundamental issue regarding lexical representation concerns the relationship between knowledge of word forms and word syntax (the grammatical properties of words). One question is whether word form and word syntax are independently represented. And, if they are, what is the processing relationship between these components of word knowledge in the course of lexical selection?

With regard to a possible distinction between word form and word syntax, the critical evidence has been the reports of individuals who display *intact knowledge of a word's grammatical properties despite being unable to recover the phonological form of the word.* A particularly clear example of this pattern is the case of Dante, reported by Badecker, Miozzo, and Zanuttini (1995; see also Henaff Gonon, Bruckert, & Michel, 1989; Miozzo & Caramazza, 1997; Shapiro & Caramazza, 2003a; Vigliocco, Vinson, Martin, & Garrett, 1999). In one experiment Dante was asked to produce 200 single spoken words in picture naming and sentence completion tasks. He was able to correctly name only 56% of these items. For each of the 88 items he was unable to name, he was asked (at the time at which he was unable to name the item) to make a number of forced-choice judgments designed to evaluate his access to the word's grammatical and phonological properties. Specifically he was asked to make forced-choice judgments about grammatical gender (masculine/feminine), word length, first letter, last letter and rhyming (e.g, does it rhyme with word X or word Y). As indicated in Table 3, Dante was 98% accurate with gender judgments but his performance was no different from chance on the judgments that concerned the form of the word. That is, Dante was able to access a word's syntax although he was unable to recover its phonology. His inability to access word phonology was indicated both by his inability to name the word, and his inability to make above-chance judgments regarding form features. Furthermore, the authors determined that the failure in making judgments regarding phonological form could not be attributed to lack of understanding of the tasks themselves as Dante was accurate in making these same phonological judgments for words that he could name.[1]

This pattern of performance clearly indicates that word syntax and word form are represented with sufficient neural independence that they can be selectively affected by neurological damage. This evidence of the independent representation of word form and word syntax quite naturally leads to the question of the processing relationship between the two. The current debate on this topic can be described as "the lemma dilemma."

Table 3. *Dante's accuracy on forced-choice questions on trials where he could not generate the target word (N = 88) (data combined from picture-naming and sentence completion tasks)*

	% correct
Grammatical gender	98
Word length	50
First letter	53
Last letter	47
Rhyming word	48

There are two major positions on the question. The position of Levelt and colleagues as well as others (Dell, 1986, 1990; Garrett, 1980; Kempen & Huijbers, 1983; Levelt, 1989, 1992; Levelt, Roelofs, & Meyer, 1999; Roelofs, 1992; Roelofs, Meyer, & Levelt, 1998) is that there is an independent level of lexical representations, referred to as lemmas, that represent or are linked to grammatical features. According to this position, lemmas are abstract, amodal representations that include or provide access to a word's grammatical features. Furthermore, and central to the claim, is the proposal that lemmas are the gatekeepers to a word's form and, as such, must be accessed prior to retrieval of the spoken (or written form) (Figure 2a). Within the cognitive neuropsychological literature the notion of abstract, lexical-grammatical representations is supported by evidence that certain individuals suffer from difficulties that are post-semantic yet pre-formal. For, example, there are the cases where a morphological deficit affects all input and output modalities in a very similar manner (Badecker, Rapp, & Caramazza, 1995). This can be explained by assuming that a (disrupted) morphological process operates over lexical representations that are shared across input and output, spoken and written modalities. The fact that these representations are shared across modalities indicates that they are abstract and amodal. Also thought to be

[1] Furthermore, the pattern reported in Table 3 was also observed for the subset of items for which grammatical gender cannot be predicted by the final segment of the word (i.e., nouns ending in/o/that are feminine and nouns ending in/a/that are masculine; nouns ending in/e/,/i/, and/u/that can be either masculine or feminine).

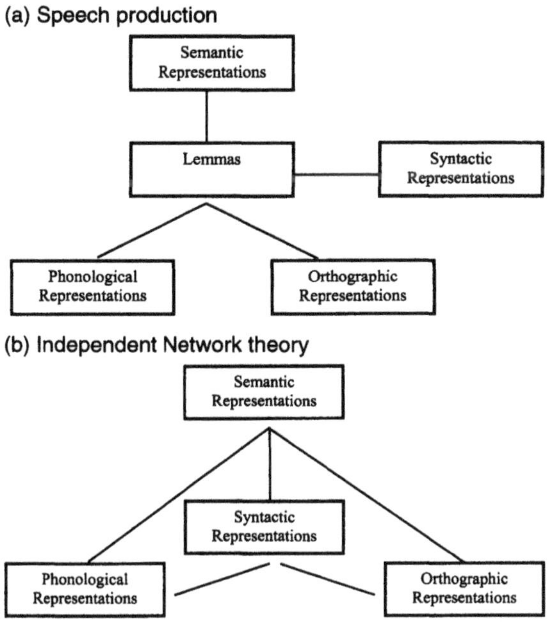

Figure 2. *Schematic of lemma theories of (a) speech production and (b) the Independent Network theory of Caramazza (1997b).*

supportive of the lemma proposal are cases in which morphological processing of both regularly (e.g., walked) and irregularly inflected forms (e.g., went) is affected (see below; see also Allen & Badecker, 1999, for evidence and arguments from the psycholinguistic literature). The rationale in these cases is that for regular and irregular forms to be similarly affected they must share a common and presumably abstract, amodal lexical representation.

However, even prior to 1984, there was scepticism regarding the notion of modality-neutral lexical representations (Allport & Funnell, 1981; Butterworth, 1983). This scepticism has continued and Caramazza (1997) and Caramazza and Miozzo (1997) have more recently claimed that an additional amodal lexical representational level is unnecessary. They have argued that the empirical facts can be understood without positing lemma representations. They propose, instead, that a word's grammatical features are linked to its form and that, in contrast to the lemma position, word syntax is accessible either from form or (depending on the type of grammatical feature) from semantics (Figure 2b).

There are two major disputed questions in this debate: First, whether or not there is an amodal, *lexical* level of representation that links to both word form and syntax. Second, whether word syntax must be accessed prior to word form. Although in the next section we discuss some additional evidence that is relevant to this debate, a full review of the arguments and relevant evidence is beyond the scope of this paper; instead, we refer the interested reader to additional papers (e.g., Caramazza, 1997; Caramazza & Miozzo, 1997, 1998; Levelt et al., 1999; Rapp & Caramazza, 2002; Roelofs et al., 1998).

Regardless of the eventual resolution of these questions, what is clear from the evidence is that word form and word syntax are independently represented. Thus, the findings we have reviewed concerning the basic organisation of the architecture indicate a fairly robust consensus that the word production system consists (at a minimum) of independent semantic, syntactic, phonological (and orthographic) components.

The organisation of the speech (phonological) output lexicon

In addition to the progress that has been made in understanding the independent components of the lexical system, there have also been significant advances specifically in understanding the organisation and representational content of the phonological output lexicon itself.

Grammatical category distinctions

A number of cases of naming difficulties that disproportionately affect one grammatical category (nouns, verbs, or function words) have been reported. These deficits have manifested themselves in both comprehension and production, or selectively in comprehension or production, and within production in both written and spoken naming or selectively in spoken or written naming (see Rapp & Caramazza, 2002, for a review). These patterns clearly indicate that grammatical category plays a role at some point in the word production process. However, the persistent challenges have been: (1) to determine if these selective deficits are truly grammatical rather than artifactual, and (2) if grammatical, to establish the level/s in spoken naming process at which grammatical category distinctions are represented.

With regard to the issue of the grammatical nature of the deficits, there have been a number of proposals that attribute the reported deficits to nongrammatical factors that are often correlated with grammatical category. It has been suggested that what may actually be relevant is some semantic variable such as abstractness/imageability (see Bird, Howard, & Franklin, 2000; Moss, Tyler, Durrant-Peatfield, & Bunn, 1998; but see Shapiro & Caramazza, 2001, for a critical commentary).

There have, however, been a number of lines of evidence that at the least, not all cases of apparent grammatical category deficits can be explained by semantic factors. Specifically, noun/verb dissociations have been documented even when factors such as abstractness have been controlled across grammatical categories (e.g., Berndt, Haendiges, Burton, & Mitchum, 2001, 2002). Additional evidence against a strictly semantic account are the reports of category-specific morphological deficits (Laiacona & Caramazza, 2004; Shapiro & Caramazza, 2001, 2003b; Shapiro, Shelton, & Caramazza, 2000; Tsapkini, Jarema, & Kehayia, 2001). For example, JC (Shapiro & Caramazza, 2001) had more difficulty producing the plural of nouns (*guide* → *guides*) than the third person of their verb homophones.[2] The fact that the grammatical category difficulty was specifically morphological makes a semantic account of the grammatical category dissociation unlikely.

With regard to the question of the level of processing at which grammatical category distinctions are represented, one possibility is that grammatical category is an organising feature at a central, amodal level of representation (such as the lemma level) that is shared in spoken and written output, and possibly also for comprehension and production. Such a level would most likely play a key role in sentence production and morphology. Another possibility is that grammatical category distinctions are modality-specific and represented at the level of phonological (and orthographic) form, either exclusively, or in addition to being represented at a central, amodal level.

One of the most compelling lines of evidence indicating that grammatical category organisation is both nonsemantic and active beyond a central, amodal level are the reports of grammatical category deficits that are modality specific. In these cases there is a selective deficit in producing words of one grammatical category and the deficit is restricted to either the spoken or written modality. Caramazza and Hillis (1991) reported two such cases: one exhibited selective difficulty in producing spoken verbs versus spoken nouns but had no particular difficulty with written verbs or nouns; the other case had difficulty producing written verbs versus nouns, with sparing of spoken verbs and nouns (for other cases of modality-specific noun/verb deficits, see also Baxter & Warrington, 1985; Berndt & Haendiges, 2000; Hillis & Caramazza, 1995b; Rapp & Caramazza, 1998).

In addition to cases such as these there are also cases of *single* individuals who exhibit a double dissociation of grammatical category by modality. A number of these have exhibited difficulty with the open class vocabulary in spoken production and the closed class vocabulary in written production (Assal, Buttet, & Jolivet, 1981; Bub & Kertesz, 1982; Coslett, Gonzales-Rothi, & Heilman, 1984; Lecours & Rouillon, 1976; Lhermitte & Derouesne, 1974; Patterson & Shewell, 1987; Rapp et al., 1997), prompting their characterisation as "oral Wernicke vs. written Broca" (Assal et al., 1981). One of the most striking dissociations of grammatical category

[2] Note, this was also the case even for nonwords—"this is a wug; these are ___" was less difficult than "these people wug, this person ___." The reverse pattern was exhibited by JR (Shapiro & Caramazza, 2003b).

by modality is that of KSR (Rapp & Caramazza, 2002; see also Hillis & Caramazza, 1995b) who exhibited a double dissociation of nouns/verbs by modality. As indicated in Table 4, in single word picture naming tasks, KSR had more difficulty producing spoken nouns than verbs and more difficulty producing written verbs than nouns. Examples of his responses when asked to say or write a sentence are shown in Figure 3, where it can be seen that, for example, in response to a picture of a girl pushing a wagon he writes "the girl is actions a wagon," but he says "The girl is holding the/b aI g/."

The pattern of modality-specific, grammatical category impairment is compelling because the integrity of the grammatical category in one modality indicates that the deficit cannot be an artifact of some semantic variable. Furthermore, the pattern also indicates that some grammatical category distinction must originate at a post-semantic, modality-specific level of processing. Typically these deficits are interpreted as revealing that the phonological and orthographic lexicons are organised in such a manner that neurological damage can selectively affect the retrieval of words from one grammatical category. This conclusion is not, of course, inconsistent with an architecture in which earlier levels of representation are *also* organised in a manner that respects grammatical category distinctions.

The relevance of these data to the lemma dilemma is that they are problematic for the view that grammatical category distinctions are present only at a modality-independent level of representation, as has been suggested by certain lemma-based accounts (Dell, 1990; Levelt, 1989; Levelt et al., 1999; Roelofs, 1992; Roelofs et al., 1998). However, they may be accommodated within a lemma-based account if

Table 4. *Distribution of KSR's responses on spoken and written naming of nouns and verbs*

	Spoken noun	*Spoken verb*	*Written noun*	*Written verb*
Correct	71	89	93	55
Other word	10	6	1	14
Nonword	17	2	3	4
Omission	1	0	1	7
Morphological	1	2	2	11
Other word + morphological	0	0	0	7

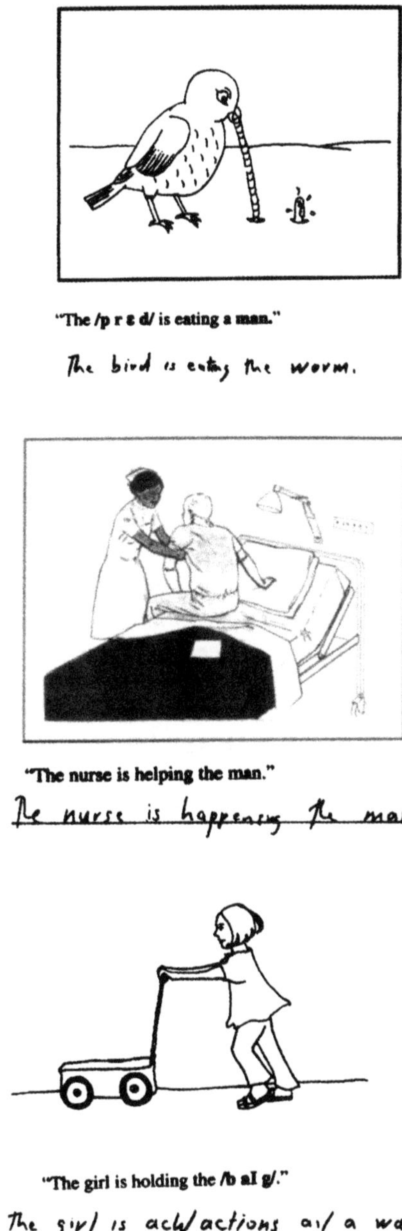

Figure 3. *KSR's written and spoken naming of target pictures.*

the modality-specific, grammatical category distinctions are represented in the connections between the lemma and form levels. Nonetheless, Rapp and Caramazza (2002) suggest that more detailed aspects of KSR's performance represent a challenge for the lemma-based accounts (see also Caramazza & Miozzo, 1998).

Lexical category distinctions

In addition to post-semantic grammatical category distinctions, there have been claims of additional post-semantic category-specific deficits that, presumably, reveal the organisation of the phonological lexicon. These have involved a number of categories, including: abstract/concrete words (Franklin, Howard, & Patterson, 1995), semantic categories (e.g., body parts, fruits and vegetables, colours, etc.; Beauvois, 1982; Dennis, 1976; Hart, Berndt, & Caramazza, 1985), letter names (Goodglass, Wingfield, Hide, & Theurkauf, 1986), number names (McCloskey, Sokol, & Goodman, 1986), and proper nouns (see below). Among these, perhaps the strongest case has been made for the proper/common noun distinction. At any rate, since the critical pattern of evidence is essentially the same regardless of the category, proper nouns will serve as a representative case. These lexical category distinctions are assumed to represent a further differentiation of the noun component of the phonological lexicon (see Figure 5 on p. 57).

The critical evidence takes the form of *selective difficulty in naming proper but not common nouns in the face of intact comprehension of proper nouns.* This pattern is exemplified by the case of PC reported by Semenza and Zettin (1988; see also Lucchelli & DeRenzi, 1992; McKenna & Warrington, 1978; Semenza & Zettin, 1989; Warrington & McCarthy, 1987). PC was 100% ($n = 303$) correct in his naming of pictures, real objects, and naming to definition of items from the categories of vegetables, fruits, body parts, colours, letters, transportation, pasta, furniture, and numbers, as well as adjectives and verbs. In contrast, his accuracy in naming proper names (people, cities, rivers, countries, mountains) in response to picture stimuli, maps, or definitions was extremely poor, with an accuracy of only 2% ($n = 119$). Also contrasting with his poor naming of proper nouns was the observation that his comprehension of the names and pictures was apparently intact (97% correct, $n = 119$). For example, in response to a picture of the then Italian prime minister, although PC was unable to name him, he correctly said: "he is the first socialist holding this position in our country."

The fact that comprehension is intact indicates that the naming deficit does not arise at the semantic level, revealing a differentiation between proper and common nouns either in the organisation of the phonological lexicon itself, or in the processes involved in accessing proper and common names from the phonological lexicon. One consistent concern with this interpretation has been the possibility that proper nouns are more vulnerable to damage than common nouns, not because they are independently represented, but simply because they are lower in frequency. However, the fact that PC, for example, was able to name very infrequent common nouns but no proper nouns (even frequent ones) renders such an account unlikely. The evidence that would most readily speak to this concern would be cases of selective sparing of proper nouns. Such cases have been reported (Cipolotti, 2000; Cipolotti, McNeil, & Warrington, 1993; McKenna & Warrington, 1978; Schmidt & Buchanan, 2004; Semenza & Sgaramella, 1993), although they all have been somewhat problematic as they have involved only extremely impaired individuals who usually could be tested only in the written modality (see Schmidt, Buchanan, & Semenza, 2003, for a review). Despite these limitations, although BWN (Schmidt et al., 2003) could only produce written responses, he was 100% correct with proper nouns but only 50% correct with common nouns. With common nouns, he produced either semantic errors (*clown* → *man*) or omissions, despite communicating that he knew their meaning.

If the proper/common noun dissociations indicate a representational distinction at the level of the phonological output lexicon, then we would expect (as in the case of post-semantic, grammatical category deficits) to observe modality-specific deficits affecting proper and common nouns. One such case was recently reported by Cipolotti (2000). This individual showed proper name superiority (just for country names) in the spoken modality (100% for country names vs. 30% for objects) but not the written modality (100% correct on both country names and objects). Such a pattern supports the differentiation of proper vs common nouns at the level of the phonological lexicon. As was the case for grammatical category organisation, this does not, however, preclude the differentiation of common and proper nouns at higher levels such as within the semantic system and, indeed, there have been cases exhibiting selective impairment of conceptual knowledge for proper names that support this (Lyons, Hanley, & Kay, 2002; Miceli, Capasso, Daniele, Esposito, Magarelli, & Tamaiuolo, 2000).

Given the quite robust evidence for a distinction between proper and common nouns at the level of the phonological lexicon one can, quite naturally, wonder what purpose it would serve. Thus, whereas the specification of grammatical category at the level of form may play a role in sentence production and productive morphological processes, the functional role of a proper/common noun distinction is less obvious. It has been suggested (Semenza & Zettin, 1989) that the distinction may have its origins in differences in the learning of the two categories of words. Specifically, Semenza and Zettin (1989) pointed out that unlike common nouns, proper nouns are referring expressions that are arbitrary in that they apply only to a specific referent and do not imply any particular set of semantic attributes. Some support for the relevance of this fact is that several individuals with selective difficulties with proper names also had difficulty in learning arbitrary paired associates (Hittmaier-Delazer, Denes, Semenza, & Mantovan, 1994; Lucchelli & De Renzi, 1992; Semenza & Zettin,1989; but see Saetti, Marangolo, DeRenzi, Rinalidi, & Lattanzi, 1999). Clearly, however, the underlying basis for lexical category distinctions at the level of the phonological lexicon requires further investigation.

Morphological decomposition

Are morphologically complex words stored in memory as whole word representations or in terms of their constituent morphemes? This single question has dominated psycholinguistic work on the mental lexicon and, fortunately, it is an issue regarding which cognitive neuropsychological evidence has been particularly informative.

There are a number of possible distinctions that can be considered and that add to the complexity of the question. First, there are the possible distinctions between levels of representation. The question of morphological composition certainly refers to the representation of morphologically complex words at the level of phonological form. However, if one assumes an abstract level of lexical representation such as the lemma, the question can also refer to this representational level as well. Another distinction is that between regular and irregular morphology. Compositionality at the level of form is not equally plausible for all morphologically complex words. In English, for example, although there is a highly regular compositional pattern that characterises the past tense of the vast majority of verbs (e.g., walk–walked), there are also the more idiosyncratic patterns of the so-called irregular verbs (e.g., tell–told; is–was; hit–hit), which render them less obvious candidates for morphological composition at the level of form. The nature of the distinction between regular/irregular morphological patterns has been the focus of particularly intense debate over the past 20 years (for recent reviews, see Marslen-Wilson & Tyler, 1998; McClelland & Patterson, 2002; Pinker & Ullman, 2002). Finally there is the distinction between inflectional and derivational morphology: What may be true of inflectional morphology need not be the case for derivational morphology. With regard to questions of composition/decomposition, these three distinctions are largely independent of one another. That is to say, evidence for decomposition for one category does not necessarily have implications regarding another. As a result, a rather diverse set of proposals has been put forward. Rather than attempting to review this very considerable body of work, we focus here on those patterns for which the evidence of compositionality is clearest, namely for *regularly inflected words at the level of phonological form.*

The general pattern that strongly supports the claim of decomposed phonological representations of inflected words is the following: *morphological errors in spoken production in the context of intact comprehension of morphological contrasts.* Intact comprehension assures that the morphological deficit in fact arises at the level of the phonological output lexicon and not at a more central level of morphological representation and processing. The third element of this pattern is *evidence ruling out nonmorphological (semantic or form-based) interpretations of the errors* (e.g., Pillon, de Partz, Raison, & Seron, 1991).

There have been a number of different performance configurations that generally fit this pattern and that have supported the notion of morphological decomposition at the level of lexical phonological form. Inflected neologisms constitute one such case. Certain individuals have been reported who produce neologisms for the stem of a word that is otherwise appropriately inflected (e.g., "he's really *knawling* over me" (Buckingham & Kertesz, 1976); "she /*wiksəz*/" (Butterworth & Howard, 1987); "tuto il *ternessico* che mi aspetta" Semenza, Butterworth, Panzeri, & Ferreri, 1990; see also Buckingham, 1981; Caplan, Keller, & Locke, 1972). Although these errors are extremely compelling, one of the difficulties has been in clearly establishing that the errors do not represent a phonological deformation of a whole-word form that diminishes towards the end of the word. Furthermore, it has often been difficult to evaluate comprehension of morphological contrasts in these cases (e.g., Semenza et al., 1990, reported individuals with significant comprehension impairments; similar comprehension impairments were found for 2/5 cases reviewed by Butterworth & Howard, 1987).

An especially compelling pattern of performance that has been informative with regard to the question of morphological decomposition at the level of phonological form is the production of morphologically illegal combinations of stem and affix (e.g., *blackness* → *blackage*). Such combinations are surely not stored in the lexicon and must, therefore, be the result of morphologically based compositional processes. FS (Miceli & Caramazza, 1988) produced errors of this type, for example, *resisteva* (he was resisting) was produced as *resistire* (correct stem with the infinitival form for verbs of the 3rd conjugation), as did cases reported in Semenza et al. (1990) (e.g., *fratellanza* [brotherhood] → *fratellismo*) (see also Laine, Niemi, Koivuselkä-Sallinen, & Hyönä, 1995). One case that clearly presents all of the elements of the critical pattern identified above is that of SJD, reported by Badecker and Caramazza (1991). In spontaneous speech and oral reading, SJD produced morphologically illegal errors such as *poorest* read as *poorless, youthful as youthly, discussing* as *discussionly.* Although SJD did produce some phonological errors, a phonological basis for the morphologically illegal errors was ruled out because SJD produced morphological errors only for inflected forms (e.g., links, teas) and not for homophonic unaffixed forms (e.g., lynx, tease) (see Table 5). Furthermore, a semantic or input locus for these errors was ruled out because many of the illegal morphological combinations were accompanied by clearly adequate definitions (e.g., *cloudless* → *cloudness, it means if*

Table 5. *Percentage of total responses in SJD's reading aloud of matched sets of affixed/unaffixed homophones (examples of each potential error type on each list are shown in parentheses)*

	Affixed homophones (e.g., TEAS)	*Unaffixed homophones (e.g., TEASE)*
Correct	50	85
Morphological error (bowled → bowl; lynx → link)	42	0
Phonological error (frays → prays; bread → breast)	8	15

the sun is clear, with no clouds at all). Finally, additional evidence of a form-based locus of impairment was that regularly inflected forms were affected (60% correct) while irregularly inflected were not (92%) and, in fact, these behaved similarly to uninflected forms (90%). This implicates a level of representation—such as phonological form—where regularly and irregularly inflected forms are most likely to be represented in a distinct manner.

The evidence of decomposed phonological forms implies that there are morphological processes that manipulate morpheme-sized phonological representations in composing inflected forms. Whether these morphological processes are themselves modality-specific or whether they are amodal and simply manipulate modality-specific morphemic representations is unclear from the available data. In either case, it would be predicted that there might be cases of modality-specific morphological deficits; that is, we should expect to find cases in which the patterns reported above are present in either the written or spoken modality with intact morphological composition in the other modality. There is some evidence that this may indeed be the case. Berndt and Haendiges (2000) described an individual with selective difficulties in producing written verbs who produced morphological errors in writing but never in speaking (see also the data in Table 4 above; Rapp & Caramazza, 2002).

It is important to be clear that the finding of form-level morphological decomposition is not at odds with, nor does it preclude, there being compositional morphological processes operating at a more abstract level. In fact there are a number of lines of evidence that indicate that this may indeed be the case (see Allen & Badecker, 2001, for a review of evidence from spoken production; see also Marslen-Wilson & Tyler, 1998, 2005, for reviews of research in comprehension and production). Badecker (1997; see also Badecker & Caramazza, 1987) reported the case of FM, who produced a large number of morphological errors and significantly higher error rates on both regularly (e.g., *asked*) and irregularly (e.g., *ate)* inflected verbs compared to uninflected verbs (e.g., *ask, eat*). The fact that both regular and irregular forms were similarly affected (in contrast to the pattern exhibited by SJD described above) suggests that the deficit was at a level at which both are similarly represented. This would seem to exclude the phonological level. Furthermore, Badecker (1997) argued that a simple semantic account of these errors is ruled out by asymmetries in FM's productions. In particular, he produced many errors where an inflected form was replaced by its corresponding base form (e.g., *asked* → *ask*), but few errors where the reverse occurred (e.g., *ask* → *asked*). If FM's errors were based purely on semantic similarity, there should be no such asymmetry; the semantic distance involved in both errors is identical. Instead his errors are apparently influenced by the compositional structure of inflected forms, whether regular or irregular. This points to an abstract level of representation where morphological processes deal with abstract morphosyntactic structures in a manner that is "blind" to differences in surface form (e.g., [talk] + past is handled similarly to [eat] + past). (For other lines of neuropsychological evidence that support a level of morphological representation that is form-independent, see Laine et al., 1995.) Along somewhat similar lines there is the evidence that morphological representations and processes may be shared across modalities. This includes individuals with deficits that affect the processing of both regularly and irregularly inflected forms relative to uninflected forms, in both comprehension and production, across written and spoken modalities (Badecker et al., 1995). It should be noted, however, that the neuropsychological evidence for a strictly abstract and/or amodal level of morphological representation is scarce and not without its limitations. Important in this regard is the fact that there have been no reports of individuals who make morphological errors who do not also have phonological deficits (Miceli, Capasso, & Caramazza, 2004), suggesting an especially close link between morphology and form.

In summary, with respect to regularly inflected forms, there is clear support for morphological decomposition at the level of the speech production lexicon. Other patterns of cognitive neuropsychological evidence suggest additional levels of morphological representation, although the neuropsychological evidence is more controversial on this point. The overall picture may be consistent with a distinction between decomposed lexical phonological representations on the one hand and morphological processes that deal with abstract morphosyntactic structures on the other. This type of distinction would seem to map naturally onto the lemma/lexeme (form) distinction that has been proposed; but, as we have indicated earlier, this conclusion has been vigorously contested (Caramazza, 1997). Clearly, the resolution of this set of intimately inter-related issues concerning the syntactic and morphological nature of lexical representation and processing will be one of the major challenges facing future cognitive neuropsychological research.

Activation dynamics

The spoken word production architecture developed to this point has been largely a static one, as there has been little discussion of the temporal attributes of processing. However, the issue of activation dynamics is clearly an important one in the context of spoken word production and in this section we focus on the progress that has been made in understanding the roles of feedback and cascading activation in spoken word production.

Feedback and cascading activation

The debate on interactivity in spoken word production has been dominated by two sets of positions—the discrete and the interactive. While there are a number of variants within these two sets of positions, we take the proposal of Levelt and colleagues (Levelt, Schriefers, Vorberg, Meyer, Pechmann, & Havinga, 1991a; Levelt et al., 1999) to be representative of the highly discrete view and that of Dell and colleagues (Dell, 1986; Dell, Schwartz, Martin, Saffran, & Gagnon, 1997) to be representative of the highly interactive view. Both positions assume the general two-stage framework depicted in Figure 1, with Stage 1 referred to as *lexical selection* and Stage 2 as *phonological encoding*.

According to the highly discrete position, processing proceeds in a strictly feedforward direction, with the selection of an item at each level (e.g., semantic, lexical and phonological) taking place before activation is passed on to the subsequent level (Levelt et al., 1991a). Within such an architecture (see Figure 4a), Stage 1 begins when semantic information regarding the target produces activation of the target and its semantically related competitors at the semantic and lexical levels. This stage of lexical selection ends when a single lexical unit is selected; competing lexical units are not allowed to pass on their activation to the phoneme level. Then, during Stage 2, only the phonemes for the selected lexical unit are activated and selected.[3]

According to an interactive position (see Figure 4b), Stage 1 begins (as in the discrete architecture) when semantic information regarding the target produces activation of the target and its semantically related competitors, and Stage 1 continues as all of the activated lexical units pass on activation to the phoneme level. Furthermore, activation throughout Stages 1 and 2 involves not only a forward flow of activation but also a backward flow between the phonological and lexical levels as well as between the lexical and semantic levels. Stage 1 ends with the selection of the most active lexical unit; however, within this framework, selection means only that the activation level of the selected unit is raised above that of its competitors; competitors are allowed to pass on their activation. During

[3] As discussed above, Levelt and colleagues' position with regard to lexical representation is that there are two levels of lexical representation-lemmas and lexemes, prior to the phoneme level. They assume that only a single selected lemma will activate its corresponding lexeme, and only this lexeme can pass on activation to the phoneme level. Despite this additional level/stage, it is not obvious that this changes any of the predictions we will discuss here.

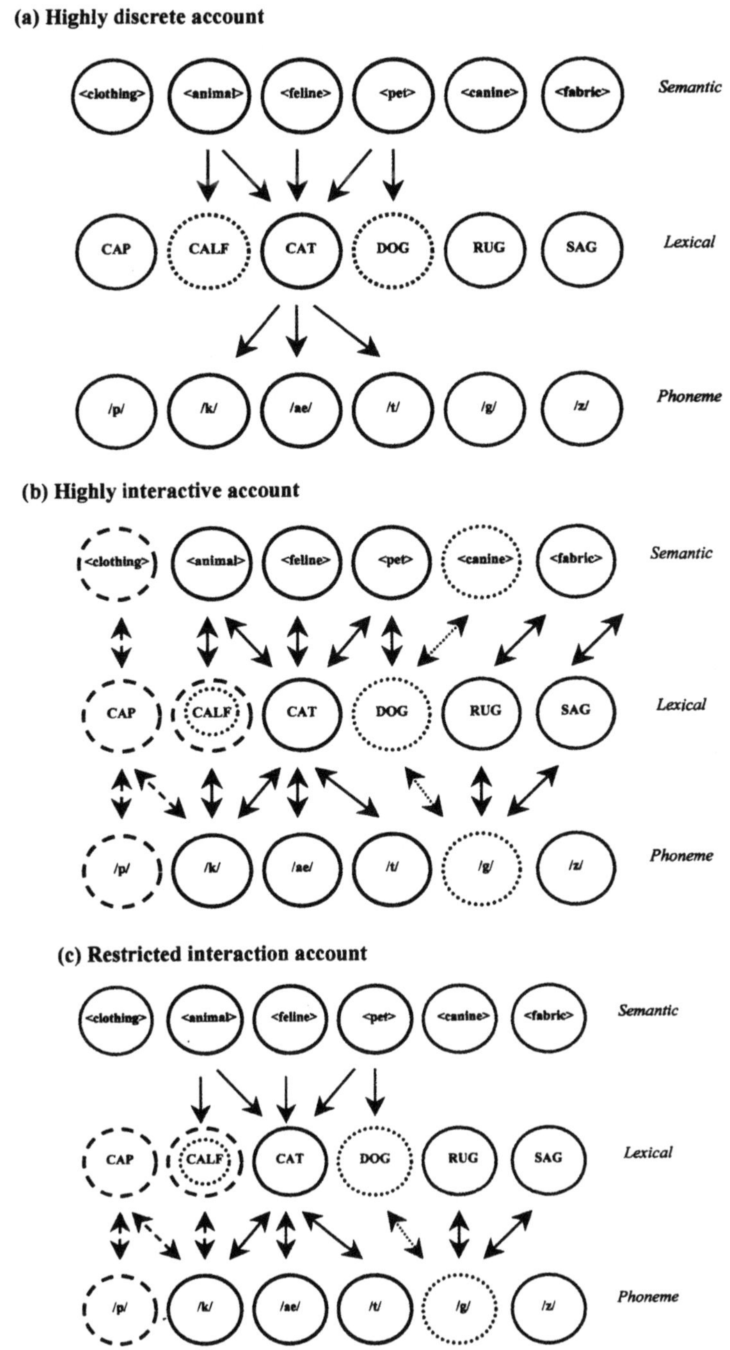

Figure 4. *(a) Highly discrete, (b) highly interactive, and (c) restricted interaction accounts of spoken word production for the target word "cat". Dotted lines and units show activation due to semantic overlap with the target; dashed lines and units show activation due to phonological overlap with the target. (Concentric circle denote units activated by both semantic and phonological overlap. Greyed-out units are not strongly activated by the target.)*

Stage 2, processing at all levels continues until the end of the stage, at which time the most active phoneme units are selected.

These two positions are similar in terms of the representational types they assume and their commitment to a two-stage framework. In addition, they share the assumption that both the target and its semantic competitors are active during Stage 1 lexical selection. They differ primarily in that the interactive position assumes cascading activation and feedback throughout the entire process. Prominent among the various lines of evidence that have been considered in trying to adjudicate between these positions have been analyses of mixed errors and form-based errors.

Speakers sometimes produce a word that is related in meaning to a target word (e.g., *shirt* → *skirt*). A number of analyses of spontaneous and experimentally induced speech errors produced by both neurologically intact and neurologically injured individuals have indicated that these semantic errors show a higher degree of phonological similarity to the intended word than would be predicted by a highly discrete account (Blanken, 1998; Brédart & Valentine, 1992; Dell & Reich, 1981; Dell et al., 1997; Harley, 1984; Kulke & Blanken, 2001; Martin, Gagnon, Schwartz, Dell, & Saffran, 1996a; Martin, Weisberg, & Saffran, 1989; Rapp & Goldrick, 2000; but see Best, 1996; del Viso, Igoa, & García-Albea, 1991; Igoa, 1996; Levelt, 1983, 1992; Nickels, 1995). Similarly, analyses of both lexical (e.g., *mitten* → *muffin*) and nonlexical (e.g., *trumpet* → "*chirpet*") form-based errors have indicated that lexical form-based errors occur at rates greater than would be expected in a highly discrete system (Baars, Motley, & MacKay, 1975; Best, 1996; Dell, 1986, 1990; Dell & Reich, 1981; Gagnon, Schwartz, Martin, Dell, & Saffran, 1997b; Harley, 1984; Humphreys, 2002; Nooteboom, 2003, 2004; Stemberger, 1985; but see del Viso et al., 1991; Garrett, 1976; Nickels & Howard, 1995). This latter finding is referred to as the "lexical bias effect" as it suggests that production system is biased to produce word outcomes.

Both mixed error and lexical bias effects are thought to require at least some form of feedback. Interactive theories account for lexical bias as follows: As activation passes from the lexical representation of a target (CAT) to its phonemes (/k//ae//t/), feedback connections send activation from these phonemes back to all lexical units that share phonemes with the target, including form-related neighbours of the target (e.g., HAT, BAT, MAT, RAT). These, in turn, activate their constituent phonemes, including those that are not shared with the target (/h/for HAT). These then reactivate their lexical level representations, creating "positive feedback loops" (Dell, 1986). Nonword responses (e.g., GAT) do not benefit from this type of support and, for that reason, when a disruption in processing occurs, the phonemes of the form-related neighbours of CAT will more successfully compete for selection than the phonemes of nonwords (i.e.,/h/will be a stronger competitor than/g/for the onset position).

With regard to the mixed error effect, the interactive architecture accounts for it by assuming that the feedback connections (from phonology to the lexical level and also from the lexical level to semantics) allow for interaction between semantic and phonological processes. Because of this, the mixed neighbours of a target (RAT) will be more active than other competitors that are either only semantically (DOG) or only phonologically (HAT) related to the target. As a result, all other things being equal,[4] if an error arises in the course of lexical selection, a mixed neighbour is a more likely error than a semantic or phonological neighbor.

Mixed error and lexical bias effects cannot be readily accounted for within highly discrete architectures and their proponents have presented a number of arguments challenging the validity of these effects in neurologically intact individuals (e.g., attributing effects to speaker's monitoring

[4] For example, the probabilities need to take into account the numbers of neighbours of the various types (see Rapp & Goldrick, 2000, for further discussion).

of their speech; Baars et al., 1975; Levelt, 1983, 1992; Levelt et al., 1999; Levelt, Schriefers, Vorberg, Meyer, Pechmann, & Havinga, 1991b; Nooteboom, 2003; Roelofs, 2004a, 2004b). It is beyond the scope of this paper to review and evaluate these arguments (see Rapp & Goldrick, 2000, 2004). We instead focus our discussion on the evidence from aphasic production that has been brought to bear on the question of interactivity in spoken word production.

There have been a number of analyses of aphasic errors that have attempted to determine whether or not mixed errors (Blanken, 1998; Dell et al., 1997b; Kulke & Blanken, 2001; Martin et al., 1996; Rapp & Goldrick, 2000) or form-based lexical errors (Best, 1996; Gagnon et al., 1997) occur at rates higher than would be expected by chance in a discrete architecture. Dell and colleagues (Dell et al., 1997b; see also Martin, Dell, Saffran, & Schwartz, 1994; Martin, Saffran, & Dell, 1996b; Schwartz & Brecher, 2000; Schwartz, Wilshire, Gagnon, & Polansky, 2004) used simulations to test the hypothesis that a wide range of patterns of spoken naming deficits could be accounted for within a highly interactive architecture. They showed that the fit between observed and simulated patterns was substantially better than the fit obtained for randomly generated patterns of errors. This success indicated that the evidence was generally consistent with the interactive two-stage account. In addition to the claims Dell and colleagues made regarding activation dynamics, they also made two other significant claims regarding the nature of the damage that gives rise to word naming deficits. First, they specifically argued that the fit between observed and simulated data was achieved by assuming that spoken naming deficits arise from global damage affecting all levels of the spoken production system (the globality assumption). Second, they further proposed that damage takes one of two forms, affecting either representational integrity (increased decay rates of the nodes throughout the system) or information transmission (noise on the connections between representational levels). Of these claims, the globality assumption has generated the most controversy and has been weakened by a number of challenges (Caramazza, Papagno, & Ruml, 2000; Cuetos, Aguado, & Caramazza, 2000; Dell, Lawler, Harris, & Gordon, 2004; Foygel & Dell, 2000; Rapp & Goldrick, 2000; Ruml & Caramazza, 2000; Ruml, Caramazza, Shelton, & Chialant, 2000; see Dell, Schwartz, Martin, Saffran, & Gagnon, 2000, for a reply to some of these challenges). In contrast, the proposal that the specific nature of the damage (i.e., whether it affects representations, the connections between them, the rate of activation) may produce different effects is one which has also been put forward in different forms by a number of investigators (e.g., access/storage deficits, see Crutch & Warrington, 2001; Warrington & Shallice, 1979), and is a topic that will continue to be the focus of numerous research efforts.

Rapp and Goldrick (2000, 2004; Goldrick & Rapp, 2002) followed up on the work of Dell and colleagues. Rather than examining if the existing data are simply *consistent* with the highly interactive architecture depicted in Figure 4b, this work sought to determine the specific architectural features (e.g., feedback, cascading activation) that are *required* to account for a set of critical performance patterns. Through a series of computer simulation studies this work examined the predictions of theories that varied with regard to the degree of interactivity that was assumed. They examined simulations instantiating both highly discrete and interactive architectures, and also architectures of intermediate interactivity. Those with intermediate levels of interactivity included a two-stage architecture that assumed cascading activation but lacked feedback, and one that incorporated cascading activation and feedback but in which the feedback was limited. Specifically, in the latter architecture (referred to as the Restricted Interactivity Account, or RIA) there was feedback from the phonological to the lexical level, but not from the lexical level back up to semantics (Figure 4c). After a extensive series of analyses, Rapp and Goldrick concluded that of all the architectures they examined, the RIA provided the best fit to the critical patterns of both the normal and aphasic data (but see Roelofs, 2004a, 2004b; Ruml et al., 2000). They claimed that with regard to the architecture of spoken word production "the important generalization is that although interaction is necessary, it is also true that interactivity is problematic as it increases beyond some optimal point" (p. 491).[5]

In addition to its theoretical implications, the work on activation dynamics also serves to underscore two more general points. One is the realisation that there is no atheoretical method for computing chance, rather that chance is simply the rate at which something would occur in some theory that does not include the feature of interest. For example, in the case at hand, chance is the rate at which mixed errors and form-related lexical errors would be predicted by a theory that does not include feedback. Once these rates are established, they can be compared to the observed rates. If they are at odds with one another, then the data represent a challenge to the theory that lacks the feature of interest. The second point is the increasing relevance of computer simulation to the development and testing of theories of spoken word production (see below, as well as Harley, 1993, 1995; Harley & MacAndrew, 1995; Laine, Tikkala, & Juhola, 1998; Plaut & Shallice, 1993; Wright & Ahmad, 1997). It is not surprising that questions of activation dynamics have led to extensive simulation work because the introduction of mechanisms such as feedback greatly increases the complexity of a theory. Given this, computer simulations can serve as an invaluable tool for clarifying the consequences of introducing activation dynamics into a theory and, therefore, the predictions of the different theoretical positions.

Summary: 1984–2004

In the above sections we have reviewed six questions on which, in our view, clear and significant progress has been made over the last 20 years of cognitive neuropsychological research on spoken word production. As indicated earlier, there are many more exciting questions that have been investigated and important findings that have been reported than we have discussed; we have limited ourselves to highlighting the most reliable and robust of these that have had implications for fundamental aspects of our understanding of spoken word production.

Our review indicates that, arguably, in the last 20 years cognitive neuropsychology has made its strongest contributions to questions concerning the organisation and content of the phonological lexicon. These can be summarised schematically in Figure 5. The evidence reveals an internally complex, long-term memory system that encodes morpheme-based phonological representations that are organised in a manner that respects grammatical and (certain) lexical categories. Furthermore, research reveals that this lexicon is dynamic, that lexical items compete for selection with other items that are concurrently active, and that both top-down (semantic–word) and bottom up (phoneme–word) constraints are brought to bear on this competition. Presumably, these characteristics allow for the effective selection and composition of word forms that are required for sentence production.

Methodological points

A number of methodological observations emerge from this review. One concerns the sometimes critical role played by *written* spelling data in elucidating questions of *spoken* word production. An examination of the integrity of written language production often allows us to determine if effects of interest observed in spoken word production arise at modality-specific levels of representation and processing (e.g., the phonological lexicon) or at modality-independent levels (syntactic or semantic levels). When an effect is present in spoken production but absent in

[5] This conclusion is assumed to hold across a range of simulation implementations, e.g., whether the representations in the system are localist or distributed or whether the system is implemented as an attractor network or in some other class of activation spreading architecture.

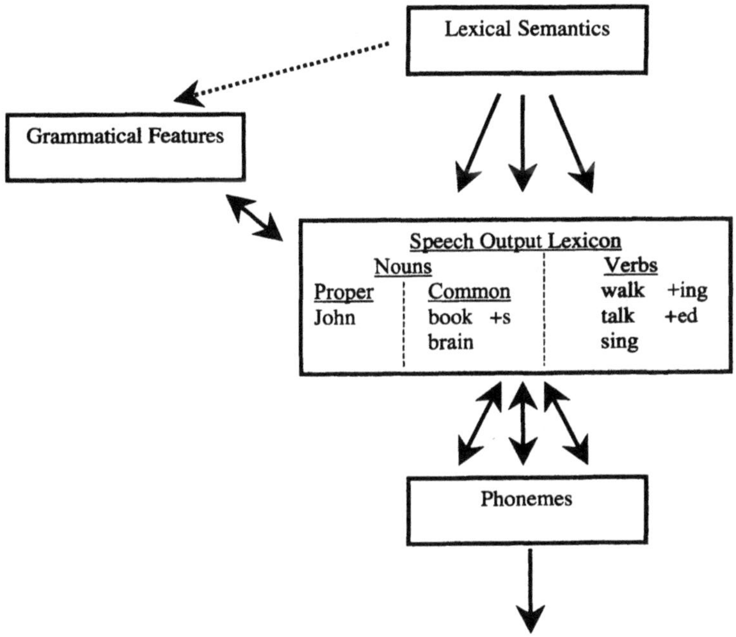

Figure 5. *Framework for current theories of speech production, incorporating findings reviewed in the article. Multiple arrows between processing components denote cascading activation; double-headed arrows indicate feedback between components. Dashed line indicates uncertainty regarding relationship between lexical semantic representations and grammatical features.*

written production, a case can be made for the modality-specific locus of the effect. In the cases reviewed above we see spelling data playing a critical role in the determination of post-semantic grammatical category distinctions as well as in the understanding that semantic errors can arise from disrupted access to the phonological lexicon from intact semantic representations.

Another point concerns the role of clinical categories and syndromes. Consistent with the insights of the cognitive neuropsychology pioneers of the 1970s and early 1980s, the progress that we have reported has not relied on clinical or syndrome characterisations of the individuals and/or their performance patterns. Instead, performance has been evaluated and interpreted relative to existing theories of intact language processing. This approach appears to have been highly productive, providing insights into both the content and organisation of the unimpaired spoken word production system, and an understanding of the spoken word production deficits themselves.

Finally, it is worth noting that although dissociations and double dissociations have played an important role in the advances we have reported, this is not the only type of evidence that has been brought to bear on the questions of interest. For example, on the question of the separability of lexical semantic and form representations, although one element of the critical pattern was, indeed, the dissociation between word comprehension and spoken word production, the other critical element concerned *the types of errors* produced in spoken naming. Namely, it was the fact that the errors were semantic errors that was critical to establishing a lexical rather than post-lexical locus of impairment. Another example concerns the work on morphological decomposition. Here, most critical was the type of error that was produced, namely the illegal combinations of stems and affixes (e.g., blackage, youthly). The argument was that these illegal combinations could not have been stored in the phonological lexicon and that, therefore, they must have been the product of compositional processes operating over morpheme-sized representations.

In sum, the last 20 years have been fruitful ones both with regard to the number of empirical findings with strong theoretical implications, as well as in terms of our understanding of a number of methodological issues. These advances provide reasonably firm foundations on which to construct an increasingly deeper and more detailed understanding of spoken word production. In the next sections, we discuss topics on which relatively less progress has been made and which, we anticipate, may occupy our research efforts in the upcoming years.

Spoken word production: Circa 2004 and beyond

If we consider Figure 5 as a summary of the current state of theorising, a number of deficiencies are immediately evident. First, it appears that progress has been made largely in our understanding of word selection, with considerably less progress having been made in understanding subsequent phonological processing stages. Second, the relationship between lexical processing and sentence processing is not indicated. Third, the relationship between word production and comprehension (one of the issues raised by Ellis, 1985) is not specified. Finally, there has been virtually no specification of the computational/representational machinery that allows words to produced *in real time*. That is, not only are various aspects of activation dynamics (e.g., competition, inhibition, decay, buffering) underspecified, crucially, the representation of time itself (ordering, timing, and duration) is strikingly absent. We briefly discuss each of these topics, identifying the opportunities and challenges faced by cognitive neuropsychological research in these areas.

Phonological processing

Subsequent to word selection, there are a number of sound-based processing stages including (at a minimum): phonological encoding and buffering, articulatory planning, and motor execution. Given the pervasiveness of spoken production difficulties following left-hemisphere damage, it is quite alarming that there has been relatively little cognitive neuropsychological research on these topics. This is not to say that there have not been a number of excellent papers; however, these have been scarce relative to the number of opportunities available to study deficits arising at these levels, as well as relative to the progress that has been made in the neighbouring linguistic disciplines of phonology and phonetics.

There is a fairly broad consensus that there is a distinction between two basic types of phonological processes—sometimes referred to as lexical and post-lexical. Thus, it is generally assumed that a lexical phonological process (or set of processes) recovers the largely arbitrary lexical phonological representations from long-term memory. These representations are often assumed to be "abstract" in that they lack at least some of the predictable aspects of phonological structure (but see Bybee, 2001; Crompton, 1982). A subsequent post-lexical process (or set of processes) elaborates these lexical phonological representations to produce (more) fully-specified post-lexical phonological representations that contain the information necessary to engage subsequent articulatory and motor processes. Despite general agreement on this broad distinction, there is little agreement regarding the specific content of lexical and post-lexical phonological representations and processes.

To date much of the work directed at understanding the nature of phonological representations and the forces operating in the course of spoken word production has been influenced by linguistic work on markedness. Markedness refers to the typological distribution of sound structure; marked structures are found in few languages, while unmarked structures are found in many languages. If these notions are relevant for phonological processing, marked phonological structures might be expected to be more difficult to process than unmarked structures. For example, it has been proposed (e.g., Clements, 1990) that segments within particular syllable positions (e.g., consonants within a syllable onset) are ordered in a systematic manner with certain orderings being more marked than others—a principle referred to as sonority. Following on from this, Romani and Calabrese (1998) and Romani, Olson, Semenza, and Granà (2002) reported that the sonority principle accounted for the pattern of errors observed in impaired spoken production and specifically concluded that sonority exerted an influence on post-lexical processing (i.e., articulatory planning). A preference for less marked structures has been generally found to be the case in a number of studies since the seminal work of Blumstein (1973), who studied the conversational production of a group of English-speaking aphasic individuals (see also Béland, 1990; Béland & Favreau, 1991; Béland, Paradis, & Bois, 1993; Carter, Gerken, & Holland, 1998; Christman, 1994; Code & Ball, 1994; Den Ouden, 2002; Kohn, Melvold, & Smith, 1995; Nespoulous, Joanette, Béland, Caplan, & Lecours, 1984; Nespoulous, Joanette, Ska, Caplan, & Lecours, 1987; Nespoulous & Moreau, 1997, 1998; but see Favreau, Nespolous, & Lecours, 1990; and for case studies, see Béland & Paradis, 1997; Kohn & Smith, 1994; Romani & Calabrese, 1998; Romani et al., 2002). While these studies all point to the relevance of the notions of markedness somewhere within speech production, they are limited by a lack of detailed information regarding the level at which these effects arise. This is because, in addition to their production deficits, many of the individuals in these studies suffered from comprehension deficits (e.g., nearly half of the individuals studied in Den Ouden (2002) or (sometimes subtle) deficits to articulatory processing (see Blumstein, 1998, for a review).

In fact, the differences of opinion regarding the organisation of the spoken production system not only concern the detailed content of phonological representations but also the level at which the various aspects of phonological representation are specified. Some researchers posit an early specification of featural, syllabic, and prosodic information at the lexical level, others posit a later post-lexical or even articulatory specification of this information, and yet others propose that different aspects of phonological information are represented at different levels. Cognitive neuropsychological research provides the opportunity to use selective deficits affecting specific processes to develop a deeper understanding of the representational and processing distinctions respected by the phonological machinery. In doing so it may also contribute to what currently may well be the most controversial issue in linguistic theories of sound structure—the distinction between phonology and phonetics. The distinction between the categorical, discrete, and abstract descriptions of the phonology and the continuous, graded variables traditionally associated with phonetics (Hale & Reiss, 2000; Keating, 1988; Pierrehumbert, 1990) has recently been vigorously debated (e.g., Ohala, 1990; Pierrehumbert, Beckmann, & Ladd, 2000) and alternative positions put forward. In this context, the challenge for cognitive neuropsychological work (as it has been for theoretical linguistics) is to identify the level at which the phenomena of interest (e.g., errors) arise (Goldrick & Rapp, 2004). To date this has been difficult because, among other things, the representational types supporting phonological, phonetic, and articulatory processes are typically assumed to be similar along a number of dimensions.

As a consequence of the difficulties involved in attributing deficits to particular levels of representation, researchers have reached different conclusions regarding the level at which particular aspects of phonological structure are represented. For example, Béland, Caplan, and Nespoulous (1990) and Kohn and Smith (1994) came to different conclusions regarding the level at which syllabic structure is specified. Similarly, while Romani et al. (2002) claimed that features are specified during post-lexical processing, Kohn et al. (1995) claimed that (marked) features are specified at the lexical level.

In sum, the situation in 2004 is not unlike that faced by Ellis, who in 1985 was concerned with the relative paucity of research in this area. Future theoretical work must do more to contribute to a more precise understanding of the level/s at which featural, syllabic, and prosodic information is represented and processed in the course of spoken, word production (see Goldrick & Rapp, 2004). Another important issue concerns the phonological level/s at which *lexical* variables such as grammatical category, lexical frequency, and neighbourhood density are relevant. Some theories assume fairly restricted early representation of lexical variables while others posit a more widespread representation or influence of lexical factors at post-lexical and even articulatory levels. Finally, also important are questions of activation dynamics (similar to those raised earlier) regarding the extent to which processing is highly interactive or modular in this part of the spoken production system.

The relationship between spoken word production and comprehension

This question is still a holdover from the set of issues identified by Ellis (1985) and is a part of the far broader question concerning the relationship between perception and action in a variety of domains (e.g., nonlinguistic actions, written language, etc.). Within the domain of spoken word production, this question consists of a number of subquestions, such as: (a) Are common mechanisms (lexicons, buffer, etc.) used in comprehension and production? (b) Does the feedback connectivity in spoken production correspond to the feed-forward connectivity in comprehension? (c) How does monitoring of spoken word production operate and what is its relationship to comprehension? Although considerable work has been carried out on many of these topics (e.g., Dell et al., 1997b; Howard, 1995; Martin & Saffran, 2002; Nickels & Howard, 1995; Romani, 1992; Shallice, Rumiati, & Zadini, 2000), we are far from having clear answers.

The challenge for cognitive neuropsychological work is to derive clear predictions that discriminate between the shared and independent systems views. The principal strategy thus far has been to document if production and comprehension deficits occur in association with one another or if they dissociate. The primary difficulty has been that, at least at a general level, dissociations and associations can be accommodated by both shared and independent systems views. Dissociations are accounted for by a shared systems view by assuming that, at least in one modality, the deficit affects access to the representations of interest. Associations are accounted for by an independent systems view as the accidental result of neural damage affecting multiple components. One possibility is that progress in developing a more detailed understanding of the phonological processes involved in production (see the previous section) will provide a more substantive basis from which to formulate hypotheses that will allow us to investigate and understand the relationship between comprehension and production. That is, as we understand production better, we will be in a stronger position to test whether or not the same representations and processes are involved in comprehension.

Speaking words in sentences: Grammatical and morphological processes

Research efforts on spoken word production and sentence production have proceeded fairly independently of one another. This has had the advantage that it has allowed us to establish some terra firma in the two domains, providing the theoretical and evidential scaffolding required to support interaction and integration of these domains. Given these advances, the general question now before us is: How are lexical selection, phonological encoding, and articulatory planning affected by sentential context?

There is evidence that lexical selection and encoding processes are affected by taking place within a sentential context or merely by occurring in the context of a string of words (e.g., Caramazza & Hillis, 1989; McCarthy & Kartsounis, 2000; Nespoulous et al., 1988; Schwartz & Hodgson, 2002; Wilshire & McCarthy, 2002). For example, Schwartz and Hodgson asked MP to name the same set of pictures in two contexts. One was standard picture naming, with a single picture and a single response required (e.g., picture of cat → "A cat"). In the second context, two pictures were presented side by side, and MP was required to name them both in a single utterance (e.g., cat, ear → "A cat and an ear"). Her accuracy in the first context was relatively high (92%), but it was dramatically decreased in the second context (42%). Note that many of her errors were not simply reversals (e.g., "An ear and a cat"), but productions of the incorrect lexical item (e.g., "A pie and a fan") These results indicate that lexical selection can be influenced by the spoken language context.

Given the role of morphology and grammatical features in sentential syntax, work in this area should be particularly helpful in shedding light on the various debates concerning the morphological and grammatical representation and processing of words that have been highlighted in previous sections. Furthermore, the temporal relationships between processes sensitive to the grammatical, morphological, and phonological aspects of words also require clarification. Contrasting spoken word production in contexts in which grammatical and morphological processes are most likely to be engaged with single word production will surely provide important insights into the spoken production system.

Activation dynamics: Competition, inhibition, decay, buffering

Although most theories characterise speech production as involving activation flow among various representational types, there is a striking lack of specificity regarding the means by which this activation is regulated and controlled. This is a crucial question, because competition among activated representations plays a significant role in speech production. During lexical selection, multiple semantic competitors are activated (e.g., during the processing of "cat," "dog", and "rat" are also activated). When producing sequences of words or sounds, the representations of sounds and words to be produced or that have already been produced may all be simultaneously active (as shown by anticipation and perseveration errors). Therefore, a critical set of issues concerns the mechanisms that mediate this competition.

One relevant mechanism is the selection process—how is it that a single activated representation comes to dominate processing? One selection mechanism that has been proposed involves enhancement of the most active representation. In some theories, the most active unit's activation is greatly increased at certain "selection" points. This activation advantage allows the selected unit to dominate processing. One way that this enhancement can be achieved is through an outside mechanism that simply adds activation to the "winning" unit (Dell, 1986, 1988; Dell et al., 1997b; Goldrick & Rapp, 2002; MacKay, 1987; Rapp & Goldrick, 2000). Another enhancement method is a competitive process by which active representations deactivate competitors to a degree that is proportional to their own activation strength. That is, the more active a representation, the more it can drive down (inhibit) the activation of its competitors (see Dell & O'Seaghdha, 1994, for a review). This is often implemented in language production theories using lateral connections among units of a similar representational type (e.g., Berg & Schade, 1992; Cutting & Ferreira, 1999; Harley, 1993; Meyer & Gordon, 1985; Schade & Berg, 1992; Stemberger, 1985). Note that a similar process occurs in attractor-based systems (e.g., Plaut & Shallice, 1993). Here, since different representations compete for realisation over a single set of units, the activation of one representation necessarily blocks the activation of another (see Page (2000), for further discussion of the relationship between lateral inhibition and attractors). Another selection-related mechanism involves "gating" activation flow—units are not allowed to pass on activation to other units until they meet some response criterion (e.g., a threshold of activation: Dell et al., 2004; Laine et al., 1998;

or a relative activation level that is sufficiently greater than competitors: Levelt et al., 1999; Roelofs, 1992).

While inhibition is an active process, intimately related to selection, other theories have adopted a more passive mechanism to drive down competitor activation. These theories posit that all activation levels constantly decay towards resting levels; units can only maintain activity if they receive outside input (Dell, 1986, 1988; Dell et al., 1997b; Harley, 1993; Martin et al., 1994). A number of studies have suggested that a pathological increase in decay can account for spoken production deficits (Dell et al., 1997b; Martin, 1996; Martin et al., 1994; Martin & Saffran, 1992; Schwartz & Brecher, 2000). One problem for this proposal is that other studies have argued that the patterns of performance attributed to excessive decay can be accounted for by other forms of damage (Foygel & Dell, 2000; Wright & Ahmad, 1997).

Thus, although many studies have invoked disruption to selection mechanisms to account for spoken production deficits, specific questions about these selection mechanisms (e.g., enhancement, inhibition, or decay) have not received much attention.

Another issue regarding activation dynamics is the role of buffering processes. Buffering comes into play when interacting processes function on different time scales. For example, in planning a sentence, a plan for a phrase might be activated and this phrase may need to be maintained active while each component lexical item is retrieved. Similarly, when a lexical item is retrieved, it may need to remain active to guide post-selection processing of its phonological components. A small number of case studies (see Shallice et al., 2000, for a review) have attributed production deficits to impairment to a phonological buffering process. Although these studies support the presence of such a buffer, considerable work remains to be done in specifying the precise nature of temporal ordering mechanisms (see below) and the structure of representations that are buffered. Furthermore, there are a number of reasons to think that there are buffers or buffering processes operating at multiple levels in the system. That is, speech production behaviour requires that activation be maintained at various points in processing; existing theoretical and empirical studies have done little to resolve how this is accomplished.

Representing time

The precise orchestration of events over time is an essential aspect of producing spoken words. It is, therefore, imperative for the time dimension to be more fully integrated into our theories of spoken word production. Specifically, the temporal dimension is an essential component of mechanisms and representations involved in ordering (e.g., to distinguish "cat" and "tack," segment order must be respected), timing (e.g., to correctly articulate voiced and voiceless stop consonants, the relative timing of consonant release and vocal fold vibration must be controlled), and duration (e.g., to signal obstruent voicing/devoicing in word-final position, the length of the previous vowel must be controlled). Fortunately, there is both theoretical and computational work that can contribute to cognitive neuropsychological efforts to bring patterns of impaired performance to bear on this important aspect of spoken production.

Recent theoretical work in phonology and phonetics directly tackles the problem of incorporating the temporal dimension into the representational formalism. This research includes proposals such as those of Browman and Goldstein (1992) in articulatory or gestural phonology. In their approach, the temporal dimension forms a part of categorical phonological representations themselves, providing an interface with the more graded, continuous representations of phonetics. As can be seen in Figure 6, in this theoretical framework the duration of gestures is specified and (although not depicted) the temporal coordination of these gestures is specified as well. Introducing this temporal information into phonological representations has extended the descriptive and explanatory power of linguistic theories (Browman & Goldstein, 1992; for

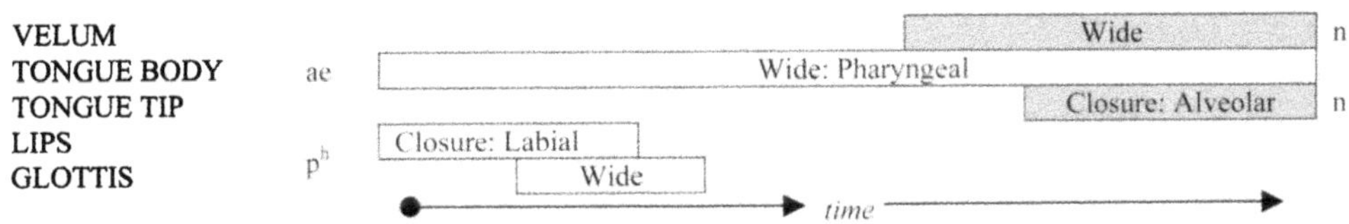

Figure 6. *Schematic articulatory phonological representation (gestural score) of "pan." Articulators are shown on the left-hand side. Letters on the left and the right show the association between articulatory gestures and elements in the segmental transcriptions. For each articulator, the bar represents the time during which the articulator is active. Labels within the bar refer to the degree of constriction; for some articulators, location is specified following a colon. Wide indicates that the degree of constriction is low, while closure indicates a high degree of constriction.*

recent applications, see Davidson, 2004; Gafos, 2002; Hall, 2003).

Computational work on the questions of time and serial production has developed in a number of directions. The shared objective of the various approaches is to understand how information is represented and processed to allow for the production of learned, temporally ordered sequences such as spoken words (see Lashley, 1951, for a seminal discussion of this issue).

One line of work has focused on developing more sophisticated versions of older chaining mechanisms used for encoding order. In a chain-based representation, the production of one element (e.g., a phoneme) triggers the production of the following element in the sequence, by virtue of being linked to it. Recurrent network simulations represent recent work along these general lines. In these networks, learning involves encoding the relationship between a distributed representation of an element (e.g., a set of phonemic features) and a distributed representation of the previous element and/or learning context (Dell, Juliano, & Govindjee, 1993; Elman, 1990; Jordan, 1986). Once learning has taken place, the activation of an element provides the context for the activation of the subsequent element in the sequence, and this process continues successively until the end of the sequence. These networks can reproduce a number of salient findings that have been reported for slips of the tongue produced by neurologically intact individuals, including such things as the preservation of phonotactic regularity and consonant/vowel status in substitutions, etc. (see Anderson, Milostan, & Cottrell, 1998; Dell et al., 1993, for discussions of the strengths and limitations of this approach). These approaches have not yet, however, been applied to spoken word production in aphasia.

Another direction taken to understanding the ordering question has been the computational instantiation of slot and filler mechanisms. This work contrasts with the chaining approach in assuming a fundamental distinction between content and structure. Information regarding ordering is represented in a structural frame, while the elements to be ordered are independently represented. For example, in producing the form of the word "creed"/k r i d/, both a frame specifying a monosyllabic word with a complex onset, nucleus, and simple coda and the phonological content (the component phonemes) of the word are retrieved from memory. Subsequently, the phonemes are linked to their respective syllabic positions via some "slot filling" process. Errors may arise at various points; for example, in the retrieval of the frame or the content, in the course of slot filling, or in readout from the filled slots (e.g., Dell, 1986, 1988; Dell, Burger, & Svec, 1997a; Hartsuiker, 2001; Levelt, 1989; Levelt et al., 1999; MacKay, 1972, 1987; Meijer, 1994; Shattuck-Hufnagel, 1979, 1992; Stemberger, 1985). (Note that these approaches use different methods for binding frame and content; see Dell, Ferreira, & Bock, 1999; Levelt et al., 1999.) This approach has been successfully applied to slips of the tongue and, to a more limited extent, to data from aphasia as well (e.g., Pate, Saffran, & Martin, 1987; Schwartz, Saffran, Bloch, & Dell, 1994; Wilshire, 2002;

Wilshire & McCarthy, 1996). The slot and filler approach has been able to handle certain of the phenomena not well accommodated by recurrent networks, but faces its own set of challenges (for a review, see Dell et al., 1997a).

Finally, there are a number of computational approaches in which order, timing, and duration are represented through the association of the elements to be ordered (e.g., phonemes) with timing units that have intrinsic temporal characteristics (e.g., oscillators). It is this direct and explicit incorporation of timing elements into the production process that distinguishes this approach from the previous two. In describing the basic logic of the approach, Brown, Preece, and Hulme (2000) use a clock analogy such that the hour, minute, and second hands of a clock are analogous to slow, intermediate, and fast oscillators. During learning, the clock starts and, as time passes, each phoneme is associated with (linked to) a particular configuration of the hands. Then, at the time of retrieval, the clock is started and its associated elements are produced as time unfolds (for specific applications to speech production, see Harris, 2002; Hartley & Houghton, 1996; Vousden, Brown, & Harley, 2000). This approach has a number of advantages over the previous ones, although it too suffers from its own set of limitations.

In sum, to date questions regarding the representation and processing of temporal ordering and duration have scarcely been addressed in cognitive neuropsychological work on spoken word production. However, advances in theoretical linguistics and computational theories of speech production provide a number of frameworks within which to pursue this complex, yet critical, dimension of spoken word production.

Conclusions

The last 20 years have been a testament to the vitality and productiveness of the cognitive neuropsychological approach in the domain of spoken word production. We have seen clear progress made on a number of macro and micro structural issues. This work has revealed a dynamic, yet internally structured system that is instantiated in the brain in a manner that allows for the fairly selective damage to individual components of meaning, form, as well as grammatical and lexical properties. The next 20 years will require that we build on the architectural and representational foundations of the preceding years in order to develop far more detailed and computationally explicit theories of spoken language processing. Such theories will help us to understand the real-time transitions from categorical to continuous representations that allow us to fluently produce words both in isolation and in sentences. It is difficult to imagine that computer simulation will not play an important role in the theory testing and development that will be required to make progress on these questions. Furthermore, it is essential that we continue to actively work for the broader integration of cognitive neuropsychological evidence into theory development in cognitive science, psychology, linguistics, and neuroscience.

PrEview proof published online 2 September 2005

REFERENCES

Allen, M., & Badecker, W. (1999). Stem homograph inhibition and stem allomorphy: Representing and processing inflected forms in a multi-level lexical system. *Journal of Memory and Language, 41*, 105–123.

Allen, M., & Badecker, W. (2001). Morphology: The internal structure of words. In B. Rapp (Ed.), *The handbook of cognitive neuropsychology: What deficits reveal about the human mind* (pp. 211–232). Philadelphia, PA: Academic Press.

Allport, D. A. (1983). Language and cognition. In R. Harris (Ed.), *Approaches to language* (pp. 61–94). Oxford: Pergamon Press.

Allport, D. A., & Funnell, E. (1981). Components of the mental lexicon. *Philosophical Transactions of the Royal Society (London), B295*, 397–410.

Anderson, K., Milostan, J., & Cottrell, G. W. (1998). Assessing the contribution of representation to results. In M. A. Gernsbacher & S. J. Derry (Eds.), *Proceedings of the 20th annual conference of the Cognitive Science Society* (pp. 48–53). Hillsdale, NJ: Lawrence Erlbaum Associates Inc.

Assal, G., Buttet, J., & Jolivet, R. (1981). Dissociation in aphasia: A case report. *Brain and Language, 13*, 223–240.

Baars, B. J., Motley, J. T., & MacKay, D. (1975). Output editing for lexical status from artificially elicited slips of the tongue. *Journal of Verbal Learning and Verbal Behavior, 14*, 382–391.

Badecker, W. (1997). Levels of morphological deficit: Indications from inflectional regularity. *Brain and Language, 60*, 360–380.

Badecker, W., & Caramazza, A. (1987). The analysis of morphological errors in a case of acquired dyslexia. *Brain and Language, 32*, 360–380.

Badecker, W., & Caramazza, A. (1991). Morphological composition in the lexical output system. *Cognitive Neuropsychology, 8*, 335–367.

Badecker, W., Miozzo, M., & Zanuttini, R. (1995). The two-stage model of lexical retrieval: Evidence from a case of anomia with selective preservation of grammatical gender. *Cognition, 57*, 193–216.

Badecker, W., Rapp, B., & Caramazza, A. (1995). A modality-neutral lexical deficit affecting morpho-syntactic representations. *Brain and Language, 51*, 83–84.

Basso, A., Taborelli, A., & Vignolo, L. A. (1978). Dissociated disorders of speaking and writing in aphasia. *Journal of Neurology, Neurosurgery and Psychiatry, 41*, 556–563.

Battistella, E. L. (1996). *The logic of markedness*. Oxford: Oxford University Press.

Baxter, D. M., & Warrington, E. K. (1985). Category specific phonological dysgraphia. *Neuropsychologia, 23*, 653–666.

Beauvois, M. F. (1982). Optic aphasia: A process of interaction between vision and language. *Philosophical Transactions of the Royal Society (London), B 298*, 33–47.

Béland, R. (1990). Vowel epenthesis in aphasia. In J.-L. Nespoulous & P. Villiard (Eds.), *Morphology, phonology, and aphasia* (pp. 235–252). New York: Springer-Verlag.

Béland, R., Caplan, D., & Nespoulous, J.-L. (1990). The role of abstract phonological representations in word production: Evidence from phonemic paraphasias. *Journal of Neurolinguistics, 5*, 125–164.

Béland, R., & Favreau, Y. (1991). On the special status of coronals in aphasia. In C. Paradis & J.-F. Prunet (Eds.), *Phonetics and phonology, vol. 2: The special status of coronals* (pp. 201–221). San Diego: Academic Press.

Béland, R., & Paradis, C. (1997). Principled syllabic dissolution in a primary progressive aphasia case. *Aphasiology, 11*, 1171–1196.

Béland, R., Paradis, C., & Bois, M. (1993). Constraints and repairs in aphasic speech: A group study. *Canadian Journal of Linguistics, 38*, 279–302.

Berg, T., & Schade, U. (1992). The role of inhibition in a spreading-activation model of language production I: The psycholinguistic perspective. *Journal of Psycholinguistic Research, 21*, 405–434.

Berndt, R. S., & Haendiges, A. (2000). Grammatical class in word and sentence production: Evidence from an aphasic patient. *Journal of Memory and Language, 43*, 249–273.

Berndt, R. S., Haendiges, A., Burton, M., & Mitchum, C. (2001). Grammatical class and imageability in aphasic word production: Their effects are independent. *Journal of Neurolinguistics, 15*, 353–371.

Berndt, R., Haendiges, A., Burton, M., & Mitchum, C. (2002). Grammatical class and imageability in aphasic word production Their effects are independent. *Journal of Neurolinguistics, 15*, 353–371.

Best, W. (1996). When racquets are baskets but baskets are biscuits, where do the words come from? A single case study of formal paraphasic errors in aphasia. *Cognitive Neuropsychology, 13*, 443–480.

Bird, H., Howard, D., & Franklin, S. (2000). Why is a verb like an inanimate object? Grammatical category and semantic category deficits. *Brain and Language, 72*, 246–309.

Blanken, G. (1998). Lexicalisation in speech production: Evidence from form-related word substitutions in aphasia. *Cognitive Neuropsychology, 15*, 321–360.

Blumstein, S. (1973). *A phonological investigation of aphasic speech*. The Hague: Mouton.

Blumstein, S. (1998). Phonological aspects of aphasia. In M. Sarno (Ed.), *Acquired aphasia* (3rd ed., pp. 157–185). San Diego: Academic Press.

Brédart, S., & Valentine, T. (1992). From Monroe to Moreau: An analysis of face naming errors. *Cognition, 45*, 187–223.

Browman, C. P., & Goldstein, L. (1992). Articulatory phonology: An overview. *Phonetica, 49*, 155–180.

Brown, G. D. A., Preece, T., & Hulme, C. (2000). Oscillator-based memory for serial order. *Psychological Review, 107*, 127–181.

Bub, D., Cancelliere, A., & Kertesz, A. (1985). Whole-word and analytic translation of spelling-to-sound in a nonsemantic reader. In K. E. Patterson, J. C. Marshall, & M. Coltheart (Eds.), *Surface dyslexia: Neuropsychological and cognitive studies of phonological reading* (pp. 15–34). Hove, UK: Lawrence Erlbaum Associates Ltd.

Bub, D., & Kertesz, A. (1982). Evidence for lexicographic processing in a patient with preserved written over oral single word naming. *Brain, 105*, 697–717.

Buckingham, H. (1981). Where do neologisms come from? In J. Brown (Ed.), *Jargonaphasia* (pp. 39–62). New York: Academic Press.

Buckingham, H., & Kertesz, A. (1976). *Neologistic jargon aphasia*. Amsterdam: Swets & Zeitlinger.

Butterworth, B. (1983). Lexical representation. In B. Butterworth (Ed.), *Language production* (Vol. 2, pp. 257–294). London: Academic Press.

Butterworth, B., & Howard, D. (1987). Paragrammatisms. *Cognition, 26*, 1–37.

Bybee, J. L. (2001). *Phonology and language use*. Cambridge: Cambridge University Press.

Caplan, D., Keller, L., & Locke, S. (1972). Inflection of neologisms in aphasia. *Brain, 95*, 169–172.

Caramazza, A. (1984). The logic of neuropsychological research and the problem of patient classifications in aphasia. *Brain and Language, 21*, 9–20.

Caramazza, A. (1986). On drawing inferences about the structure of normal cognitive systems from the analysis of impaired performance: The case for single-patient studies. *Brain and Cognition, 5*, 41–66.

Caramazza, A. (1997). How many levels of processing are there in lexical access? *Cognitive Neuropsychology, 14*, 177–208.

Caramazza, A., & Hillis, A. E. (1989). The disruption of sentence production: Some dissociations. *Brain and Language, 36*, 625–650.

Caramazza, A., & Hillis, A. E. (1990). Where do semantic errors come from? *Cortex, 26*, 95–122.

Caramazza, A., & Hillis, A. E. (1991). Lexical organisation of nouns and verbs in the brain. *Nature, 349*, 788–790.

Caramazza, A., & Miozzo, M. (1997). The relation between syntactic and phonological knowledge in lexical access: Evidence from the "tip-of-the-tongue" phenomenon. *Cognition, 64*, 309- 343.

Caramazza, A., & Miozzo, M. (1998). More is not always better: A response to Roelofs, Meyer, and Levelt. *Cognition, 69*, 231–241.

Caramazza, A., Papagno, C., & Ruml, W. (2000). The selective impairment of phonological processing in speech production. *Brain and Language, 75*, 428–450.

Caramazza, A., & Shapiro, K. (2004). The representation of grammatical knowledge in the brain. In L. Jenkins (Ed.), *Variation and universals in biolinguistics*. Amsterdam: Elsevier.

Carter, A., Gerken, L., & Holland, A. (1998, November). *Markedness and syllable structure in aphasic word production*. Paper presented at the Academy of Aphasia annual meeting, Santa Fe, NM. [Abstract in *Brain and Language, 65*, 202–203.]

Christman, S. S. (1994). Target-related neologism formation in jargonaphasia. *Brain and Language, 46*, 109–128.

Cipolotti, L. (2000). Sparing of country and nationality names in a case modality-specific oral output impairment: Implications for theories of speech production. *Cognitive Neuropsychology, 17*, 709–729.

Cipolotti, L., McNeil, J. E., & Warrington, E. K. (1993). Spared written naming of proper nouns: A case report. *Memory, 1*, 289–311.

Cipolotti, L., & Warrington, E. K. (1995). Semantic memory and reading abilities: A case report. *Journal of the International Neuropsychological Society, 1*, 104–110.

Clements, G. N. (1990). The role of the sonority cycle in core syllabification. In M. Beckman & J. Kingston (Eds.), *Papers in laboratory phonology I: Between the grammar and physics of speech* (pp. 283–333). Cambridge: Cambridge University Press.

Code, C., & Ball, M. J. (1994). Syllabification in aphasic recurring utterances: Contributions of sonority theory. *Journal of Neurolinguistics, 8*, 257–265.

Coltheart, M., Masterson, J., Byng, S., Prior, M., & Riddoch, J. (1983). Surface dyslexia. *Quarterly Journal of Experimental Psychology, 35A*, 469–495.

Coslett, H. B. (1991). Read but not write "idea": Evidence for a third reading mechanism. *Brain and Language, 40*, 425–443.

Coslett, H. B., Gonzales-Rothi, L. J., & Heilman, K. M. (1984). Reading: Selective sparing of closed-class words in Wernicke's aphasia. *Neurology, 34*, 1038–1045.

Crompton, A. (1982). Syllables and segments in speech production. In A. Cutler (Ed.), *Slips of the tongue and language production* (pp. 109–162). Berlin: Mouton.

Crutch, S. J., & Warrington, E. K. (2001). Refractory dyslexia: Evidence of multiple task-specific phonological output stores. *Brain, 124*, 1533–1543.

Cuetos, F., Aguado, G., & Caramazza, A. (2000). Dissociation of semantic and phonological errors in naming. *Brain and Language, 75*, 451–460.

Cutting, J. C., & Ferreira, V. S. (1999). Semantic and phonological information flow in the production lexicon. *Journal of Experimental Psychology: Learning, Memory, and Cognition, 25*, 318–344.

Davidson, L. (2004). *Consonant cluster phonotactics in a gesturally based grammar*. Manuscript submitted for publication.

Dell, G. S. (1986). A spreading activation theory of retrieval in sentence production. *Psychological Review, 93*, 283–321.

Dell, G. S. (1988). The retrieval of phonological forms in production: Tests of predictions from a connectionist model. *Journal of Memory and Language, 27*, 124–142.

Dell, G. S. (1990). Effects of frequency and vocabulary type on phonological speech errors. *Language and Cognitive Processes, 4*, 313–349.

Dell, G. S., Burger, L. K., & Svec, W. R. (1997a). Language production and serial order: A functional analysis and a model. *Psychological Review, 104*, 123–147.

Dell, G. S., Ferreira, V., & Bock, K. (1999). Binding, attention, and exchanges. *Behavioral and Brain Sciences, 22*, 41–42.

Dell, G. S., Juliano, C., & Govindjee, A. (1993). Structure and content in language production: A theory of frame constraints in phonological speech errors. *Cognitive Science, 17*, 149–195.

Dell, G. S., Lawler, E. N., Harris, H. D., & Gordon, J. K. (2004). Models of errors of omission in aphasic naming. *Cognitive Neuropsychology, 21*, 125–145.

Dell, G. S., & O'Seaghdha, P. G. (1994). Inhibition in interactive activation models of linguistic selection and sequencing. In D. Dagenbach & T. H. Carr (Eds.), *Inhibitory processes in attention, memory, and language* (pp. 409–453). San Diego: Academic Press.

Dell, G. S., & Reich, P. A. (1981). Stages in sentence production: An analysis of speech error data. *Journal of Verbal Learning and Verbal Behavior, 20*, 611–629.

Dell, G. S., Schwartz, M. F., Martin, N., Saffran, E. M., & Gagnon, D. A. (1997b). Lexical access in aphasic and nonaphasic speakers. *Psychological Review, 104*, 801–838.

Dell, G. S., Schwartz, M. F., Martin, N., Saffran, E. M., & Gagnon, D. A. (2000). The role of computational models in neuropsychological investigations of language: Reply to Ruml & Caramazza (2000). *Psychological Review, 107*, 635–645.

Del Viso, S., Igoa, J. M., & García-Albea, J. E. (1991). On the autonomy of phonological encoding: Evidence from slips of the tongue in Spanish. *Journal of Psycholinguistic Research, 20*, 161–185.

Dennis, M. (1976). Dissociated naming and locating of body parts after anterior temporal lobe resection: An experimental case study. *Brain and Language, 3*, 147–163.

Den Ouden, D.-B. (2002). *Phonology in aphasia: Syllables and segments in level-specific deficits.* Doctoral dissertation, Rijksuniversiteit Groningen, The Netherlands.

Ellis, A. W. (1985). The production of spoken words: A cognitive neuropsychological perspective. In A. W. Ellis (Ed.), *Progress in the psychology of language, Vol. 2* (pp. 107–145). Hove, UK: Lawrence Erlbaum Associates Ltd.

Ellis, A. W. (1987). Intimations of modularity, or, the modularity of mind: Doing cognitive neuropsychology without syndromes. In M. Coltheart, G. Sartori, & R. Job (Eds.), *The cognitive neuropsychology of language* (pp. 397–480). Hove, UK: Lawrence Erlbaum Associates Ltd.

Elman, J. L. (1990). Finding structure in time. *Cognitive Science, 14*, 213–252.

Favreau, Y., Nespoulous, J.-L., & Lecours, A. R. (1990). Syllable structure and lexical frequency effects in the phonemic errors of four aphasics. *Journal of Neurolinguistics, 5*, 165–187.

Forster, K. (1976). Accessing the mental lexicon. In R. J. Wales & E. Walker (Eds.), *New approaches to language mechanisms* (pp. 257–287). Amsterdam: North-Holland.

Foygel, D., & Dell, G. S. (2000). Models of impaired lexical access in speech production. *Journal of Memory and Language, 43*, 182–216.

Franklin, S., Howard, D., & Patterson, K. (1995). Abstract word anomia. *Cognitive Neuropsychology, 12*, 549–566.

Funnell, E. (1983). Phonological processes in reading: New evidence from acquired dyslexia. *British Journal of Psychology, 74*, 159–180.

Gafos, A. I. (2002). A grammar of gestural coordination. *Natural Language and Linguistic Theory, 20*, 269–337.

Gagnon, D. A., Schwartz, M. F., Martin, N., Dell, G. S., & Saffran, E. M. (1997). The origins of formal paraphasias in aphasics' picture naming. *Brain and Language, 59*, 450–472.

Garrett, M. F. (1976). Syntactic processes in sentence production. In R. J. Wales & E. Walker (Eds.),

New approaches to language mechanisms (pp. 231–255). Amsterdam: North-Holland.

Garrett, M. F. (1980). Levels of processing in sentence production. In B. Butterworth (Ed.), *Language production (Vol. I): Speech and talk* (pp. 177–220). New York: Academic Press.

Goldrick, M., & Rapp, B. (2002). A restricted interaction account (RIA) of spoken word production: The best of both worlds. *Aphasiology, 16*, 20–55.

Goldrick, M., & Rapp, B. (2004). *Lexical and post-lexical phonological representations in spoken production*. Manuscript submitted for publication.

Goodglass, H., Wingfield, A., Hide, M. R., & Theurkauf, J. C. (1986). Category-specific dissociations in naming and recognition by aphasic patients. *Cortex, 22*, 87–102.

Hale, M., & Reiss, C. (2000). Phonology as cognition. In N. Burton-Roberts, P. Carr, & G. Docherty (Eds.), *Phonological knowledge: Conceptual and empirical issues* (pp. 161–184). Oxford: Oxford University Press.

Hall, N. (2003). *Gestures and segments: Vowel intrusion as overlap*. Doctoral dissertation, University of Massachusetts, Amherst. ROA-637–0104, Rutgers Optimality Archive, http://roa.rutgers.edu/.

Harley, T. A. (1984). A critique of top-down independent levels models of speech production: Evidence from non-plan-internal speech errors. *Cognitive Science, 8*, 191–219.

Harley, T. A. (1993). Phonological activation of semantic competitors during lexical access in speech production. *Language and Cognitive Processes, 8*, 291–309.

Harley, T. A., (1995). Connectionist models of anomia: A comment on Nickels. *Language and Cognitive Processes, 10*, 47–58.

Harley, T. A., & MacAndrew, S. B. G. (1995). Interactive models of lexicalization: Some constraints from speech error, picture naming, and neuropsychological data. In J. P. Levy, D. Bairaktaris, & J. A. Bullinaria (Eds.), *Connectionist models of memory and language* (p. 311–331). London: UCL Press.

Harris, H. D. (2002). Holographic reduced representations for oscillator recall: A model of phonological production. In W. D. Gray & C. D. Schunn (Eds.), *Proceedings of the 24th annual meeting of the Cognitive Science Society*. Hillsdale, NJ: Lawrence Erlbaum Associates Inc.

Hart, J., Berndt, R. S., & Caramazza, A. (1985). Category-specific naming deficit following cerebral infarction. *Nature, 316*, 439–440.

Hartley, T., & Houghton, G. (1996). A linguistically constrained model of short-term memory for nonwords. *Journal of Memory and Language, 35*, 1–31.

Hartsuiker, R. J. (2001). The addition bias in Dutch and Spanish phonological speech errors: The role of structural context. *Language and Cognitive Processes, 17*, 61–96.

Hayes, B. (1999). Phonetically-driven phonology: The role of Optimality Theory and inductive grounding. In M. Darnell, E. Moravscik, M. Noonan, F. Newmeyer, & K. Wheatly (Eds.), *Functionalism and formalism in linguistics, volume I: General papers* (pp. 243–285). Amsterdam: John Benjamins.

Henaff Gonon, M., Bruckert, R., & Michel, F. (1989). Lexicalization in an anomic patient. *Neuropsychologia, 27*, 391–407.

Hillis, A. E., & Caramazza, A. (1991). Mechanisms for accessing lexical representations for output: Evidence from a category-specific semantic deficit. *Brain and Language, 40*, 106–144.

Hillis, A. E., & Caramazza, A. (1995a). Converging evidence for the interaction of semantic and phonological information in accessing lexical information for spoken output. *Cognitive Neuropsychology, 12*, 187–227.

Hillis, A. E., & Caramazza, A. (1995b). Representation of grammatical knowledge in the brain. *Journal of Cognitive Neuroscience, 7*, 396–407.

Hittmair-Delazer, M., Denes, G., Semenza, C., & Mantovan, M. C. (1994). Anomia for people's names. *Neuropsychologia, 32*, 465–476.

Howard, D. (1995). Lexical anomia: Or the case of the missing lexical entries. *Quarterly Journal of Experimental Psychology, 48A*, 999–1023.

Humphreys, K. R. (2002). *Lexical bias in speech errors*. Doctoral dissertation, University of Illinois at Urbana-Champaign.

Igoa, J. M. (1996). The relationship between conceptualization and formulation processes in sentence production: Some evidence from Spanish. In M. Carreiras, J. García-Albea, & N. Sebastián-Galles (Eds.), *Language processing in Spanish* (pp. 305–351). Hillsdale, NJ: Lawrence Erlbaum Associates Inc.

Jordan, M. I. (1986). *Serial order: A parallel distributed processing approach*. Institute for Cognitive Science Technical Report 8604. La Jolla, CA: University of California at San Diego. [Reprinted in J. W. Donahoe & V. P. Dorsel (Eds.), (1997). *Neural-network models of cognition: Biobehavioral foundations* (pp. 221–277). Amsterdam: Elsevier Science Press.]

Keating, P. A. (1988). The phonology–phonetics interface. In F. Newmeyer (Ed.), *Linguistics: The Cambridge survey*, Vol. 1 (pp. 281–302). Cambridge: Cambridge University Press.

Kempen, G., & Huijbers, P. (1983). The lexicalization process in sentence production and naming: Indirect elicitation of words. *Cognition, 14*, 185–209.

Kohn, S. E., Melvold, J., & Smith, K. L. (1995). Consonant harmony as a compensatory mechanism in fluent aphasic speech. *Cortex, 31*, 747–756.

Kohn, S. E., & Smith, K. L. (1994). Distinctions between two phonological output deficits. *Applied Psycholinguistics, 15*, 75–95.

Kulke, F., & Blanken, G. (2001). Phonological and syntactic influences on semantic misnamings in aphasia. *Aphasiology, 15*, 3–15.

Laiacona, M., & Caramazza, A. (2004). The noun/verb dissociation in language production: Varieties of causes. *Cognitive Neuropsychology, 21*, 103–123.

Laine, M., Niemi, J., Koivuselkä-Sallinen, P., & Hyönä, J. (1995). Morphological processing of polymorphemic nouns in a highly inflected language. *Cognitive Neuropsychology, 12*, 457–502.

Laine, M., Tikkala, A., & Juhola, M. (1998). Modelling anomia by the discrete two-stage word production architecture. *Journal of Neurolinguistics, 11*, 275–294.

Lambon Ralph, M. A., Ellis, A. W., & Franklin, S. (1995). Semantic loss without surface dyslexia. *Neurocase, 1*, 363–369.

Lambon Ralph, M. A., Ellis, A. W., & Sage, K. (1998). Word meaning blindness revisited. *Cognitive Neuropsychology, 15*, 389–400.

Lashley, K. S. (1951). The problem of serial order in behavior. In L. A. Jeffress (Ed.), *Cerebral mechanisms in behavior: The Hixon symposium* (pp. 112–136). New York: John Wiley.

Lecours, A. R., & Rouillon, F. (1976). Neurolinguistic analysis of jargonaphasia and jargonagraphia. In H. Whitaker & H. Whitaker (Eds.), *Studies in neurolinguistics*, Vol. 2 (pp. 96–144). New York: Academic Press.

Levelt, W. J. M. (1983). Monitoring and self-repair in speech. *Cognition, 14*, 41–104.

Levelt, W. J. M. (1989). *Speaking: From intention to articulation*. Cambridge, MA: MIT Press.

Levelt, W. J. M. (1992). Accessing words in speech production: Stages, processes, and representations. *Cognition, 42*, 1–22.

Levelt, W. J. M., Roelofs, A., & Meyer, A. S. (1999). A theory of lexical access in speech production. *Behavioral and Brain Sciences, 22*, 1–75.

Levelt, W. J. M., Schriefers, H., Vorberg, D., Meyer, A. S., Pechmann, T., & Havinga, J. (1991a). The time course of lexical access in speech production: A study of picture naming. *Psychological Review, 98*, 122–142.

Levelt, W. J. M., Schriefers, H., Vorberg, D., Meyer, A. S., Pechmann, T., & Havinga, J. (1991b). Normal and deviant lexical processing: Reply to Dell and O'Seaghdha (1991). *Psychological Review, 98*, 615–618.

Lhermitte, F., & Derouesne, J. (1974). Paraphasies et jargonaphasie dans le langage oral avec conservation du langage écrit. *Revue Neurologique, 130*, 21–38.

Lucchelli, F., & DeRenzi, E. (1992). Proper name anomia. *Cortex, 28*, 21–30.

Lyons, F., Hanley, J. R., & Kay, J. (2002). Anomia for common names and geographical names with preserved retrieval of names of people: A semantic memory disorder. *Cortex, 38*, 23–35.

MacKay, D. G. (1972). The structure of words and syllables: Evidence from errors in speech. *Cognitive Psychology, 3*, 210–227.

MacKay, D. G. (1987). *The organisation of perception and action: A theory for language and other cognitive skills*. New York: Springer-Verlag.

Marin, O. S. M., Saffran, E. M., & Schwartz, M. F. (1976). Dissociations of language in aphasia: Implications for normal function. *Annals of the New York Academy of Sciences, 280*, 868–884.

Marshall, J. C. (1986). The description and interpretation of aphasic language disorder. *Neuropsychologia, 24*, 5–24.

Marslen-Wilson, W., & Tyler, L. K. (1998). Rules, representations and the English past tense. *Trends in Cognitive Sciences, 2*, 428–435.

Marslen-Wilson, W., & Tyler, L. K. (2005). The lexicon, grammar, and past tense: Dissociation revisited. In M. Tomasello & D. Slobin (Eds.), *Beyond nature-nuture: Essays in honour of Elizabeth Bates*. Hillsdale, NJ: Lawrence Erlbaum Associates Inc.

Martin, N. (1996). Models of deep dysphasia. *Neurocase, 2*, 73–80.

Martin, N., Dell, G. S., Saffran, E. M., & Schwartz, M. F. (1994). Origins of paraphasias in deep dysphasia: Testing the consequences of a decay impairment to an interactive spreading activation model of lexical retrieval. *Brain and Language, 47*, 609–660.

Martin, N., Gagnon, D. A., Schwartz, M. F., Dell, G. S., & Saffran, E. M. (1996a). Phonological facilitation of semantic errors in normal and aphasic speakers. *Language and Cognitive Processes, 11*, 257–282.

Martin, N., & Saffran, E. M. (1992). A computational study of deep dysphasia: Evidence from a single case study. *Brain & Language, 43*, 240–274.

Martin, N., & Saffran, E. M. (2002). The relationship of input and output phonological processing: An evaluation of models and evidence to support them. *Aphasiology, 16*, 107–150.

Martin, N., Saffran, E. M., & Dell, G. S. (1996b). Recovery in deep dysphasia: Evidence for a relation between auditory-verbal STM capacity and lexical errors in repetition. *Brain and Language, 52*, 83–113.

Martin, N., Weisberg, R. W., & Saffran, E. M. (1989). Variables influencing the occurence of naming errors: Implications for models of lexical retrieval. *Journal of Memory and Language, 28*, 462–485.

McCarthy, R. A., & Kartsounis, L. D. (2000). Wobbly words: Refractory anomia with preserved semantics. *Neurocase, 6*, 487–497.

McCarthy, R. A., & Warrington, E. K. (1986). Phonological reading: Phenomena and paradoxes. *Cortex, 22*, 359–380.

McClelland, J. L., & Patterson, K. (2002). Rules or connections in past-tense inflections: What does the evidence rule out? *Trends in Cognitive Sciences, 6*, 465–472.

McCloskey, M. Sokol, S., & Goodman, R. A. (1986). Cognitive processes in verbal-number production: Inferences from the performance of brain-damaged subjects. *Journal of Experimental Psychology: General, 115*, 307–330.

McKenna, P., & Warrington, E. K. (1978). Category-specific naming preservation: A single case study. *Journal of Neurology, Neurosurgery, and Psychiatry, 41*, 571–574.

Meijer, P. J. A. (1994). Towards a new model of phonological encoding. In A. Ram & K. Eiselt (Eds.), *Proceedings of the 16th annual conference of the Cognitive Science Society* (pp. 619–623). Hillsdale, NJ: Lawrence Erlbaum Associates Inc.

Meyer, D. E., & Gordon, P. C. (1985). Speech production: Motor programming of phonetic features. *Journal of Memory and Language, 24*, 3–26.

Miceli, G., Benvegnú, B., Capasso, R., & Caramazza, A. (1997). The independence of phonological and orthographic lexical forms: Evidence from aphasia. *Cognitive Neuropsychology, 14*, 35–70.

Miceli, G., Capasso, R., & Caramazza, A. (2004). The relationships between morphological and phonological errors in aphasic speech: Data from a word repetition task. *Neuropsychologia, 42*, 273–287.

Miceli, G., Capasso, R., Daniele, A., Esposito, T., Magarelli, M., & Tamaiuolo, F. (2000). Selective deficit for people's names following left temporal damage: An impairment of domain-specific conceptual knowledge. *Cognitive Neuropsychology*, 17, 489–516.

Miceli, G., & Caramazza, A. (1988). Dissociation of inflectional and derivational morphology. *Brain and Language, 35*, 24–65.

Miozzo, M., & Caramazza, A. (1997). On knowing the auxiliary of a verb that cannot be named: Evidence for the independence of grammatical and phonological aspects of lexical knowledge. *Journal of Cognitive Neuroscience, 9*, 160–166.

Moss, H. E., Tyler, L. K., Durrant-Peatfield, M., & Bunn, E. M. (1998). Two eyes of a see through: Impaired and intact semantic knowledge in a case of selective deficit for living things. *Neurocase*, 4, 291–310.

Nespoulous, J. L., Dordain, M., Perron, C., Ska, B., Bub, D., Caplan, D., Mehler, J., & Lecours, A. R. (1988). Agrammatism in sentence production without comprehension deficits: Reduced availability of syntactic structures and or grammatical morphemes? A case study. *Brain and Language, 33*, 273–295.

Nespoulous, J.-L., Joanette, Y., Béland, R., Caplan, D., & Lecours, A. R. (1984). Phonological disturbances in aphasia: Is there a "markedness effect" in aphasic phonetic errors? In F. C. Rose (Ed.), *Advances in neurology, Vol. 42: Progress in aphasiology* (pp. 203–214). New York: Raven Press.

Nespoulous, J.-L., Joanette, Y., Ska, B., Caplan, D., & Lecours, A. R. (1987). Production deficits in Broca's and conduction aphasia: Repetition versus reading. In E. Keller & M. Gopnik (Eds.), *Motor and sensory processes of language* (pp. 53–81). Hillsdale, NJ: Lawrence Erlbaum Associates Inc.

Nespoulous, J.-L., & Moreau, N. (1997). Repair strategies and consonantal cluster production in Broca's aphasia. In Y. Lebrun (Ed.), *From the brain to the mouth: Acquired dysarthria and dysfluency in adults* (pp. 71–80.). Dordrecht, The Netherlands: Kluwer.

Nespoulous, J.-L., & Moreau, N. (1998). "Repair strategies" and the production of segmental errors in aphasia: Epentheses vs. syncopes in consonantal clusters. In E. G. Visch-Brink & R. Bastiaanse

(Eds.), *Linguistic levels in aphasiology* (pp. 133–145). San Diego: Singular.

Nickels, L. (1992). The autocue? Self-generated phonemic cues in the treatment of a disorder of reading and naming. *Cognitive Neuropsychology, 9*, 307–317.

Nickels, L. (1995). Getting it right? Using aphasic naming errors to evaluate theoretical models of spoken word production. *Language and Cognitive Processes, 10*, 13–45.

Nickels, L., & Howard, D. (1995). Phonological errors in aphasic naming: Comprehension, monitoring, and lexicality. *Cortex, 31*, 209–237.

Nooteboom, S. G. (2004). Listening to oneself: Monitoring speech production. In R. Hartsuiker, Y. Bastiaanse, A. Postma, & F. Wijnen (Eds.), *Phonological encoding and monitoring in normal and pathological speech*. Hove, UK: Psychology Press.

Nooteboom, S. G. (2003). *Phoneme-to-word feedback and the origin of lexical bias in phonological speech errors*. Manuscript submitted for publication.

Ohala, J. J. (1983). The origin of sound patterns in vocal tract constraints. In P. MacNeilage (Ed.), *The production of speech* (pp. 189–216). New York: Springer-Verlag.

Ohala, J. J. (1990). There is no interface between phonology and phonetics: A personal view. *Journal of Phonetics, 18*, 153–171.

Page, M. (2000). Connectionist modelling in psychology: A latent manifesto. *Behavioral and Brain Sciences, 23*, 443–467.

Pate, D. S., Saffran, E. M., & Martin, N. (1987). Specifying the nature of the production impairment in a conduction aphasic: A case study. *Language and Cognitive Processes, 2*, 43–84.

Patterson, K., & Shewell, C. (1987). Speak and spell: Dissociations and word-class effects. In M. Coltheart, G. Sartori, & R. Job (Eds.), *The cognitive neuropsychology of language* (pp. 273–294). Hove, UK: Lawrence Erlbaum Associates Ltd.

Pierrehumbert, J. B. P. (1990). Phonological and phonetic representation. *Journal of Phonetics, 18*, 375–394.

Pierrehumbert, J. (2002). Word-specific phonetics. In C. Gussenhoven & N. Warner (Eds.), *Papers in laboratory phonology VII* (pp. 101–140). Berlin: Mouton de Gruyter.

Pierrehumbert, J. B. P., Beckman, M. E., & Ladd, D. R. (2000). Conceptual foundations of phonology as a laboratory science. In N. Burton-Roberts, P. Carr, & G. Docherty (Eds.), *Phonological knowledge: Conceptual and empirical issues* (pp. 273–304). Oxford: Oxford University Press.

Pillon, A., de Partz, M.-P., Raison, A.-M., & Seron, X. (1991). L'orange, c'est le frutier de l'orangine: A case of morphological impairment? *Language and Cognitive Processes, 6*, 137–167.

Pinker, S., & Ullman, M. T. (2002). The past and future of the past tense. *Trends in Cognitive Sciences, 6*, 456–463.

Plaut, D. C., & Shallice, T. (1993). Deep dyslexia: A case study of connectionist neuropsychology. *Cognitive Neuropsychology, 10*, 377–500.

Rapp, B., Benzing, L., & Caramazza, A. (1997). The autonomy of lexical orthography. *Cognitive Neuropsychology, 14*, 71–104.

Rapp, B., & Caramazza, A. (1998). A case of selective difficulty in writing verbs. *Neurocase, 4*, 127–140.

Rapp, B., & Caramazza, A. (2002). Selective difficulties with spoken nouns and written verbs: A single case study. *Journal of Neurolinguistics, 15*, 373–402.

Rapp, B., & Goldrick, M. (2000). Discreteness and interactivity in spoken word production. *Psychological Review, 107*, 460–499.

Rapp, B., & Goldrick, M. (2004). Feedback by any other name is still interactivity: A reply to Roelofs' comment on Rapp & Goldrick (2000). *Psychological Review, 111*, 573–578.

Raymer, A. M., & Berndt, R. S. (1996). Reading lexically without semantics: Evidence from patients with probable Alzheimer's disease. *Journal of the International Neuropsychological Society, 2*, 340–349.

Roelofs, A. (1992). A spreading-activation theory of lemma retrieval in speaking. *Cognition, 42*, 107–142.

Roelofs, A. (2004a). Error biases in spoken word planning and monitoring by aphasic and nonaphasic speakers: Comment on Rapp and Goldrick (2000). *Psychological Review, 111*, 561–572.

Roelofs, A. (2004b). Comprehension-based versus production-internal feedback in planning spoken words: A rejoinder to Rapp and Goldrick (2004). *Psychological Review, 111*, 579–580.

Roelofs, A., Meyer, A. S., & Levelt, W. J. M. (1998). A case for the lemma-lexeme distinction in models of speaking: Comment on Caramazza and Miozzo (1997). *Cognition, 69*, 219–230.

Romani, C. (1992). Are the distinct input and output buffers? Evidence from an aphasic patient with an impaired output buffer. *Language and Cognitive Processes, 7*, 131–162.

Romani, C., & Calabrese, A. (1998). Syllabic constraints on the phonological errors of an aphasic patient. *Brain and Language, 64*, 83–121.

Romani, C., Olson, A., Semenza, C., & Granà, A. (2002). Patterns of phonological errors as a function of a phonological versus an articulatory locus of impairment. *Cortex, 38*, 541–567.

Ruml, W., & Caramazza, A. (2000). An evaluation of a computational model of lexical access: Comment on Dell et al. (1997). *Psychological Review, 107*, 609–634.

Ruml, W., Caramazza, A., Shelton, J. R., & Chialant, D. (2000). Testing assumptions in computational theories of aphasia. *Journal of Memory and Language, 43*, 217–248.

Saetti, M. C., Marangolo, P., DeRenzi, E., Rinalid, M. C., & Lattanzi, E. (1999). The nature of the disorder underlying the inability to retrieve proper names. *Cortex, 35*, 675–685.

Saffran, E. M. (1982). Neuropsychological approaches to the study of language. *British Journal of Psychology, 73*, 317–337.

Sartori, G., Masterson, J., & Job, R. (1987). Direct route reading and the locus of lexical decision. In M. Coltheart, G. Sartori, & R. Job (Eds.), *Cognitive neuropsychology of language* (pp. 59–77). Hove, UK: Lawrence Erlbaum Associates Ltd.

Schade, U., & Berg, T. (1992). The role of inhibition in a spreading-activation model of language production II: The simulation perspective. *Journal of Psycholinguistic Research, 21*, 435–462.

Schmidt, D., & Buchanan, L. (2004). Selective sparing of written production of proper nouns and dates in aphasia. *Brain and Cognition, 55*, 406–408.

Schmidt, D., Buchanan, L., & Semenza, C. (2003, October). *Proper nouns and dates in aphasia.* International Mental Lexicon Research Group Meeting.

Schwartz, M. F. (1984). What the classical aphasia categories can't do for us, and why. *Brain and Language, 21*, 3–8.

Schwartz, M. F., & Brecher, A. (2000). A model-driven analysis of severity, response characteristics, and partial recovery in aphasics' picture naming. *Brain and Language, 73*, 62–91.

Schwartz, M. F., & Hodgson, C. (2002). A new multiword naming deficit: Evidence and interpretation. *Cognitive Neuropsychology, 19*, 263–288.

Schwartz, M. F., Saffran, E. M., & Marin, O. S. M. (1980). Fractionating the reading process in dementia: Evidence for word-specific print-to-sound associations. In M. Coltheart, K. E. Patterson, & J. C. Marshall (Eds.), *Deep dyslexia* (pp. 259–269). London: Routledge & Kegan Paul.

Schwartz, M. F., Saffran, E. M., Bloch, D. E., & Dell, G. S. (1994). Disordered speech production in aphasic and normal speakers. *Brain and Language, 47*, 52–88.

Schwartz, M. F., Wilshire, C. E., Gagnon, D. A., & Polansky, M. (2004). Origins of nonword phonological errors in aphasic picture naming. *Cognitive Neuropsychology, 21*, 159–186.

Semenza, C., Butterworth, B., Panzeri, M., & Ferreri, T. (1990). Word formation: New evidence from aphasia. *Neuropsychologia, 28*, 499–502.

Semenza, C., & Sgaramella, T. (1993). Production of proper names: A clinical study of the effects of phonemic cueing. *Memory, 1*, 265–228.

Semenza, C., & Zettin, M. (1988). Generating proper names: A case of selective inability. *Cognitive Neuropsychology, 5*, 711–772.

Semenza, C., & Zettin, M. (1989). Evidence from aphasia for the role of proper names as pure referring expressions. *Nature, 342*, 678–679.

Shallice, T. (1979). Case-study approach in neuropsychological research. *Journal of Clinical Neuropsychology, 1*, 183–211.

Shallice, T., Rumiati, R. I., & Zadini, A. (2000). The selective impairment of the phonological output buffer. *Cognitive Neuropsychology, 17*, 517–546.

Shallice, T., Warrington, E. K., & McCarthy, R. (1983). Reading without semantics. *Quarterly Journal of Experimental Psychology, 35A*, 111–138.

Shapiro, K., & Caramazza, A. (2001). Sometimes a noun is just a noun: Comments on Bird, Howard, and Franklin (2000). *Brain and Language, 76*, 202–212.

Shapiro, K., & Caramazza, A. (2003a). Looming a loom: Evidence for independent access to grammatical and phonological properties in verb retrieval. *Journal of Neurolinguistics, 16*, 85–112.

Shapiro, K., & Caramazza, A. (2003b). Grammatical processing of nouns and verbs in left frontal cortex? *Neuropsychologia, 41*, 1189–1198.

Shapiro, K., Shelton, J., & Caramazza, A. (2000). Grammatical class in lexical production and morphological processing: evidence from a case of fluent aphasia. *Cognitive Neuropsychology, 17*, 665–682.

Shattuck-Hufnagel, S. (1979). Speech errors as evidence for a serial-ordering mechanism in sentence production. In W. E. Cooper & E. C. T. Walker (Eds.), *Sentence processing: Psycholinguistic studies presented to Merrill Garrett* (pp. 295–341). Hillsdale, NJ: Lawrence Erlbaum Associates Inc.

Shattuck-Hufnagel, S. (1992). The role of word structure in segmental serial ordering. *Cognition, 42*, 213–259.

Stemberger, J. P. (1985). An interactive activation model of language production. In A. W. Ellis (Ed.), *Progress in the psychology of language, Vol. 1* (pp. 143–186). Hillsdale, NJ: Lawrence Erlbaum Associates Inc.

Tsapkini, K., Jarema, G., & Kehayia, E. (2001). A morphological processing deficit in verbs but not in nouns: A case study in a highly inflected language. *Journal of Neurolinguistics, 15*, 265–288.

Vigliocco, G., Vinson, D. P., Martin, R. C., & Garrett, M. F. (1999). Is "count" and "mass" information available when the noun Is not? An investigation of tip of the tongue states and anomia. *Journal of Memory and Language, 40*, 534–558.

Vousden, J. I., Brown, G. D. A., & Harley, T. A. (2000). Serial control of phonology in speech production: A hierarchical model. *Cognitive Psychology, 41*, 101–175.

Warrington, E. K., & McCarthy, R. A. (1987). Categories of knowledge: Further fractionations and an attempted integration. *Brain, 110*, 1273–1296.

Warrington, E. K., & Shallice, T. (1979). Semantic access dyslexia. *Brain, 102*, 43–63.

Wilshire, C. E. (2002). Where do aphasic phonological errors come from? Evidence from movement errors in picture naming. *Aphasiology, 16*, 169–197.

Wilshire, C. E., & McCarthy, R. A. (1996). Experimental investigations of an impairment in phonological encoding. *Cognitive Neuropsychology, 13*, 1059–1098.

Wilshire, C. E., & McCarthy, R. A. (2002). Evidence for a context-sensitive word retrieval disorder in a case of nonfluent aphasia. *Cognitive Neuropsychology, 19*, 165–186.

Wright, J. F., & Ahmad, K. (1997). The connectionist simulation of aphasic naming. *Brain and Language, 59*, 367–389.

Wu, D. H., Martin, R. C., & Damian, M. F. (2002). A third route for reading? Implications from a case of phonological dyslexia. *Neurocase, 8*, 274–295.

COGNITIVE NEUROPSYCHOLOGY, 2006, 23 (1), 74–95

The neuropsychology of sentence processing: Where do we stand?

Randi C. Martin
Rice University, Houston, TX, USA

In the early 1980s, sentence comprehension deficits were attributed to a loss of syntactic knowledge in agrammatic Broca's aphasics and to a short-term memory deficit in conduction aphasics. Findings in the remainder of the decade called both of these claims into question and presented general difficulties for the group study approach. Results from case studies support the representational independence of syntactic and semantic information but the interaction of these knowledge sources during processing. Working memory is still considered to provide critical constraints on sentence comprehension, but the capacity involved appears to be largely independent of the phonological storage involved in word list recall. Current computational approaches to sentence comprehension provide the means of accounting for the interaction of multiple sources of information and working memory requirements, but have yet to be tested against neuropsychological findings.

Part of the upsurge of interest in neuropsychological investigations of cognitive processes during the 1970s arose from findings that demonstrated striking deficits in sentence comprehension among patients who could understand the individual words in the sentence. A seminal study by Caramazza and Zurif (1976) showed that patients classified as Broca's aphasics or as conduction aphasics had difficulty matching a sentence to a picture for a sentence like "The tiger that the lion chased was yellow." Their incorrect picture choices were virtually always those that depicted a reversal of agent and patient. They did not choose pictures with a lexical distractor, that is, pictures that substituted an incorrect noun or verb. Wernicke's aphasics also had difficulty with the task but were equally likely to choose lexical distractor pictures as those depicting a reversal of agent and patient. Thus, the sentence comprehension difficulty of the Wernicke's aphasics might be attributed to difficulty in understanding individual word meanings, whereas the pattern for the Broca's and conduction aphasics pointed to a deficit in processes specific to understanding a sentence.

Much of the subsequent work focused on Broca's aphasics, whose speech is described as "agrammatic," that is, marked by simplified grammatical structure and the omission of function words and inflectional markers (Goodglass & Kaplan, 1972). A study by Schwartz, Saffran, and Marin (1980) showed that Broca's aphasics demonstrated a syntactic comprehension deficit not only for the complex centre-embedded relative clause sentences used by Caramazza and Zurif (1976) but also for simple reversible active and passive sentences (e.g., "The boy chased the dog" and "The boy was chased by the dog"). The co-occurrence of agrammatic speech and deficits in sentence comprehension related to syntactic

Correspondence should be addressed to Randi C. Martin, Psychology, MS-25, Rice University, PO Box 1892, Houston, TX 77251-1892, USA (Email: rmartin@rice.edu).

The writing of this manuscript was supported by NIH grant DC-00218 to Rice University. The author would like to thank Meredith Knight for her help with technical aspects of this manuscript.

DOI:10.1080/02643290500179987

processing led to the formulation of the syntactic deficit hypothesis (Berndt & Caramazza, 1980), that is, to the claim that Broca's aphasics had a general deficit in syntactic processing that was the source of both their production and comprehension patterns. This hypothesis had strong appeal to a number of constituencies in psychology and linguistics because, if true, the hypothesis would support the existence of a syntactic processing module, independent from semantics, that was drawn on in both comprehension and production. Given the Broca's aphasics' putative lesion localisation, this syntactic processing module was presumably localised to left inferior frontal cortex.

While the syntactic deficit hypothesis seemed to provide an elegant means of accounting for both the comprehension and production problems of Broca's aphasics, even the original findings of Caramazza and Zurif (1976) were problematic for the notion that there was a single syntactic processing module underlying production and comprehension. Specifically, the conduction aphasics in that study showed comprehension deficits indistinguishable from those of the Broca's aphasics but did not produce agrammatic speech. A possible explanation for the seeming dissociation came from other case studies of conduction aphasia that demonstrated verbal short-term memory deficits for these patients, particularly in phonological short-term memory. Consequently, the claim was made that the comprehension deficit of conduction aphasics derived from their short-term memory deficit (Caramazza, Basili, Koller, & Berndt, 1981; Friedrich, Martin, & Kemper, 1985). The argument was that phonological short-term memory was necessary for holding words in their correct order and for maintaining the representations of function words and inflections that had little semantic content but were critical for syntactic processing.

A number of reports from around the time of the founding of the journal *Cognitive Neuropsychology* provided a challenge to the syntactic deficit hypothesis and to the hypothesis that a short-term memory deficit might be the source of sentence comprehension deficits in patients who were not agrammatic speakers. The first section below provides a brief overview of these findings. The next section provides an update on these issues from more recent studies. The following section provides evidence from case studies regarding the structure of the language processing system and the working memory system that supports it. The final section deals with current theoretical issues and potential directions for future research.

STATUS OF SYNTACTIC DEFICIT HYPOTHESIS AND RELATED SHORT-TERM MEMORY DEFICIT HYPOTHESIS 1982–1989

Syntactic deficit hypothesis

In addition to the difficulties in accounting for the syntactic comprehension deficits of conduction aphasics, other problems arose for the syntactic deficit hypothesis shortly after its proposal (see Berndt, 1991, 1998; R. Martin, 2000, for overviews of this literature). Large-scale group studies of syntactic comprehension deficits in aphasia found a similar rank ordering of difficulty across different syntactic structures for all groups (i.e., Broca's, conduction, transcortical, Wernicke's), with differences between the groups being due mainly to differences in the overall level of impairment (Caplan & Hildebrandt, 1988; Naeser et al., 1987). Such studies might be criticised on the grounds that these groups' sentence comprehension difficulties could arise from different sources, despite the similarity in pattern. For example, more syntactically complex sentences (e.g., datives or relative clause sentences) would necessarily include more nouns than simple actives and passives, and if patients had difficulty comprehending these nouns, poorer performance on comprehension could result for sentences with more nouns. Also, short-term memory deficits could potentially account for greater difficulty with more complex sentences that had more words to be retained. It should be noted, however, that a case study of a mild Wernicke's aphasic found poor syntactic comprehension on simple active and passive sentences, even though the patient did not have agrammatic speech, poor noun comprehension, or a short-term memory deficit (R. Martin & Blossom-Stach, 1986).

Dissociations in the opposite direction were also observed—that is, patients were reported who produced agrammatic speech but who did not have syntactic comprehension deficits (Kolk, Van Grunsven, & Keyser, 1985; Miceli, Mazzucchi, Menn, & Goodglass, 1983; Nespoulous et al., 1988). Kolk and Van Grunsven (1985) reported the sentence–picture matching performance for 11 agrammatic speakers for reversible active and passive sentences (see Figure 1). Two of these patients scored 100% correct on both sentence types. The worst performance was observed for one patient who scored below 50% correct on both structures (where 50% would be chance). Other patients' scores fell in a continuum between these extremes, with some showing better performance on actives than passives but others showing similar levels of performance on both. Thus, not only did some of these patients show preserved comprehension, but it was also clear that patients did not demonstrate all-or-none deficits in syntactic processing. One might hypothesise that the variation in degree of deficit could be related to the degree of grammatical disruption in production. However, a study of four patients with some degree of agrammatism in speech found no match between the degree of production deficit and the degree of syntactic comprehension deficit (R.C. Martin, Wetzel, Blossom-Stach, & Feher, 1989).

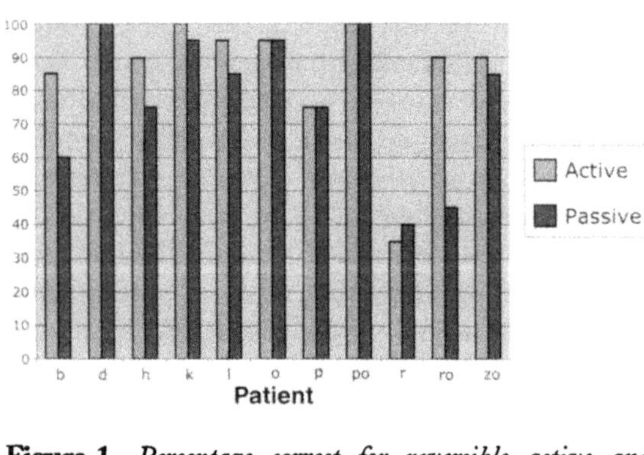

Figure 1. *Percentage correct for reversible active and passive sentences for agrammatic patients from Kolk and Van Grunsven (1985).*

A serious challenge to the syntactic deficit hypothesis was also presented by findings from a study by Linebarger, Schwartz, and Saffran (1983). These researchers examined grammaticality judgments on a variety of sentence constructions for patients who produced agrammatic speech as well as syntactic comprehension deficits on sentence–picture matching. Some examples of incorrect sentences are shown below:

1. I hope you to go to the store now.
2. Which are you going to give records to Louise?
3. Is the boy is having a good time?
4. This job was expected Frank to get.

The four patients tested in this study performed remarkably well on these grammaticality judgments, with mean performance on many of the constructions ranging from 80–90% correct. This was the case even though these patients scored at or below chance on simple passives, and some scored at chance on simple actives on sentence–picture matching. As can be seen in these examples, the ungrammaticality of the sentences often hinged on detecting errors in function words or inflections—the specific items thought to be most affected in agrammatic speech. A number of subsequent studies have replicated the findings of good performance on grammaticality judgments together with impaired performance on comprehension (e.g., Lu et al., 2000; Lukatela, Crain, & Shankweiler, 1988).

Some researchers have criticised studies employing grammaticality judgments on the grounds that these judgments require "off-line" metalinguistic skills, and thus draw on processes unlike those involved in natural language comprehension (Zurif & Grodzinsky, 1983). However, studies of grammaticality judgments that have used online processing techniques have also demonstrated preserved performance (Shankweiler, Crain, Gorrell, & Tuller, 1989; Wulfeck, 1988). Saffran and colleagues proposed the "mapping deficit" hypothesis to explain this dissociation (Linebarger, 1990; Saffran & Schwartz, 1988). According to this hypothesis, these patients are able to parse a sentence (i.e., analyse its syntactic structure) but are unable to carry out the mapping between grammatical roles (e.g., subject, object) and thematic roles (e.g., agent, theme) that is implied by this structural information.

It should be noted that some findings from early German aphasiologists had anticipated these later results that have caused difficulty for the syntactic deficit hypothesis. De Bleser (1987) and Howard (1985) published translations (Isserlin, 1922/1985) and commentary on some of these writings in *Cognitive Neuropsychology*. These early researchers reported aphasic patients who demonstrated agrammatic speech in production but preserved syntactic comprehension as well as patients with poor syntactic comprehension but preserved ability to detect grammatical errors in sentences.

Short-term memory deficits as the source of syntactic comprehension deficits

Caramazza et al. (1981) and Friedrich et al. (1985) reported the short-term memory and sentence comprehension abilities of two patients who fit the classification of conduction aphasia. Both had very reduced digit and word spans of one to two items. They performed poorly even on short-term recognition probe tasks, thus ruling out an output deficit as the source of the short-term memory deficit. Although the patient reported by Friedrich et al. (patient EA) had some difficulty in speech perception (Friedrich, Glenn, & Marin, 1984), this impairment was quite mild and could not account for her severe short-term memory deficit (R.C. Martin & Breedin, 1992). Both patients showed deficits in syntactic comprehension for reversible active and passive sentences and for centre-embedded subject and object relative clause sentences. A later study that included EA (R.C. Martin, 1987) showed that her comprehension of reversible passive sentences had improved to about 90% correct (across simple passives and relative clause sentences with a main clause passive), but she continued to have difficulty with centre-embedded relative clauses that had either an embedded passive or an object relative construction.

Vallar and Baddeley (1984a) reported the sentence comprehension abilities of a patient PV, who had a relatively pure deficit in verbal short-term memory. In a separate study (Vallar & Baddeley, 1984b), these authors demonstrated that PV had a deficit specifically in maintaining phonological information but had good phonological processing abilities. PV performed well on a syntactic battery (Parisi & Pizzamiglio, 1970); however, this battery did not include the complex centre-embedded subject and object relative clause sentences that were tested in the Carramazza et al. (1981) and R.C. Martin (1987) studies. Thus, it is difficult to determine the extent to which PV's syntactic comprehension may have been better than that of the other two patients.

Vallar and Baddeley's (1984a) main results with regard to sentence processing came from a task employing sentence anomaly judgments. On this task, PV performed well on short sentences or on longer sentences in which maintenance of word order was not crucial. However, PV did display comprehension problems when tested on anomaly detection in longer sentences where the maintenance of word order was critical. An example of such a sentence was "One could reasonably claim that sailors are often lived on by ships of various kind." Vallar and Baddeley argued that PV's poor performance resulted from having to maintain exact word order information for sentences that exceeded her "sentence span," which was found to be about 5–6 words.

While Vallar and Baddeley (1984b) refer to the need to retain word order, it is clear that a sentence like the example above requires the processing of syntactic structure—here, the passive structure. For their long sentences, for which word order was argued not to be critical (e.g., "Lettuce is the kind of person that one rarely meets in a schoolroom"), the anomaly could be detected on the basis of semantic incompatibility of the content words in the main clause. Thus, syntactic processing was not necessary to detect the error. Unfortunately, these authors apparently did not include short sentences in which syntactic processing was required (e.g., "Sailors are often lived on by ships"), thus one cannot determine whether length was a critical factor. Even though a few passives were included in the Parisi and Pizzamiglio (1970) battery, these passives were reversible (e.g., "The dog was chased by the boy"), and thus there were no strong semantic constraints in the fit of the nouns to thematic roles about the verb. As will be discussed later with respect to a more recent study by Saffran, Schwartz, and Linebarger (1998), patients' ability to process various syntactic structures can be compromised when there are strong semantic factors that lead to the assignment of thematic roles to nouns (as in sailors living on ships rather than vice versa) that contradict those indicated by the syntax. As suggested by Caplan, Vanier, and Baker (1986), it is possible that normal subjects handle such sentences by reviewing a phonological record of the sentence when they detect a discrepancy between a pragmatically based interpretation and one derived from full processing of the syntactic structure. For PV, such a review would be impossible because of her restricted phonological capacity.

From the above outcomes, it is clear that there was some evidence from this time period that patients with a phonological short-term memory deficit might have difficulty with the comprehension of long and complex sentences, though the exact role that phonological storage played in these deficits was not clear. As discussed by R.C. Martin (1987), the theoretical basis for the connection between phonological storage and syntactic processing had not been plainly laid out in prior studies. One notion was that a limited phonological capacity implied that sentence processing was carried out over a narrow window in terms of the words considered simultaneously (Caramazza et al., 1981), thus causing difficulty in constructing the appropriate syntactic structure. How this would give rise to difficulty in understanding passive sentences is not transparent. If patients can understand "the boy" as they hear it, followed by comprehension of "was chased," it is unclear why comprehension would fail for a sentence such as "The boy was chased by the dog." One might predict difficulties with comprehending sentences such as "The boy that carried the girl had red hair," if the patients mistakenly associated the girl with red hair (rather than the boy) because "girl" is closer than "boy" to the "had red hair" phrase. However, R.C. Martin (1987) found that patient EA, shown to have a phonological short-term memory deficit, scored 100% correct in interpreting the main clause in such sentences.

R.C. Martin (1987) found that EA did have difficulty comprehending centre-embedded relative clauses, but only when the embedded clause was either a passive or an object relative form. (She did not have difficulty when the main clause, rather than the embedded clause, was a passive.) This led to the "downstream buffer" hypothesis—that is, the hypothesis that phonological storage was important in holding subsequent parts of a sentence if processing of an immediately preceding section was difficult and incomplete by the time the subsequent parts were perceived. When processing of the difficult section was complete, the remainder of the sentence being held in the phonological buffer could be addressed. EA, because of her very restricted phonological capacity, had to either give up on processing the embedded clause and attend to the predicate of the main clause or complete processing of the embedded clause and lose the end of the sentence. This explanation assumed that EA's processing of the embedded clause was normal. However, a study of a developmental case, RE, with a phonological STM deficit (Butterworth, Campbell, & Howard, 1986) showed that she had no difficulty comprehending centre-embedded object relative clause sentences from the Test for Reception of Grammar (Bishop, 1983) (RE also performed at a normal level on a variety of other long and complex sentences such as "The bus is preceded by the train which the triangle is below"). Consequently, it seems that EA may have had some difficulty in processing these complex embedded clauses, and that, rather than this being an outcome of her phonological storage deficit per se, it was the interaction of her syntactic processing deficit and her short-term memory deficit that gave rise to her pattern of performance.

RECENT HISTORY OF STUDIES OF SYNTACTIC PROCESSING

A fixation on agrammatism

As discussed earlier, considerable evidence accumulated in the 1980s that not all patients with agrammatic speech had syntactic comprehension deficits. Subsequent studies confirmed the considerable variation in comprehension performance. Berndt, Mitchum, and Haendiges (1996) carried out a review of studies of Broca's aphasics that had assessed their comprehension of reversible active and passive sentences, with a total of 64 data sets from 42 patients being reviewed. They showed that approximately one third of the patients did well on both structures, about one third scored at chance on both structures, and the other third showed above-chance performance on actives and at-chance performance on passives. In addition to data showing substantial variation in comprehension impairment, other data called into question the consistency of the speech production pattern of those diagnosed as having agrammatic speech. Saffran, Berndt, and Schwartz (1989) showed that morphological (e.g., function word omissions) and structural (e.g., reduced sentence length) aspects of agrammatism could dissociate, as some patients showed preserved morphological production but reduced structure (see Rochon, Saffran, Berndt, & Schwartz, 2000, for an update on these findings). A few cases showing the reverse pattern have also been reported (e.g., Miceli et al., 1983; Nespoulous et al., 1988; Thompson, Fix, & Gitelman, 2002). Even among those patients showing morphological deficits, the specific problems that were noted have been found to vary considerably across patients (Miceli, Silveri, Romani, & Caramazza, 1989). Consequently, even the determination of what constitutes "agrammatism" in production is not clear-cut (Miceli, 1999).

Despite these problems with the syndrome approach, a number of researchers have persisted in putting forward hypotheses concerning the nature of the syntactic deficit in agrammatism. Some of these hypotheses take the approach that there is a disruption in some type of linguistic knowledge, whereas others hypothesise a disruption in syntactic processing mechanisms. Among the proposed linguistic deficit hypotheses are a deficit with certain types of moved elements (termed the trace deletion hypothesis—Grodzinsky, 1990), a deficit for structures in which there is a double dependency (Mauner, Fromkin, & Cornell, 1993), and a deficit with the closed-class vocabulary (Jarema & Friederici, 1994). Among the hypothesised processing deficits are slowed syntactic parsing (Haarmann & Kolk, 1991), slowed reactivation of noun phrases (Swinney, Zurif, & Nicol, 1989), and a delay in integration of lexical information (Swaab, Brown, & Hagoort, 1997). (These lists are not exhaustive.) Given the general problems with the syndrome approach, there seems to be little point in delving into the fine details of each of these different perspectives. However, some consideration will be given to the trace deletion hypothesis because it has had considerable currency in the literature.

Grodzinsky and colleagues (Grodzinsky, 1990, 1995; Grodzinsky, Pinango, Zurif, & Drai, 1999) have postulated that Broca's aphasics have difficulty in comprehension only for sentences that, according to linguistic theory, have certain moved elements, specifically, complete phrases. For example, in passive sentences (e.g., "The girl was carried by the boy"), the object of the verb ("the girl") is assumed to have been moved out of its deep structure position following the verb to a position before the verb, leaving behind a trace (see Radford, 1988). The linkage of this noun phrase to its trace is assumed to be critical for comprehension. These traces are assumed to be missing in the representation of syntactic structure for Broca's aphasics, resulting in poor comprehension of passives and other structures hypothesised to have moved noun phrases. One prediction of this approach is that these patients will show above-chance performance on active sentences and at-chance performance on passive sentences. Performance on actives should be good because this structure does not involve a trace. Performance on passives should be at chance because both the initial noun phrase and that in the "by" phrase should be assigned subject of the verb, with the patient randomly choosing between them. As clearly demonstrated by Berndt et al. (1996) in their review, not all agrammatic patients show this comprehension pattern. Grodzinsky et al. (1999) analysed a set of cases partially overlapping those reviewed by Berndt et al. and came to the conclusion that the data did support their claims. A critical component of their approach was to argue that data had to be analysed at the group level rather than at the single case level. However, this is clearly untenable if the claim is that *all* Broca's aphasics show a certain comprehension pattern. Their approach is rather like arguing that one can support the claim that *all* Norwegians have an IQ of 100 by showing that the group mean for a sample does not differ from 100. Caramazza, Capitani, Rey, and Berndt (2001) provided a thorough critique of the Grodzinsky et al. approach, pointing out the logical and statistical flaws in the reasoning.

Other studies that have set out to test specific predictions of the trace deletion hypothesis have obtained disconfirming results. Druks and Marshall (1994) reported one agrammatic patient who actually showed the reverse of the predicted pattern on actives and passives, scoring above chance on passives and at chance on actives. R.C. Martin (1987) found that agrammatic patients did not show below-chance performance on truncated passives, which might be predicted, given that there is no "by" phrase. Druks and Marshall examined comprehension in four Hebrew-speaking agrammatic patients, exploiting the fact that it is possible to have passives without a trace in Hebrew. They found that two patients scored above chance on actives, on passives with trace, and on passives without trace, whereas the other two patients scored above chance on actives but at chance on passives either with or without trace. In fact, Badecker, Nathan, and Caramazza (1991) pointed out over a decade ago that no single case had been reported that actually fit the trace deletion hypothesis by showing comprehension deficits for all sentence types with traces and good comprehension for those without. The same appears to be the case today. The patient reported by Hildebrandt, Caplan, and Evans (1987) came close, but did not have difficulty with relatively simple sentences assumed to have a trace of a moved noun phrase (e.g., passives and cleft object sentences).

Grodzinsky and colleagues account for the good performance of some patients in some grammaticality judgment studies on the grounds that the sentence types included in these studies did not test for sensitivity of the legality of moved elements. However, some of the sentence types that were used in the Linebarger et al. (1983) study did involve the processing of moved elements, and patients performed well on these sentence types (see Linebarger, 1990, for discussion).

A recent study by Grodzinsky and Finkel (1998) demonstrated difficulties for Broca's aphasics on grammaticality judgments for sentences having traces of complete phrases. However, Wernicke's aphasics also had difficulties with these sentences and there was no significant interaction between group and performance on sentences with and without traces. Moreover, this study did not report comprehension data for the same patients, so one cannot determine whether there was any correspondence between the accuracy of grammaticality judgments and performance on comprehension of sentences with moved elements.

Recently, Wilson and Saygun (2004) followed up on the Grodzinsky and Finkel study. They argued that the sentences with trace in Grodzinsky and Finkel seemed intuitively to be quite difficult compared to those without traces. Thus, they constructed easy and hard sentences with traces and easy and hard sentences without traces (with ease or difficulty verified by control subject performance). They tested six Broca's, eight Wernicke's, and eight anomic aphasics on the grammaticality judgments for these sentences and assessed their performance on sentence comprehension for actives and passives. The order of difficulty for the age-matched control subjects was that trace/hard sentences were the most difficult (and more difficult than other/hard), whereas trace/easy sentences were the easiest (and easier than the other/easy sentences). All of the patient groups performed more poorly than controls, but the Broca's and Wernicke's aphasics showed the same pattern as controls across sentence types. Only the comparison of the anomics vs controls revealed an interaction between group and sentence type; however, the pattern for the anomics was difficult to interpret. When the Broca's aphasics were divided into those who showed the sentence comprehension pattern predicted by the trace deletion hypothesis (i.e., above chance on actives and at chance on passives) and those who did not, no difference in performance on the grammaticality judgments was found. In fact, it should be noted that only two of the Broca's aphasics showed the predicted pattern, along with two Wernicke's and one anomic patient.

When relating patient performance on the grammaticality judgments to lesion site, Wilson and Saygun (2004) found that damage to a posterior temporal region was most highly associated with impaired performance on grammaticality judgments, whether scored in terms of overall accuracy or performance on sentences with trace. These authors suggest that the brain areas involved in syntactic processing are distributed widely across the left perisylvian region. They also noted that they found considerable impairment in performance on grammaticality judgments for all groups, in contrast to most previous findings. The source of this difficulty is unclear, though the sentence types were substantially different from those in most previous studies.

Case study evidence for the independence of semantic and syntactic information

The original findings of Caramazza and Zurif (1976) generated a great deal of excitement because they appeared to demonstrate the independence of syntactic and semantic knowledge by showing that syntactic knowledge could be selectively disrupted. However, several studies have shown that at least some agrammatic Broca's aphasics do not have difficulty with grammaticality judgments on a variety of structures. Thus, many patients do not show a complete deficit in syntactic processing (and some actually do quite well even on comprehension tests). Although Wilson and Saygun's (2004) study did find difficulties with grammaticality judgments for their patients, their results suggested that a posterior temporal region was critical for such judgments; however, a similar area has been implicated in lexical and semantic processing (Hillis et al., 2001). Thus, patients who performed poorly in the Wilson and Saygun study on sentence comprehension and grammaticality judgments may have also performed poorly on tests of lexical or semantic knowledge.

To determine whether one may, in fact, see selective disruptions or selective preservations of syntax, it is necessary to look to single case studies where all the various aspects of language processing have been documented thoroughly. As reviewed in the next sections, some such studies have been carried out, with results supporting the independence of representations of syntax and semantics.

Preservation of syntax with impaired semantics. Some striking evidence for the independence of syntax and semantics has come from the examination of cases with Alzheimer's or semantic dementia or with progressive aphasics who have severe disruptions of semantic knowledge but show evidence of preserved syntactic abilities (Breedin & Saffran, 1999; Hodges, Patterson, & Tyler, 1994; Kempler, Curtiss, & Jackson, 1987; Schwartz & Chawluk, 1990; Schwartz, Marin, & Saffran, 1979). For example, Hodges et al. (1994) tested a semantic dementia case, patient PP, on a variety of semantic and syntactic tests. PP was impaired on various semantic tasks including naming, picture–word matching, and attribute judgments, with performance approaching chance levels as the dementia progressed. On a sentence processing task involving word monitoring, which is assumed to tap "online" processing (see Tyler, 1992), PP showed a normal pattern of sensitivity to grammatical structure and grammatical violations, but, unlike controls, was unaffected by the semantic properties of the sentences.

In their study of another semantic dementia case (DM), Breedin and Saffran (1999) showed that this patient's performance on a grammaticality test tapping a wide range of grammatical structures remained at the same high level (95%—within normal range) during a time period in which performance on semantic tasks declined dramatically. Using a sentence enactment task, they further demonstrated that DM had a preserved ability to assign thematic roles based on sentence structure. In this task, DM heard a sentence and then saw an enactment of an action that either matched the sentence or reversed the roles of agent and patient. The patient was then asked to point to one of the participants in the action. The two participants for each sentence were chosen to be ones that the patient could not discriminate on single word comprehension tests (e.g., lion and tiger). So, for example, the experimenter might demonstrate a tiger biting a lion, and on a matching trial the patient would hear "The lion was bitten by the tiger. Show me tiger." On a nonmatching trial the patient would hear "The tiger was bitten by the lion. Show me lion." The sentences included subject (e.g., "That's the lion that bit the tiger") and object cleft (e.g., That's the lion that the tiger bit") constructions in addition to active and passive sentences. On the matching trials, DM performed flawlessly on all of these structures. On the nonmatching trials, the patient routinely chose the wrong animal, indicating that he was making his choice on the basis of the syntactic information in the sentence rather than on lexical-semantics. That is, the evidence indicated that the patient had analysed the structure of the sentence to determine, for example, that in a passive sentence such as "The tiger was bitten by the lion" that "lion" was the agent. The patient then had to be able to analyse the enactment sufficiently to determine what entity was carrying out the action and point to that entity. An earlier study along these lines by Schwartz and Chawluk (1990) reported similar findings. Thus, these results indicate that some patients with severe semantic deficits can process grammatical structure (including the long-distance dependencies between the relative clause and the head noun in cleft object sentences), and map grammatical structure onto thematic roles even though the semantics of single words may be severely disrupted.

Preservation of semantics with impaired syntax. Reported cases demonstrating the reverse dissociation are rarer, perhaps because of the difficulty of using online measures to demonstrate impaired knowledge of grammatical structure together with preserved semantic processing of sentences. That is, it is necessary to demonstrate a sensitivity to semantic aspects of sentences together with an insensitivity to grammatical aspects of sentences, ideally using measures that do not require conscious reflection on the part of the patient. Poor performance on overt judgments of grammaticality might plausibly be attributed to poor metalinguistic abilities, and thus would not necessarily constitute evidence for a disruption of syntactic knowledge. A dissociation between syntax and semantics using online test performance was obtained by Ostrin and Tyler (1995) for a patient JG, who had a left temporo-parietal lesion. JG showed a marked disruption of all syntactic abilities together with relatively preserved lexical-semantic abilities. In a standard sentence–picture matching paradigm, he showed an asyntactic comprehension pattern, but unlike the patients reported in Linebarger et al. (1983), he performed poorly on a grammaticality judgment task. Furthermore, in a series of word monitoring tasks, he showed an insensitivity to a variety of grammatical violations—violations of subcategorisation frame, violations of inflectional and derivational morphology, and, in a previous study, violations of word order (Tyler, 1992). However, like normal subjects, he showed semantic priming in a lexical decision task. This case thus showed the reverse pattern to the case of Hodges et al. (1994) (described above) on a similar set of tasks.

Interactions of lexical, semantic, and syntactic factors in comprehension

Findings from neurally intact individuals. Although the neuropsychological evidence suggests that syntax and semantics may be represented independently, there is considerable evidence from recent studies of normal subjects that these factors interact on a moment-by-moment basis during processing to determine the interpretation of sentence meaning. Much of the work with normal subjects has used syntactically ambiguous sentences to determine the point at which lexical and semantic factors have an effect on the resolution of the ambiguity. An influential model of processing, the "garden path" model, held that initial parsing decisions were made solely on the basis of heuristics related to syntactic representations (specifically, "late closure" and "minimal attachment" heuristics) (Frazier, 1987; Frazier & Rayner, 1982). Semantic and lexical factors were only used to revise a parsing decision if the application of these syntactic heuristics resulted in a semantically or syntactically anomalous representation. Recent evidence, however, suggests that lexical and semantic factors have an immediate effect on processing. For instance, for sentences with an ambiguity in noun phrase vs verb phrase attachment of a prepositional phrase (e.g., "She saw the boy with binoculars"), Spivey-Knowlton and Sedivy (1995) showed that initial decisions on attachment of the prepositional phrase were related to the type of verb—that is, there was a bias for noun phrase attachment if the verb was a perception verb like "saw," but a bias toward verb phrase attachment for an action verb like "hit." For sentences with a main verb vs reduced relative ambiguity in the interpretation of the verb (e.g., "The man *sent* the check was happy"), Trueswell, Tanenhaus, and Garnsey (1994) showed that the plausibility of the initial noun as an agent vs theme of the verb influenced initial parsing decisions. Consider, for example, the sentences:

1. The defendant examined by the lawyer was unreliable.
2. The evidence examined by the lawyer was unreliable.

A garden path effect was found for sentences like (1) in which the initial noun is plausible as the agent of the verb "examined." For such sentences, subjects are surprised by the "by the lawyer" phrase and show slower processing in this region than in the corresponding region in a nonambiguous sentence (e.g., "The defendant who was examined by the lawyer was unreliable"). No garden path effect was found for sentences like (2), in which the initial noun ("evidence") was a much more plausible theme than agent of the verb "examined."

Even the discourse or environmental context has been found to influence initial parsing decisions (Spivey & Tanenhaus, 1998; Spivey, Tanenhaus, Eberhard, & Sedivy, 2002). For example, Spivey et al. showed that if a visual scene included an apple on a towel and an apple not on a towel, then college-age participants immediately interpreted the "on the towel" phrase in a sentence such as "Put the apple on the towel in the box" as indicating which apple rather than indicating the location to put the apple. In a scene with only one apple (and that apple resting on a towel) and an empty towel, subjects were likely to interpret "on the towel" as the place to move the apple initially.

Neuropsychological findings. Two recent studies have demonstrated effects of lexical bias on sentence comprehension of aphasic patients. Gahl (2002) used a sentence anomaly task and manipulated whether the structure of each sentence matched the transitive or intransitive bias of the verb (i.e., the relative frequency with which the verb appears in either form). For example, the verb "crumble" has an intransitive bias, so the intransitive sentence "The crackers crumbled in our hands" would match the verb's bias, where the transitive sentence "The children crumbled the crackers" would have a mismatch between bias and structure. Gahl showed that a mixed group of aphasic patients performed at a higher level when the structure of the sentence matched the verb's bias for being either transitive or intransitive. In a separate study, Gahl et al. (2003) further demonstrated for a mixed group of aphasic patients that comprehension of the passive was significantly better for passive-bias verbs (e.g., "elect") than for active-bias verbs (e.g., "disturb"). In neither study was there any evident difference in the pattern for the Broca's aphasics and that for the fluent patients.[1]

Results indicating an interaction between syntax and semantics have been reported in the neuropsychological literature as well (Saffran, Schwartz, & Linebarger, 1998; Tyler, 1989). Saffran et al. tested aphasic patients who showed syntactic comprehension deficits on sentence–picture matching tasks, including five Broca's aphasics, a conduction aphasic, and a transcortical motor aphasic. A sentence anomaly task was used, in which all of the sentences had two nouns—one playing the role of agent or experiencer and one playing the role of patient. Two types of sentences were used: (1) verb constrained and (2) proposition based. In the verb-constrained sentences, one of the nouns was implausible as a filler of one thematic role of the verb, but the other noun was plausible in either role. For example, in the implausible sentence "The deer shot the hunter," a hunter can shoot or be shot, but a deer can only be shot. In the proposition-based sentences, both nouns could fill either role; however, for the implausible versions, the overall proposition was implausible. For example, in the implausible sentence "The insect ate the robin," both robins and insects can eat and be eaten, but it is implausible for something as small as an insect to eat a robin. For both sentence types, active, passive, cleft subject and cleft object versions of each of the sentences were presented.

For the control subjects, reaction times for the plausible sentences were faster for the verb-constrained sentences than for the proposition-based sentences, but for the implausible sentences, more errors were made on the verb-constrained sentences (4.7%) than on the proposition-based sentences (1.3%). The control subjects apparently had some degree of difficulty suppressing the tendency to interpret the implausible verb-constrained sentences by assigning nouns to their most semantically plausible slot, even though the syntax indicated otherwise. Recently, Ferreira (2003) reported related results showing that young normal subjects showed some difficulty in identifying the actor in implausible sentences such as "The mouse was eaten by the cheese." She noted an effect of syntactic complexity, as across experiments about 20% errors were made on passive sentences compared to less than 5% on active sentences.

The patients in the Saffran et al. (1998) study showed an exaggeration of the error effect demonstrated by the control subjects, performing much worse on the implausible verb-constrained sentences than on the implausible proposition-based sentences (see Figure 2). Even for a simple active verb-constrained sentence such as "The deer shot the hunter" the patients often said that the sentence was plausible, presumably because role assignments had been made on the basis of semantic constraints rather than on the basis of syntactic structure. In contrast, on the proposition-based sentences, with weak semantic constraints on role filling, they made only 12–17% errors on the active, passive, and cleft subject sentences. Even on the cleft object sentences, performance was above chance when considering the plausible and implausible sentences together (36% errors overall). Thus, the patients' sensitivity to syntactic structure was evident when strong semantic constraints on role filling were absent. The

[1] Gahl (2002) found no significant effect for verb bias for the Broca's aphasics, although the effect was significant for the fluent aphasics. However, there were only 5 Broca's aphasics, but 12 fluent aphasics, and thus the lack of significance for the Broca's aphasics is most likely due to a lack of power. In fact, the size of the difference between sentences that matched or did not match verb bias was at least as large for the Broca's aphasics. No direct statistical comparison was made between the results for the Broca's aphasics and the fluent aphasics.

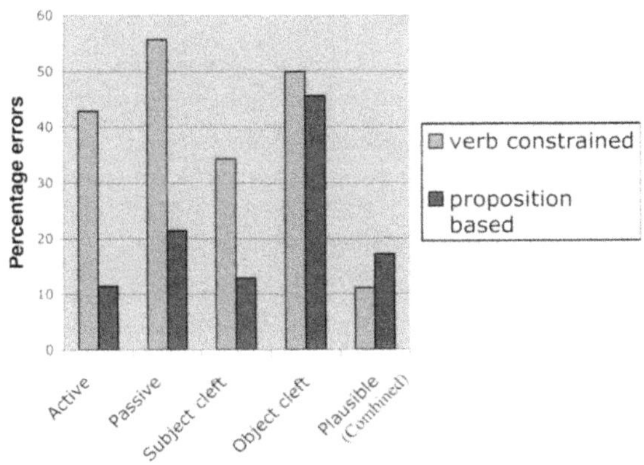

Figure 2. *Percentage correct on sentence anomaly judgments for verb-constrained and proposition-based sentences from Saffran et al. (1988) for different types of implausible sentences and plausible sentences combined across type.*

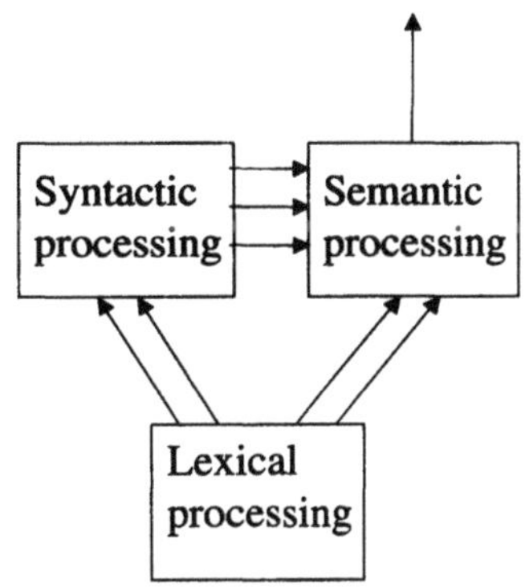

Figure 3. *Boland's (1997) concurrent model.*

results imply a weakened, though not totally disrupted, influence of syntactic structure and a stronger-than-normal role of semantic influences on sentence comprehension.

It should be noted that the results of the Saffran et al. (1998) study are not consistent with the trace deletion hypothesis, given the patients' poor performance on all sentence types (including simple actives) on the implausible verb-constrained sentences. Also, the results for the conduction aphasic and transcortical motor aphasic were similar to those obtained for the Broca's aphasics, providing further evidence that the pattern of sentence comprehension deficits observed for Broca's aphasics is not unique to this group.

Relations to current theorising

The findings reviewed earlier supported independence of the semantic and syntactic representations but processing interactions between them. One means of accounting for these findings is in terms of Boland's (1997) concurrent model of sentence processing (see Figure 3). According to this approach, syntactic and semantic representations are generated simultaneously as each word in a sentence is processed. At points of syntactic ambiguity, all possible syntactic interpretations are generated in parallel, and each is weighted according to lexical factors or frequency of the structure in the language. (For example, a higher weight would be given to a noun phrase attachment of a prepositional phrase for a perception verb and a higher weight would be given to a main verb than to a reduced relative interpretation of a verb in general, because of the infrequency of reduced relative constructions.) Semantic factors then determine which of the possible syntactic alternatives is finally selected.

A somewhat different model, termed the competitive integration model, has been proposed by Spivey and Tanenhaus (1998). In this model, there are various factors influencing the interpretation of a reduced relative/main clause ambiguity at the verb, all of which act simultaneously. As depicted in Figure 4, language frequency gives a higher weight to a main verb than to a reduced relative interpretation. However, discourse factors can affect the relative plausibility of the main verb vs reduced relative interpretation. Plausibility of the initial noun phrase as agent vs theme of the verb (not shown in model) would also have an effect. As their model was designed to account for results from reading, they also included parafoveal information beyond the verb, which could bias the interpretation. That is, the presence of a "by" following the verb would support the reduced relative interpretation. While both models assume an immediate effect of

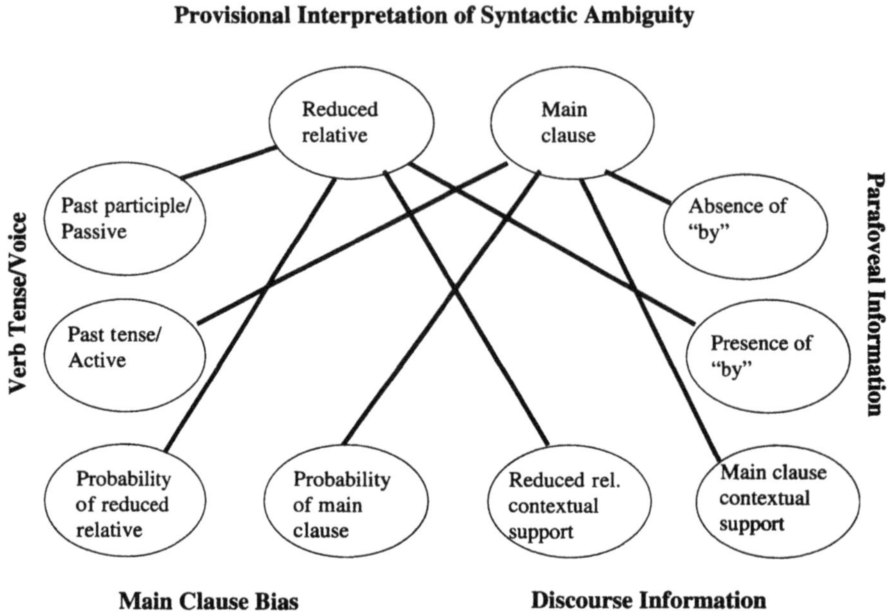

Figure 4. *Competitive integration model based on Spivey and Tanenhaus (1998), Figure 1.*

lexical, semantic, and syntactic factors, it is clear, at least in Boland's model, that there are separable syntactic and semantic modules that could be selectively affected by brain damage. In the Spivey and Tanenhaus model, this is less clear because the model depicts the analysis of only a specific syntactic ambiguity. One would have to assume some linkage between the nodes for past tense vs passive participle interpretations and syntactic interpretations for other structures if this model is to account for a selective but thoroughgoing disruption of syntactic processing.

The results from the Saffran et al. (1998) study showing an effect of verb constraint could be interpreted within the Boland (1997) or Spivey and Tanenhaus (1998) models by assuming that the weights of all syntactic representations have been reduced for the patients they tested. When there are strong semantic constraints on the assignment of nouns to thematic roles about the verb, these semantic constraints override the relatively weak outputs of the syntactic system, in which role assignments are based on syntactic structure. When there are no or only weak semantic constraints, then the results of syntactic processing can play a larger role. Similarly, the results from Gahl et al. (2003), showing better comprehension of passives for passive-bias verbs, could be explained on the grounds that a greater weight is given to a passive interpretation for a passive-bias verb than for an active-bias verb. Although early theorising about syntactic processing and representation may have led to the prediction of all-or-none loss, the assumption of a decrease in syntactic weights would allow for continuous variation in the degree of impairment, such as that reported by Kolk and Van Grunsven (1985).

For some researchers, the fact that damage was not all-or-none was taken as evidence that patient deficits should not be attributed to a loss of syntactic knowledge but instead to some type of working memory deficit (e.g., Miyake, Just, & Carpenter, 1994). As mentioned above, however, the application of the Boland (1997) or Spivey and Tanenhaus (1998) models to patient data allows for continuous variation in the strength of syntactic representations. One might question whether it is still necessary to postulate working memory deficits as the source of some comprehension deficits. Certainly some structures would seem to require some type of working memory capacity – for example when a noun that appears early in a sentence needs to be linked to verb that appears later, as in cleft object sentences such as "It was the cake that the boy liked." The models described above do not provide explicit descriptions of how such connections are made. Some recent approaches to syntactic processing give a prominent role to working memory requirements in predicting the difficulty of different syntactic structures (Gibson, 2000; Gordon, Hendrick & Johnson, 2001; Van Dyke & Lewis, 2003). Consequently, there still seems to be a need to postulate working memory involvement and to consider the potential negative consequences of working memory limitations. The next section considers recent neuropsychology evidence on the relation between syntactic processing and working memory.

RECENT HISTORY OF STUDIES RELATING SHORT-TERM MEMORY AND SENTENCE COMPREHENSION

From 1989 to the present, additional data have accumulated documenting the preservation of syntactic processing for patients with short-term memory deficits (Friedmann & Gvion, 2003; Hanten & Martin, 2001; R.C. Martin, Blossom-Stach, Yaffee, & Wetzel, 1995;[2] Waters, Caplan, & Hildebrandt, 1991). A particularly striking case was presented by Waters et al. Whereas patient BO had very restricted span and failed to show normal phonological similarity and word length effects on memory span, she showed excellent comprehension of a variety of long and syntactically complex sentences. Waters et al. attributed her short-term memory deficit to a difficulty in rehearsal, which was consistent with the patient's articulatory difficulties. However, given that the phonological similarity effect persists in normal subjects for auditory presentation under articulatory suppression (Longoni, Richardson, & Aeillo, 1993), the failure to observe this effect with BO suggests that she had difficulty with phonological storage in addition to whatever rehearsal deficit she might have had. In fact, a number of studies have shown that articulatory deficits per se do not cause a disruption of either phonological similarity or word length effects (Baddeley & Wilson, 1985; Bishop & Robson, 1989; R.C. Martin et al., 1995). Thus, BO most likely also represents a case with restricted phonological storage capacity together with excellent comprehension of syntax.

Some recent findings from McElree and colleagues (McElree, 2000; McElree, Foraker, & Dyer, 2003) with normal subjects also indicate a difference between the kind of memory involved in sentence comprehension and that involved in typical serial recall tasks. McElree and Dosher (1993) demonstrated that in word list recall tasks, item recognition involved parallel, content-addressable searches through memory, whereas recall of order information required a slow, serial search process. To the extent that the retention of order information depends on retention of phonological codes (Nimmo & Roodenrys, 2004), this slow, serial search process presumably operates on phonological representations. In contrast, during sentence processing, there is evidence that previously processed constituents can be accessed in a parallel, content-addressable fashion, even for sentences in which the order of constituents is critical (McElree et al., 2003; see also Van Dyke & Lewis, 2003).

Caplan and Waters (1999) reviewed the evidence from normal and brain-damaged populations on the relation between short-term memory, working memory, and sentence processing. Some researchers have argued that while

[2] Patient MP reported by Martin et al. (1995) showed reduced span and an absence of phonological effects for visually presented word lists, whereas his span for auditory lists was normal. His comprehension of visually presented relative clause sentences, even when presented rapidly in a word-by-word fashion, was normal.

short-term memory, as measured by span tasks, may be unrelated to sentence processing ability, measures of capacity that tap both processing and storage do relate to comprehension (Just & Carpenter, 1992). One such measure is reading span, in which subjects are asked to read a set of sentences aloud and recall the sentence-final words from each sentence at the end of the set. As discussed by Caplan and Waters (1999), however, results from a number of studies relating either working memory capacity (as from reading span) or the effect of an external load on sentence processing have failed to find an interaction between these variables and the effect of syntactic complexity. These results, along with the patient findings showing dissociations between short-term memory and sentence processing, led Caplan and Waters to postulate that there is a working memory capacity specific to the initial phases of sentence processing that is independent of the capacities tapped by short-term and working memory tasks. They divide the procedures involved in sentence processing into interpretive and post-interpretive processes. Interpretive processes include all online syntactic and semantic processes, including those involved in semantic interpretation based on the ongoing discourse. Post-interpretive processes involve using the products of interpretive processing to carry out some task, such as sentence–picture matching or enactment of the action in the sentence. Caplan and Waters argue that interpretive processing draws on a capacity specific to sentence processing, whereas post-interpretive processing draws on the capacity tapped by standard span tasks or working memory tasks.

Martin and colleagues (Hanten & Martin, 2000; R.C. Martin & He, 2004; R.C. Martin & Romani, 1994; R.C. Martin, Shelton, & Yaffee, 1994) have provided a different view on the relation between the capacities involved in span tasks and sentence processing. Specifically, they argue that span tasks tap both phonological and semantic retention (see also N. Martin & Saffran, 1997). The phonological component of span tasks is independent of the capacity involved in sentence processing. On the other hand, the semantic component does play a role in sentence comprehension in the maintenance of word meanings prior to their integration with other word meanings. Supporting this contention were results from patients with a semantic retention deficit who had difficulty detecting the semantic anomaly in sentences with several adjectives preceding a noun (e.g., "The rusty old red swimsuit") or with several nouns preceding a verb (e.g., "Rocks, trees, and shrubs grew in the back yard"). These patients did better when the adjectives followed the noun (e.g., "The swimsuit was old, red and rusty) or the nouns followed the verb (e.g., "The gardener grew shrubs, trees and rocks in the backyard") (Hanten & Martin, 2000; R.C. Martin & He, 2004; R.C. Martin & Romani, 1994). Martin and colleagues argued that when the adjectives preceded the noun, they had to be maintained as individual word meanings until the noun was processed, whereas when the adjectives followed the noun, they could be integrated with the noun as each was heard. Similarly, when the conjoined nouns preceded the verb, the role of the nouns with respect to the verb could not be determined until the verb was processed, whereas when the nouns followed the verb, their role with respect to the verb could be determined as each was heard. The patients' better performance in the "after" than the "before" conditions indicate that the patients were better able to maintain integrated semantic representations than individual word meanings. (See Haarmann, Davelaar, & Usher, 2003, for related results from normal subjects.) Martin and colleagues agree with Caplan and Waters (1999) to some extent, as both groups assume that the retention of specifically syntactic structural information is independent of both phonological and semantic capacities (R.C. Martin & Romani, 1994).

Recently, Gibson (2000) has proposed the distance-based locality theory (DLT) of sentence processing, which assumes that one major factor affecting comprehension difficulty for different syntactic structures is integration complexity. Integration complexity is determined by the distance between elements requiring integration in terms of the number of new discourse referents that intervene. (The other factor is the number of incomplete syntactic predictions that must be maintained.) This approach thus accounts for the greater difficulty of object than subject relatives on the grounds that, when processing the embedded clause verb of the object relative (for example, "carried" in "The boy that the girl carried liked the teacher"), the subject noun of the main clause ("boy") has to be integrated as the object of the embedded verb, and the embedded clause subject ("girl") intervenes between the two. In contrast, when processing the embedded clause verb for subject relatives ("The boy that carried the girl liked the teacher"), the main clause subject noun has to be integrated as the subject of the embedded verb, but no new noun phrase intervenes between the two. The greater difficulty of the sentences with prenominal than post-nominal adjectives and with subject rather than object conjoined noun phrases could also be explained in terms of this theory, if the theory is expanded to consider new adjectives in the same fashion as new discourse referents. For instance, when comprehending a noun phrase with several prenominal adjectives, the preceding adjectives would have to be reactivated to integrate with the noun after it was perceived. One would expect this integration to be more difficult as the number of prenominal adjectives increases. For post-nominal adjectives, each adjective would be integrated with the preceding noun as it was processed. This integration would cause reactivation of the noun each time an adjective was integrated with it. Thus, the noun should be maintained in an activated state. Consequently, the findings for patients and controls in the R.C. Martin and Romani study (1994; see also R.C. Martin & He, 2004) could be explained by the DLT and by the assumption that patients with a semantic short-term memory deficit would have particular difficulty with reactivation for words at greater distance.

The interpretation of R.C. Martin and Romani's (1994) findings in terms of Gibson's (2000) theory would lead to the prediction that these patients would show difficulties for more complex sentences requiring reactivation, such as centre-embedded object relatives. Given the dissociations some patients show between phonological short-term memory and the ability to process such complex structures, this reactivation does not depend, apparently, on the maintenance of a phonological form of the earlier noun. One would expect, however, that a semantic representation would have to be accessed, and thus a semantic short-term memory deficit should cause difficulties. Investigations of this prediction are currently under way.

FUTURE DIRECTIONS

Considerable data have been collected on sentence processing in aphasia, and many interesting results have been obtained. Unlike other areas of language processing, however, there have been few intensive case studies that have delineated specific deficits and used the findings to argue for or against specific theories. Although some studies have examined specific deficits with respect to, for instance, verb processing (Berndt, Haendiges, Mitchum, & Sandson, 1997; Breedin & Martin, 1996) or aspects of syntactic knowledge (see case studies in Caplan & Hildebrandt, 1988; Hildebrandt et al., 1987), the bulk of studies have been group studies. Given the general problems with the group study approach discussed here, the conclusions that can be drawn from such studies are severely limited.

Indeed, it seems that the field has been hamstrung by the emphasis on the group study approach and on agrammatism as a syndrome. Research in the area has been dominated by the question of "What characterises the syntactic comprehension deficit in agrammatism?" There is no answer to this question because there is no one coherent syndrome of agrammatism, either in production or in comprehension. Moreover, the group study approach dooms the field to the fate of all such arguments about syndromes, namely, futility. Different research groups obtain different findings, probably due to the fact that their group of agrammatic aphasics differs from someone else's group on a number of theoretically important dimensions. Findings from one research group are dismissed by other groups on the grounds that the patients probably weren't "truly" agrammatic Broca's aphasics (see Caramazza et al., 2001, for discussion). The dissension in the field caused by these types of arguments, along with a general lack of progress, is unlikely to entice new researchers into the area, unless they have a particular penchant for strident argument.

An opposing trend in the field is to claim that all patients with language disorders show basically the same pattern of sentence comprehension performance, irrespective of clinical classification or lesion site (e.g., Naeser et al., 1987). The conclusion drawn from these findings by many is that sentence processing abilities are widely distributed and, therefore, selective deficits are unlikely to occur (e.g., Dick et al., 2001; Wilson & Saygun, 2004). Such a conclusion seems unwarranted at present. As discussed by R.C. Martin (1995), and exemplified in some of the studies reported here, dramatic differences in language processing abilities across patients have been documented. Consider, for example, the dissociations between syntactic and semantic processing demonstrated in the case studies described earlier (e.g., Breedin & Saffran, 1999; Ostrin & Tyler, 1995). It is perhaps not too surprising that, in general, patients would show the same pattern of difficulty across sentence types as is shown by normal control subjects. The more complex the sentence type, the more factors that are likely to be involved in order for the correct analysis to be made. The more factors that are involved, the more likely that one or more of these will be affected for a given patient. Also, factors that are not specific to syntactic processing could increase the difficulty of processing more complex sentences. So, for example, if speech perception is somewhat impaired, such that the patient has difficulty perceiving function words or inflections that have low energy in the acoustic waveform, then comprehension that depends on these elements would be particularly impaired. (In fact, Wilson & Saygun, 2004, raise a related possibility to explain their results on grammaticality judgments.) Thus, case studies are needed that carefully examine all aspects of the patient's language processing abilities in order to determine the source of an observed sentence processing disorder.

Although it seems that thorough case studies are needed, practical considerations also weigh against researchers taking this approach. Sentence processing is obviously very complex, with many facets, and designing materials to address specific hypotheses while ruling out possible alternatives is a daunting task. In addition to the difficulties in developing tasks and materials, several studies have shown that patient performance can vary tremendously on different tasks that presumably tap the same knowledge. For example, Cupples and Inglis (1993) demonstrated that a patient who showed a syntactic comprehension deficit on sentence–picture matching showed excellent performance on the same sentence types for actor identification. A more recent study by Inglis (2003) showed that patient performance could be influenced by problem-solving abilities. Consequently, one would like to obtain online measures of processing that are not open to conscious control; however, obtaining reaction time data from brain-damaged patients is also a challenge, given possible motor control problems and the need to collect data from a large number of trials in any condition. Despite these difficulties, some interesting case studies have been carried out in the domain of sentence processing, some of which have used reaction time measures, and findings with important implications for theory have been obtained (e.g., Tyler, 1985).

In some domains, computational models have provided an excellent means for sharpening the issues and testing the predictions of theory (Coltheart, Rastle, Perry, Langdon, & Ziegler, 2001; Dell, Schwartz, Martin, Saffran, & Gagnon, 1997; Rapp & Goldrick, 2000). In the sentence processing domain, for example, some recent computational approaches have taken into account probabilistic information about the likelihood of particular syntactic structures in the language in general and with respect to particular lexical items (e.g., Jurafsky, 1996). Other models have taken into account working memory constraints, specifically incorporating findings on parallel, content-addressable search (e.g., Lewis & Nakayama, 2002). Unfortunately, although Jurafsky mentions the role that semantic constraints might play in sentence interpretation, that aspect of the model has apparently not been implemented. Of further note is that most of these models use semantic constraints only at points of syntactic ambiguity, but the data from Ferreira (2003) indicate that semantic constraints play a role for normal subjects, even with structures that are unambiguous, and certainly this is true to an even greater extent for aphasic patients. Consequently, the models would need to be revised such that semantic constraints can be powerful enough to override the output of role assignments based on syntactic structures. Many of these modelling approaches are quite new and are only in the first stages of development. Even so, attempts to account for the existing neuropsychological findings would provide critical tests of the models. Of course, in turn, attempts to model patient performance would lead to important predictions and new empirical tests regarding the nature of the patients' deficits.

PrEview proof published online 2 September 2005

REFERENCES

Baddeley, A. D., & Wilson, B. (1985). Phonological coding and short-term memory in patients without speech. *Journal of Memory and Language, 24*, 490–502.

Badecker, W., Nathan, P., & Caramazza, A. (1991). Varieties of sentence comprehension deficits: A case study. *Cortex, 27*, 311–321.

Berndt, R. S. (1991). Sentence processing in aphasia. In M. Sarno (Ed.), *Acquired aphasia* (2nd ed., pp. 223–270). San Diego: Academic Press.

Berndt, R. S. (1998). Sentence processing in aphasia. In M. Sarno (Ed.), *Acquired aphasia* (3rd ed., pp. 229–267). San Diego: Academic Press.

Berndt, R. S., & Caramazza, A. (1980). A redefinition of the syndrome of Broca's aphasia: Implications for a neuropsychological model of language. *Applied Psycholinguistics, 1*, 225–278.

Berndt, R. S., Haendiges, A., Mitchum, C., & Sandson, J. (1997). Verb retrieval in aphasia. 2: Relationship to sentence processing. *Brain and Language, 56*, 107–137.

Berndt, R. S., Mitchum, C., & Haendiges, A. (1996). Comprehension of reversible sentences in "agrammatism": A meta-analysis. *Cognition, 58*, 289–308.

Bishop, D. V. M. (1983). *TROG test for the reception of grammar*. Published by the author and available from Age and Cognitive Performance Research Centre, University of Manchester, M13 9PL.

Bishop, D., & Robson, J. (1989). Unimpaired short-term memory and rhyme judgments in congenitally speechless individuals: Implications for the notion of "articulatory coding." *Quarterly Journal of Experimental Psychology, 41A*, 12–140.

Boland, J. (1997). The relationship between syntactic and semantic processes in sentence comprehension. *Language and Cognitive Processes, 12*, 423–484.

Breedin, S., & Martin, R. (1996). Patterns of verb deficits in aphasia: An analysis of four cases. *Cognitive Neuropsychology, 13*, 51–91.

Breedin, S., & Saffran, E. (1999). Sentence processing in the face of semantic loss: A case study. *Journal of Experimental Psychology: General, 128*, 547–562.

Butterworth, B., Campbell, R., & Howard, D. (1986). The uses of short-term memory: A case study. *Quarterly Journal of Experimental Psychology, 38A*, 705–737.

Caplan, D., & Hildebrandt, N. (1988). *Disorders of syntactic comprehension*. Cambridge, MA: MIT Press.

Caplan, D., Vanier, M., & Baker, C. (1986). A case study of reproduction conduction aphasia: II. Sentence comprehension. *Cognitive Neuropsychology, 3*, 129–146.

Caplan, D., & Waters, G. (1999). Verbal working memory and sentence comprehension. *Behavioral and Brain Sciences, 22*, 77–126.

Caramazza, A., Basili, A. G., Koller, J., & Berndt, R. S. (1981). An investigation of repetition and language processing in a case of conduction aphasia. *Brain and Language, 14*, 235–271.

Caramazza, A., Capitani, E., Rey, A., & Berndt, R. S. (2001). Agrammatic Broca's aphasia is not associated with a single pattern of comprehension performance. *Brain and Language, 76*, 158–184.

Caramazza, A., & Zurif, E. (1976). Dissociation of algorithmic and heuristic processes in language comprehension: Evidence from aphasia. *Brain and Language, 3*, 572–582.

Coltheart, M., Rastle, K., Perry, C., Langdon, R., & Ziegler, J. (2001). DRC: A dual route cascaded model of visual word recognition and reading aloud. *Psychological Review, 108*, 204–256.

Cupples, L., & Inglis, A. L. (1993). When task demands induce "asyntactic" comprehension: A study of sentence interpretation in aphasia. *Cognitive Neuropsychology, 10*, 201–234.

De Bleser, R. (1987). From agrammatism to paragrammatism: German aphasiological traditions and grammatical disturbances. *Cognitive Neuropsychology, 42*, 187–256.

Dell, G., Schwartz, M., Martin, N., Saffran, E., & Gagnon, D. (1997). Lexical access in aphasic and nonaphasic speakers. *Psychological Review, 104*, 801–838.

Dick, F., Bates, E., Wulfeck, B., Utman, J. A., Dronkers, N., & Gernsbacher, M. A. (2001). Language deficits, localisation, and grammar: Evidence for a distributive model of language breakdown in aphasic patients and neurologically intact individuals. *Psychological Review, 108*, 759–788.

Druks, J., & Marshall, J. (1991). Agrammatism: An analysis and critique, with new evidence from four Hebrew-speaking aphasic patients. *Cognitive Neuropsychology, 8*, 415–433.

Ferreira, F. (2003). The misinterpretation of non-canonical sentences. *Cognitive Psychology, 47*, 164–203.

Frazier, L. (1987). Sentence processing: A tutorial review. In M. Coltheart (Ed.), *Attention and performance: Vol. XII. The psychology of reading* (pp. 559–586). Hove, UK: Lawrence Erlbaum Associates Ltd.

Frazier, L., & Rayner, K. (1982). Making and correcting errors during sentence comprehension: Eye movements in the analysis of structurally ambiguous sentence. *Cognitive Psychology, 13*, 178–210.

Friedmann, N., & Gvion, A. (2003). Sentence comprehension and working memory limitation in aphasia: A dissociation between semantic-syntactic and phonological reactivation. *Brain and Language, 86*, 23–39.

Friedrich, F., Glenn, C., & Martin, O. S. M. (1984). Interruption of phonological coding in conduction aphasia. *Brain and Language, 22*, 266–291.

Friedrich, F., Martin, R. C., & Kemper, S. (1985). Consequences of a phonological coding deficit on sentence processing. *Cognitive Neuropsychology, 2*, 385–412.

Gahl, S. (2002). Lexical biases in aphasic sentence comprehension: An experimental and corpus linguistic study. *Aphasiology, 16*, 1173–1198.

Gahl, S., Menn, L., Ramsberger, G., Jurafsky, D. S., Elder, E., Rewega, M., & Holland, A (2003). Syntactic frame and verb bias in aphasia: Plausibility judgments of undergoer-subject sentences. *Brain and Cognition, 53*, 221–228.

Gibson, E. (2000). The dependency locality theory: A distance-based theory of linguistic complexity. In A. Marantz, Y. Miyashita, & W. O'Neill (Eds.), *Image, language and brain*. Cambridge, MA: MIT Press.

Goodglass, H., & Kaplan, E. (1972). *The assessment of aphasia and related disorders*. Philadelphia: Lea & Febiger.

Gordon, P. C., Hendrick, R., & Johnson, M. (2001). Memory interference during language processing. *Cognition, 27*, 1411–1423.

Grodzinsky, Y. (1990). *Theoretical perspectives on language deficits*. Cambridge, MA: MIT Press.

Grodzinsky, Y. (1995). A restrictive theory of agrammatic comprehension. *Brain and Language, 50*, 27–51.

Grodzinsky, Y., & Finkel, L. (1998). The neurology of empty categories: Aphasics' failure to detect ungrammaticality. *Journal of Cognitive Neuroscience, 10*, 281–292.

Grodzinsky, Y., Pinango, M. M., Zurif, E., & Drai, D. (1999). The critical role of group studies in neuropsychology: Comprehension regularities in Broca's aphasia. *Brain and Language, 67*, 134–147.

Haarmann, H., Davelaar, E. J., & Usher, M. (2003). Individual differences in semantic short-term memory capacity and reading comprehension. *Journal of Memory and Language, 48*, 320–345.

Haarmann, H., & Kolk, H. (1991). Syntactic priming in Broca's aphasia: Evidence for slow activation. *Aphasiology, 5*, 247–263.

Hanten, G., & Martin, R. (2000). Contributions of phonological and semantic short-term memory to sentence processing: Evidence from two cases of closed head injury in children. *Journal of Memory and Language, 43*, 335–361.

Hanten, G., & Martin, R. (2001). A developmental short-term memory deficit: A case study. *Brain and Cognition, 45*, 164–188.

Hildebrandt, N., Caplan, D., & Evans, K. (1987). The man_i left t_i without a trace: A case study of aphasic processing of empty categories. *Cognitive Neuropsychology, 4*, 257–302.

Hillis, A. E., Kane, A., Tuffiash, E., Ulatowski, J. A., Barker, P. B., Beauchamp, N., & Wityk, R. (2001). Reperfusion of specific brain regions by raising blood pressure restores selective language functions in subacute stroke. *Brain and Language*, *79*, 495–510.

Hodges, J., Patterson, K., & Tyler, L. (1994). Loss of semantic memory: Implications for the modularity of mind. *Cognitive Neuropsychology*, *11*, 505–542.

Howard, D. (1985). Introduction to "On agrammatism" (Ueber Agrammatismus), by Max Isserlin, 1922. *Cognitive Neuropsychology*, *2*, 303–307.

Inglis, A. L. (2003). Taking expectations to task in aphasia sentence comprehension: Investigations of off-line performance. *Aphasiology*, *17*, 265–289.

Isserlin, M. (1922/1985). On agrammatism (Trans. H. Droller, with D. Howard and R. Campbell). *Cognitive Neuropsychology*, *2*, 308–345.

Jarema, G., & Friederici, A. D. (1994). Processing articles and pronouns in agrammatic aphasia: Evidence from French. *Brain and Language*, *46*, 683–694.

Jurafsky, D. (1996). A probabilistic model of lexical and syntactic access and disambiguation. *Cognitive Science*, *20*, 137–194.

Just, M., & Carpenter, P. (1992). A capacity theory of comprehension: Individual differences in working memory. *Psychological Review*, *99*, 122–149.

Kempler, D., Curtiss, S., & Jackson, C. (1987). Syntactic preservation in Alzheimer's disease. *Journal of Speech and Hearing Research*, *30*, 343–350.

Kolk, H., & Van Grunsven, M. (1985). Agrammatism as a variable phenomenon. *Cognitive Neuropsychology*, *2*, 347–384.

Kolk, H., Van Grunsven, M., & Keyser, A. (1985). On parallelism between production and comprehension in agrammatism. In M. L. Kean (Ed.), *Agrammatism*. New York: Academic Press.

Lewis, R. L., & Nakayama, M. (2002). Syntactic and positional similarity effects in the processing of Japanese embeddings. In M. Nakayama (Ed.), *Sentence processing in east asian languages*. Stanford, CA: CSLI Publications.

Linebarger, M. (1990). Neuropsychology of sentence parsing. In A. Caramazza (Ed.), *Cognitive neuropsychology and neurolinguistics: Advances in models of cognitive function and impairment*. Hillsdale, NJ: Lawrence Erlbaum Associates Inc.

Linebarger, M., Schwartz, M., & Saffran, E. (1983). Sensitivity to grammatical structure in so-called agrammatic aphasics. *Cognition*, *13*, 361–392.

Longoni, A. M., Richardson, J. T. E., & Aiello, A. (1993). Articulatory rehearsal and phonological storage in working memory. *Memory and Cognition*, *21*, 11–22.

Lu, C.-C., Bates, E., Li, P., Tzeng, O., Hung, D., Tsai, C. H., Lee, S. E., & Chung, Y.M. (2000). Judgements of grammaticality in aphasia: The special case of Chinese. *Aphasiology*, *14*, 1021–1054.

Lukatela, K., Crain, S., & Shankweiler, D. (1988). Sensitivity to inflectional morphology in agrammatism: Investigation of a highly inflected language. *Brain and Language*, *33*, 1–15.

Martin, N., & Saffran, E. M. (1997). Language and auditory-verbal short-term memory impairments: Evidence for common underlying processes. *Cognitive Neuropsychology*, *14*, 641–682.

Martin, R. C. (1987). Articulatory and phonological deficits in short-term memory and their relation to syntactic processing. *Brain and Language*, *32*, 137–158.

Martin, R. (1995). Working memory doesn't work: A critique of Miyake et al.'s capacity theory of aphasic comprehension deficits. *Cognitive Neuropsychology*, *12*, 623–636.

Martin, R. (2000). Sentence comprehension deficits. In B. Rapp (Ed.), *Handbook of cognitive neuropsychology*. Philadelphia: Psychology Press.

Martin, R., & Blossom-Stach, C. (1986). Evidence for syntactic deficits in a fluent aphasic. *Brain and Language*, *28*, 196–234.

Martin, R. C., Blossom-Stach, C., Yaffee, L., & Wetzel, F. (1995). Consequences of a central motor programming deficit for rehearsal and reading comprehension. *Quarterly Journal of Experimental Psychology*, *48A*, 536–572.

Martin, R. C., & Breedin, S. (1992). Dissociations between speech perception and phonological short-term memory. *Cognitive Neuropsychology*, *9*, 509–534.

Martin, R. C., & He, T. (2004). Semantic short-term memory and its role in sentence processing: A replication. *Brain and Language*, *89*, 76–82.

Martin, R. C., & Romani, C. (1994). Verbal working memory and sentence processing: multiple components view. *Neuropsychology*, *8*, 506–523.

Martin, R. C., Shelton, J. R., & Yaffee, L. S. (1994). Language processing and working memory: Neuropsychological evidence for separate phonological and semantic capacities. *Journal of Memory and Language*, *33*, 83–111.

Martin, R. C., Wetzel, F., Blossom-Stach, C., & Feher, E. (1989). Syntactic loss versus processing deficit: An assessment of two theories of agrammatism and syntactic comprehension deficits. *Cognition*, *32*, 157–191.

Mauner, G., Fromkin, V. A., & Cornell, T. L. (1993). Comprehension and acceptability judgments in agrammatism: Disruptions in the syntax of referential dependency. *Brain and Language*, *45*, 340–370.

McElree, B. (2000). Sentence comprehension is mediated by content-addressable memory structures. *Journal of Psycholinguistic Research*, *29*, 111–123.

McElree, B., & Dosher, B. (1993). Serial retrieval process in the recovery of order information. *Journal of Experimental Psychology: General*, *122*, 291–315.

McElree, B., Foraker, S., & Dyer, L. (2003). Memory structures that subserve sentence comprehension. *Journal of Memory and Language*, *48*, 67–91.

Miceli, G. (1999). Grammatical deficits in aphasia. In G. Denes & L. Pizzamiglio (Eds.), *Handbook of clinical and experimental neuropsychology* (pp. 245–272). Hove, UK: Psychology Press.

Miceli, G., Mazzucchi, A., Menn, L., & Goodglass, H. (1983). Contrasting cases of Italian agrammatic aphasia without comprehension disorder. *Brain and Language*, *19*, 65–97.

Miceli, G., Silveri, M. C., Romani, C., & Caramazza, A. (1989). Variations in the pattern of omissions and substitutions in grammatical morphemes in the spontaneous speech of so-called agrammatic aphasics. *Brain and Language*, *36*, 447–492.

Miller, G., & McKean, K. (1964). A chronometric study of some relations between sentences. *Quarterly Journal of Experimental Psychology*, *16*, 297–308.

Mitchum, C., Haendiges, A., & Berndt, R. (1995). Treatment of thematic mapping in sentence comprehension: Implications for normal processing. *Cognitive Neuropsychology*, *12*, 503–547.

Miyake, A., Just, M., & Carpenter, P. (1994). A capacity approach to syntactic comprehension disorder: Making normal adults perform like aphasic patients. *Cognitive Neuropsychology*, *11*, 671–717.

Naeser, M., Mazurski, P., Goodglass, H., Peraino, M., Laughlin, S., & Leaper, W. C. (1987). Auditory syntactic comprehension in nine aphasia groups (with CT scans) and children: Differences in degree but not order of difficulty observed. *Cortex*, *23*, 359–380.

Nespoulous, J. L., Dordain, M., Perron, C., Ska, B., Bub, D., Caplan, D., Mehler, J., & Lecours, A. R. (1988). Agrammatism in sentence production without comprehension deficits: Reduced availability of syntactic structures or grammatical morphemes? A case study. *Brain and Language*, *33*, 273–295.

Nimmo, L. M., & Roodenrys, S. (2004). Investigating the phonological similarity effect: Syllable structure and the position of common phonemes. *Journal of Memory and Language*, *50*, 245–258.

Ostrin, R., & Tyler, L. (1995). Dissociations of lexical function: Semantics, syntax, and morphology. *Cognitive Neuropsychology*, *12*, 345–389.

Parisi, D., & Pizzamiglio, L. (1970). Syntactic comprehension in aphasia. *Cortex*, *6*, 204–215.

Radford, A. (1988). *Transformational grammar: A first course*. New York: Cambridge University Press.

Rapp, B., & Goldrick, M. (2000). Discreteness and interactivity in spoken word production. *Psychological Review*, *107*, 460–499.

Rochon, E., Saffran, E. M., Berndt, R. S., & Schwartz, M. F. (2000). Quantitative analysis of aphasic sentence production: Further development and new data. *Brain and Language*, *72*, 193–218.

Saffran, E. M., Berndt, R. S., & Schwartz, M.F. (1989). The quantitative analysis of agrammatic production: Procedure and data. *Brain and Language*, *37*, 440–479.

Saffran, E., & Schwartz, M. (1988). "Agrammatic" comprehension it's not: Alternatives and implications. *Aphasiology*, *2*, 389–394.

Saffran, E., Schwartz, M., & Linebarger, M. (1998). Semantic influences on thematic role assignments: Evidence from normals and aphasics. *Brain and Language*, *62*, 255–297.

Schwartz, M., & Chawluk, J. (1990). Deterioration of language in progressive aphasia: A case study. In M. Schwartz (Ed.), *Modular deficits in Alzheimer-type dementia*. Cambridge, MA: MIT Press.

Schwartz, M., Marin, O. S. M., & Saffran, E. (1979). Dissociations of language function in dementia: A case study. *Brain and Language*, *7*, 277–306.

Schwartz, M., Saffran, E., & Marin, O. S. M. (1980). The word order problem in agrammatism: I. Comprehension. *Brain and Language*, *10*, 249–262.

Shankweiler, D., Crain, S., Gorrell, P., & Tuller, B. (1989). Reception of language in Broca's aphasia. *Language and Cognitive Processes*, *4*, 1–33.

Spivey, M., & Tanenhaus, M. (1998). Syntactic ambiguity resolution in discourse: Modeling the effects of referential context and lexical frequency. *Journal of Experimental Psychology: Learning, Memory, and Cognition*, *24*, 1521–1543.

Spivey, M., Tanenhaus, M., Eberhard, K., & Sedivy, J. (2002). Eye movements and spoken language comprehension: Effects of visual context on syntactic ambiguity resolution. *Cognitive Psychology*, *45*, 447–481.

Spivey-Knowlton, M., & Sedivy, J. (1995). Resolving attachment ambiguities with multiple constraints. *Cognition*, *55*, 226–267.

Swaab, T., Brown, C., & Hagoort, P. (1997). Spoken sentence comprehension in aphasia: Event-related potential evidence for a lexical integration deficit. *Journal of Cognitive Neuroscience*, *9*, 39–66.

Swinney, D., Zurif, E., & Nicol, J. (1989). The effects of focal brain damage on sentence processing: An examination of the neurological organization of a mental module. *Journal of Cognitive Neuroscience*, *1*, 25–37.

Thompson, C. K., Fix, S., & Gitelman, D. (2002). Selective impairment of morphosyntactic production in a neurological patient [Special issue: The role and neural representation of grammatical class]. *Journal of Neurolinguistics*, *15*, 189–207.

Trueswell, J., Tanenhaus, M., & Garnsey, S. (1994). Semantic influences on parsing: Use of thematic role information in syntactic ambiguity resolution. *Journal of Memory and Language*, *33*, 285–318.

Tyler, L. K. (1985). Real-time comprehension problems in agrammatism: A case study. *Brain and Language*, *26*, 259–275.

Tyler, L. K. (1989). Syntactic deficits and the construction of local phrases in spoken language comprehension. *Cognitive Neuropsychology*, *6*, 333–355.

Tyler, L. K. (1992). *Spoken language comprehension: An experimental approach to disordered and normal processing*. Cambridge, MA: MIT Press.

Vallar, G., & Baddeley, A. D. (1984a). Phonological short-term store, phonological processing and sentence comprehension: A neuropsychological case study. *Cognitive Neuropsychology*, *1*, 121–141.

Vallar, G., & Baddeley, A. D. (1984b). Fractionation of working memory: Neuropsychological evidence for a phonological short-term store. *Journal of Verbal Learning and Verbal Behavior*, *23*, 151–161.

Van Dyke, J. A., & Lewis, R. L. (2003). Distinguishing effects of structure and decay on attachment and repair: A cue-based parsing account of recovery from misanalysed ambiguities. *Journal of Memory and Language*, *49*, 285–316.

Waters, G., Caplan, D., & Hildebrandt, N. (1991). On the structure of verbal short-term memory and its functional role in sentence comprehension: Evidence from neuropsychology. *Cognitive Neuropsychology*, *8*, 81–126.

Wilson, S. M., & Saygun, A. P. (2004). Grammaticality judgments in aphasia: Deficits are not specific to syntactic structures, aphasic syndromes, or lesion sites. *Journal of Cognitive Neuroscience*, *16*, 238–252.

Wulfeck, B. (1988). Grammaticality judgments and sentence comprehension in agrammatic aphasia. *Journal of Speech and Hearing Research*, *31*, 72–81.

Zurif, E., & Grodzinsky, Y. (1983). Sensitivity to grammatical structure in agrammatic aphasics: A reply to Linebarger, Schwartz, and Saffran. *Cognition*, *15*, 207–213.

COGNITIVE NEUROPSYCHOLOGY, 2006, 23 (1), 96–109

Acquired dyslexias and the computational modelling of reading

Max Coltheart
Macquarie University, Sydney, Australia

Two current approaches to the computational modelling of reading are the connectionist triangle model approach (parallel processing; distributed representations; models developed via a connectionist training algorithm such as backpropagation) and the DRC ("dual route cascaded") model (serial processing components permitted; local representations; model architecture specified by the modeller rather than acquired by a learning algorithm). One way of testing such computational models is to lesion the computer programs that instantiate the models and study how well the impaired reading of such damaged models corresponds to the patterns of impaired reading seen in people with acquired dyslexia. This is computational cognitive neuropsychology. It has been used with both types of model in attempts to simulate acquired surface dyslexia and acquired phonological dyslexia. The results of this body of work currently favour the DRC model over connectionist models developed within the triangle model framework.

At the 1971 meeting of the International Neuropsychological Society in Engelberg, Switzerland, John Marshall and Freda Newcombe reintroduced the cognitive-neuropsychological study of reading by describing three distinct forms of acquired dyslexia—deep dyslexia, surface dyslexia and visual dyslexia—and offering theoretical interpretations of each of these in relation to an explicit information-processing model of the cognitive processes involved in reading (Marshall & Newcombe, 1973).

Their model was expressed in box-and-arrow diagrammatic form and is shown in Figure 1. The diagram asserts that there are two different processing routes via which articulation can be derived from print, i.e., via which reading aloud can be achieved. The box labels in this diagram do not make it perfectly clear exactly how each of these routes operate but the text of the paper does. Route *abdte* involves reading "via putative grapheme–phoneme correspondence rules" (Marshall & Newcombe, 1973, p. 191) and hence would generate an error when the stimulus is a word that disobeys these rules (an irregular or exception word). The other route, *abcte*, involves reading via semantics, and so could not yield a correct response when the input is a nonword. Hence this paper, along with the paper

Correspondence should be addressed to Max Coltheart, Macquarie Centre for Cognitive Science, Macquarie University, Sydney NSW, Australia 2109 (Email: max@maccs.mq.edu.au).

Versions of this paper were presented at the European Society for Cognitive Psychology meeting, Granada, 2003; the Psychonomic Society meeting, Vancouver, 2003; and the European Workshop on Cognitive Neuropsychology, Bressanone, 2004. I thank Alfonso Caramazza, Ken Forster, and Anna Woollams for useful comments on an earlier draft.

DOI:10.1080/02643290500202649

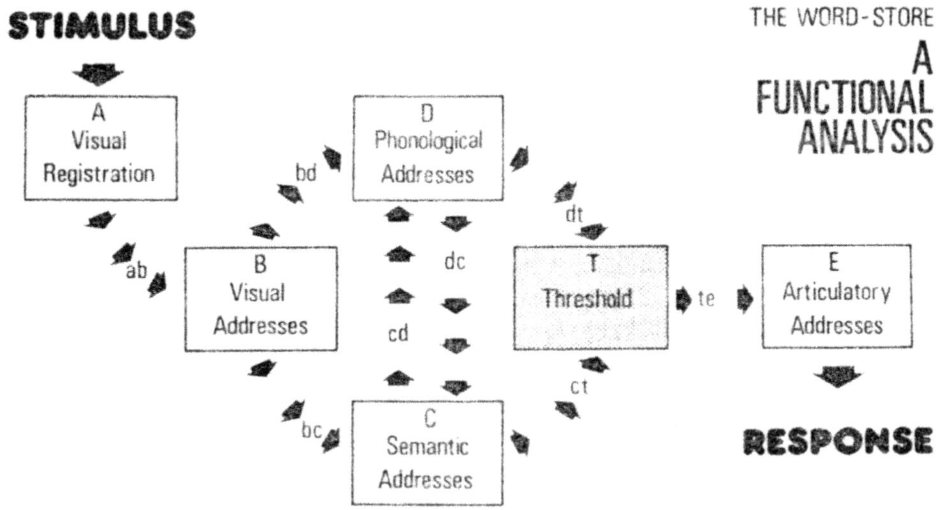

Figure 1. *A dual-route model of reading aloud (Marshall & Newcombe, 1973).*

published in the same year by Forster and Chambers (1973), introduced the dual-route model of reading aloud.

If, as a consequence of brain damage, pathway *abcte* becomes unavailable, and an irregular word is presented for reading aloud, a reading error (a regularisation error) would occur, whereas regular words or nonwords would still be read correctly. Selective impairment of the reading of irregular words with regularisation errors to these words is surface dyslexia; Marshall and Newcombe (1973) reported two such cases and interpreted them as arising from damage to pathway *abcte*.

If, instead, it is pathway *abdte* that becomes unavailable, reading aloud will depend on using semantic representations to mediate between print to speech. That will make nonword reading impossible, since of course nonwords do not have semantic representations. Marshall and Newcombe (1973; see also, 1980) suggested that this impairment would also result in semantic errors in reading aloud, i.e., would result in the form of acquired dyslexia they named deep dyslexia (either because pathway *abcte*, even when intact, cannot reliably discriminate between words that are closely similar in semantic representation, or because there is a second impairment, of the semantic system itself).

The Marshall and Newcombe paper was a seminal one in that it stimulated a great deal of cognitive-neuropsychological work on acquired dyslexia over the next decade or more, i.e., a great deal of work in which patterns of reading impairment consequent upon brain damage were interpreted in terms of some model of normal reading, and in which models of normal reading were developed, extended, and tested by using data from case studies of acquired dyslexia (for general reviews, see Coltheart, 1985; Ellis & Young, 1988; Shallice, 1988). Then, in the 1980s, there were two more crucial theoretical developments, which have both had much influence on the cognitive neuropsychology of reading: the development of the concept of distributed representation, and the introduction of computational modelling to the cognitive psychology of reading.

DISTRIBUTED REPRESENTATION

When in some cognitive system "each entity is represented by a pattern of activity distributed over many computing elements, and each computing element is involved in representing many different entities" (Hinton, McClelland, & Rumelhart, 1986, p. 77), the entities are said to have *distributed* representations. The alternative "is to use one computing element for each entity. This is called a *local* representation" (ibid.).

Prior to the introduction of the concept of distributed representation, all of the models of reading used in cognitive-neuropsychological work on acquired dyslexia posited local representations. Now there was an alternative. Furthermore, those interested in distributed representations considered that "the best psychological evidence for distributed representations is the degree to which their strengths and weaknesses match those of the human mind" (Hinton et al., 1986, p. 78) and even, crucially, that data from brain-damaged patients provided evidence that human mental representations are distributed rather than local (see, e.g., McClelland & Rumelhart, 1986, p. 527; Patterson, 1990; Rumelhart & McClelland, 1986, p. 134). The way was therefore open to develop models of reading based on distributed representations that would seek to explain acquired dyslexia as well as normal reading. The first such model of reading to incorporate distributed representations was that proposed by Seidenberg and McClelland (1989).

COMPUTATIONAL MODELLING

Models such as the Marshall-Newcombe model (see Figure 1), and all of the models of reading discussed in the reviews by Coltheart (1985), Ellis and Young (1988), and Shallice (1988) of cognitive-neuropsychological work on acquired dyslexia, were what Jacobs and Grainger (1994) referred to as *verbal models*—that is, models that are described informally or in diagrammatic form, as opposed to models that are described formally. Formal model description can be mathematical, or it can be computational. A computational model of any cognitive activity is a computer program that cannot only carry out that activity itself but does this in a way which, the modeller contends, corresponds to the way in which human beings carry out the cognitive activity in question. So, for example, a computational model of reading aloud is a computer program that is not only capable of generating some kind of phonological output representation from some kind of orthographic input representation but also does so using processing mechanisms considered to be the same as the mechanisms used by human readers as they read aloud. The first computational model of reading aloud was that proposed by Seidenberg and McClelland (1989); this model will be referred to as the SM89 model.

The general architecture of this model is shown in Figure 2. The components of the model displayed in bold are the ones that were actually implemented as a computer program by Seidenberg and McClelland. The model was trained using the backpropagation learning algorithm so that it gradually learned to produce appropriate phonological responses to orthographic inputs, i.e., to read aloud. The hollow ellipses are sets of "hidden units,'" which are needed to allow the system to learn the complex relationships that exist between the printed and spoken representations characteristic of English.

This general architecture has come to be referred to as the "triangle model" and various specific versions of it differing somewhat from

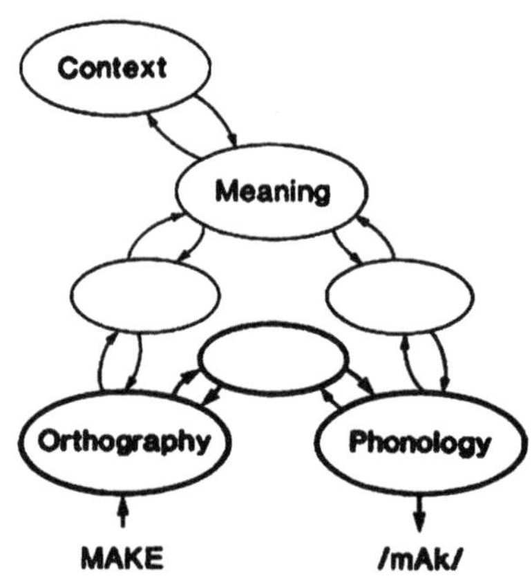

Figure 2. *The Seidenberg-McClelland model of reading aloud.*

each other have been developed—at the time of writing, seven different forms of the triangle model have been published (see Coltheart, 2005, for details of the differences between these).

ACQUIRED DYSLEXIA AND THE SEIDENBERG-McCLELLAND MODEL OF READING

One way of evaluating any model of reading, be it computational or verbal, is to consider how well the model can account for a body of empirical data collected from studies of normal readers; and Seidenberg and McClelland (1989) reported such evaluations. A second way of evaluating models of reading is to consider how well such models can account for data collected from people with acquired disorders of reading; and that was also done with the SM89 model.

Using a computational model of reading to simulate data from acquired dyslexia involves seeking ways of damaging the model that not only make its reading abnormal, but also create specific patterns of abnormal reading by the model that match specific patterns of abnormal reading seen in particular cases of acquired dyslexia. Patterson, Seidenberg, and McClelland (1989) carried out work of this kind with the SM89 model. They considered just one form of acquired dyslexia, surface dyslexia, and hence sought some way of lesioning the model so that it remained accurate at reading regular words and nonwords whilst making regularisation errors when reading irregular words. Various forms of model lesioning were explored after the model had been fully trained: setting to zero strength some proportion of the connections from orthographic input units to hidden units, setting to zero strength some proportion of the connections from hidden units to phonological output units, or setting to zero some proportion of the hidden units themselves.

The behaviour of the SM89 model when lesioned in these ways was intelligible; but the model's reading behaviour did not match the reading behaviour of patients with surface dyslexia. For example, with one set of irregular words the lesioned model's errors included 19.6% regularisation errors, while the percentage of regularisation errors made by the two surface dyslexic patients under consideration was far higher (78.8% & 81.8%, respectively). Patterson (1990) also reports an unsuccessful attempt to simulate the reading of a surface dyslexic patient KT (McCarthy & Warrington, 1986) by lesioning the Seidenberg-McClelland model.

THE NEXT TRIANGLE MODELS: PLAUT, McCLELLAND, SEIDENBERG, AND PATTERSON (1996)

Apart from the inability of the SM89 model to simulate acquired surface dyslexia, it became clear that there were facts about normal reading that this model could also not simulate, as noted by Plaut, McClelland, Seidenberg, and Patterson (1996), Plaut (1999), and others; for example, the model was far less accurate than human readers at reading nonwords (Besner, Twilley, McCann, & Seergobin, 1990). Three new triangle models aimed at remedying difficulties with the SM89 model were therefore created by Plaut et al. (1996). These new models differed from the SM89 model in a number of ways: for example, input representations were no longer distributed but local (each input unit represented a particular grapheme) and output representations were also no longer distributed but local (each output unit represented a particular phoneme).

The first of these three triangle models (a purely feedforward model) is not relevant here since it was not used in any work simulating surface dyslexia.

The second model was the same as the first except that it had feedback from phoneme units back to hidden units and was thus an attractor network. This network was lesioned in several ways in efforts to simulate acquired surface dyslexia: by removing some proportion of hidden units from the trained network, or by setting to zero some proportion of the connections between groups of units, or by adding noise to the strengths of the connections between units in the trained model. None of these methods of lesioning the model were successful in simulating the data from the surface dyslexic patient MP (Behrmann & Bub, 1992) and particularly not the data from the more severe surface dyslexic patient KT (McCarthy & Warrington, 1986). KT's nonword reading was 100% correct when his reading of low-frequency irregular words was greatly impaired (26%), and no way of lesioning the model was found that could even approximate this extreme dissociation. Any model lesion that had more than a mild effect on irregular word reading also impaired nonword reading. This problem for the model was just the same as the problem identified by Patterson (1990) for the SM89 model.

The third of the three models described in Plaut et al. (1996)—the "division of labour model"—was entirely motivated by the wish to be able to simulate surface dyslexia. This model used local representations at the input and output levels, like the other two models, but abandoned the feedback used in the second model because available computing resources were not sufficient to allow training of the model in a reasonable time if feedback connections were included. Hence the third model was a purely feedforward model, like the first, but it differed from the first model in the way it was trained.

This third model added (unimplemented) external input to the phonological output units, so as to mimic what could happen if there were an implemented semantic system that was activated by orthography and that in turn activated phonology. The training regime was set so that the input of this putative semantic pathway to the phonological output units increased monotonically over time; thus the longer the training went on, the more this model came to rely on the semantic input to phonology for correct reading of low-frequency irregular words.

So, since irregular word reading depended upon some input from the (unimplemented) semantic pathway, lesioning the model by removing input from that pathway would compromise the reading of irregular words. If this model was trained for only 400 epochs, at that point the reading of low-frequency irregular words did not depend very much on the "semantic" input; deleting that input will therefore result in mild surface dyslexia and in fact this simulated the mild surface dyslexic MP well. If the model was trained for 2000 epochs, at that point the reading of low-frequency irregular words had come to depend very much on the "semantic" input; deleting that input would therefore result in severe surface dyslexia and in fact this simulated the severe surface dyslexic KT well.

This approach to simulating surface dyslexia depends upon the claim that acquired surface dyslexia is caused by damage to the semantic system, and Plaut et al. (1996) were aware that there was already neuropsychological evidence that was difficult to reconcile with such a claim. If that claim were correct, then every patient with semantic impairment would be surface dyslexic, and Plaut et al. (1996, p. 99) acknowledged the existence of two patients whose reading of low-frequency irregular words was normal even though they suffered from a severe semantic impairment: patient WLP (Schwartz, Saffran, & Marin, 1980) and patient DRN (Cipolotti & Warrington, 1995). Three more patients with intact reading aloud of exception words in the presence of severe semantic impairment have since been reported, by Lambon Ralph, Ellis, and Franklin (1995), Gerhand (2001), and Blazely, Coltheart, and Casey (2005).

Plaut and colleagues sought to deal with this problem by suggesting that there is premorbid variation in the degree to which intact readers require semantic support for the reading aloud of exception words, with some readers requiring no support at all: It is those readers who, if they suffer semantic damage, will be the ones who will not be surface dyslexic. They acknowledge (Plaut et al., 1996, p. 99) that "this post hoc interpretation clearly requires some future, independent source of evidence." No such evidence has since emerged. Without such evidence, the triangle model account of surface dyslexia is circular and unfalsifiable. The explanation of intact irregular-word reading in the presence of semantic damage is that this person must premorbidly have been one of those readers whose irregular-word reading needed no contribution from semantics; and the only evidence that this person was indeed one of those readers is that his or her irregular-word reading is intact despite the semantic damage.

This history of attempts to interpret acquired surface dyslexia in terms of the triangle model approach has shown that all bar one attempt has been unsuccessful, and the one case where lesioning a triangle model has allowed patient data to be fitted well depends upon making a claim that is circular and that makes the model's account of surface dyslexia unfalsifiable. At present, then, the data from acquired surface dyslexia pose a serious difficulty for all of the triangle computational models of reading.

Below I consider whether a different computational model of reading, the DRC model (Coltheart, Rastle, Perry, Langdon, & Ziegler, 2001), fares any better in attempting to explain acquired surface dyslexia. Before doing so, however, I will consider how well the triangle-model approach can account for a second form of acquired dyslexia, phonological dyslexia.

PHONOLOGICAL DYSLEXIA AND THE TRIANGLE MODELS

In this form of acquired dyslexia, first described by Beauvois and Derouesné (1979; see Coltheart, 1996, for review), patients are selectively impaired at reading aloud nonwords relative to words. Plaut et al. (1996) discuss phonological dyslexia briefly and consider the possibility that an impairment within phonology itself is responsible for this form of acquired dyslexia; but they did not attempt to simulate this reading disorder by lesioning any of their models.

This was, however, subsequently done by Harm and Seidenberg (2001). Their triangle model differs from the four triangle models so far mentioned in four ways:

1. Its input units represented letters.
2. Its output units represent phonetic features.
3. Learning in the phonological units was assisted by the presence of a set of cleanup units attached to the phonological units.
4. Positional coding of orthography was relative to the vowel in the input string rather than using absolute position in the input string.

After training, the model was lesioned within phonology itself to seek to simulate phonological dyslexia (as suggested by Plaut et al., 1996). This lesioning consisted of applying multiplicative Gaussian noise just to the computations performed within the phonological system. Harm and Seidenberg (2001, p. 71) used the results of these simulations to argue for what they referred to as the *phonological impairment hypothesis*, according to which acquired phonological dyslexia "derive[s] from impaired representation and use of phonology." If that is the correct explanation of acquired phonological dyslexia, then all patients with this reading disorder will have impaired representation and use of phonology.

But this is not so. First, both Derouesné and Beauvois (1985) and Bisiacchi, Cipolotti, and Denes (1989) have reported cases of acquired phonological dyslexia in the absence of phonological impairment. The investigations of phonology in these two cases were not as detailed as one might have liked given the theoretical importance this issue has subsequently assumed; but fortunately there have been two recent studies of acquired phonological dyslexia in which phonological abilities were assessed in great detail (Caccappolo-van Vliet, Miozzo, & Stern, 2004a, 2004b). In all three cases reported in these two papers, phonological abilities were normal but nonword reading was poor. A similar result was also reported by Patterson (2000) for the phonological dyslexic CJ: "One could certainly not conclude that CJ's phonological skills in non-reading tasks were normal; but given that he *failed* to produce a correct response to about 80% of simple nonwords in reading, his roughly 80% *correct* performance in both segmentation and blending even with nonword responses suggests that his nonword reading failures cannot be attributed entirely, or even primarily, to a general phonological weakness impeding the production of unfamiliar phonological forms" (Patterson, 2000, p. 68).

Thus six patients have been reported whose phonological dyslexia cannot be explained as being due to phonological impairment. It may turn out to be possible, of course, to lesion a triangle model elsewhere than in its phonological system and to find that nonword reading is more affected than word reading. But this has not been done, and until it is this form of acquired dyslexia remains unexplained by computational models of the triangle variety. Furthermore, there are reasons to believe that selectively affecting nonword reading by damage outside the phonological component of a triangle model may not be possible. Harm and Seidenberg (2001, p. 15) report that in developmental simulations using their model, extraphonological disruption (disruption of the orthography-to-phonology pathway) affected nonword reading *less* than word reading.

This account of the interpretation of acquired phonological dyslexia in terms of the triangle model approach has shown that the model's simulation has not been successful, in the sense that the simulation predicts that all patients with acquired phonological dyslexia will have phonological impairments, and that is not true. At present, then, the data from acquired phonological dyslexia, like the data from acquired surface dyslexia, pose a serious difficulty for this particular type of computational model of reading.

ACQUIRED DYSLEXIA AND THE DRC COMPUTATIONAL MODEL OF READING

An alternative to the triangle model approach is the DRC model (Coltheart et al., 2001) shown in Figure 3. This, like the triangle models, is a computational model; unlike the triangle models, however, all of the DRC model's representations are local rather than distributed. The DRC model can be thought of as a computational realisation of the first reading model applied to the interpretation of acquired dyslexia, the Marshall-Newcombe model, shown in Figure 1 and discussed above.

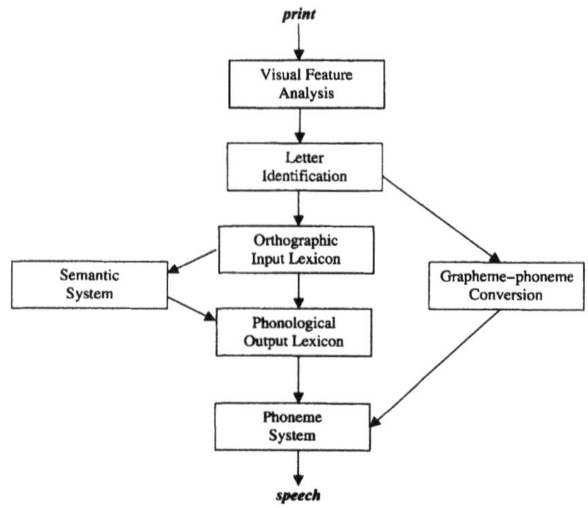

Figure 3. *The DRC model of visual word recognition and reading aloud.*

To simulate acquired surface dyslexia and acquired phonological dyslexia by lesioning the DRC model is almost *too* simple. Deletion of the grapheme–phoneme conversion system will cause nonword reading to be 0% correct while leaving irregular word reading and regular word at 100% correct: This is extreme phonological dyslexia. Appropriate damage to the lexical route after the letter identification level—deletion of the orthographic input lexicon, say—will cause irregular word read to be 0% correct while leaving nonword reading and irregular word reading at 100% correct: This is extreme surface dyslexia. Dissociations of this extremity are rarely, if ever, seen in studies of acquired dyslexia. For example, of the 11 patients with acquired phonological dyslexia described by Berndt, Haendiges, Mitchum, and Wayland (1996), the 3 whose reading of words was essentially normal in accuracy had nonword reading accuracies of 25–60% on a set of 20 nonpseudohomophonic nonwords; and none of the 6 who scored 0% on these nonwords could read words with normal accuracy. Thus not 1 of these 11 phonological dyslexics conformed to the pattern of completely abolished nonword reading with completely normal word reading. Similarly, there has been no report of a surface dyslexic patient whose irregular word reading was completely abolished while some ability to read regular words and nonwords was retained.

This raises two questions for the DRC model. The first is: Can the model simulate *partial* loss of irregular word reading or *partial* loss of nonword reading? The second is: Since the model can simulate complete dissociations of these reading abilities, does the fact that such complete dissociations have never been reported constitute an embarrassment for the model? Coltheart, Langdon, and Haller (1996) provided affirmative answers to the first of these two questions.

With respect to phonological dyslexia, Coltheart et al. (1996) showed that if one slows the rate at which the nonlexical (grapheme–phoneme rule) route of the DRC model operates, the model's nonword reading accuracy begins to decline away from 100%, though word reading remains normal. What is more, in this simulation some specific influences on nonword reading seen in cases of phonological dyslexia also occur in the reading of the DRC model when it is lesioned in this way:

1. Pseudohomophones (nonwords whose pronunciations are words, such as KOAT) yield a higher proportion of correct responses than nonpseudohomophonic nonwords.
2. This pseudohomophone advantage is larger for pseudohomophones that are orthographic neighbours of the words from which they were derived (e.g., KOAT) than for pseudohomophones for which this is not so (e.g., KOTE).

Both of these effects have been reported in cases of phonological dyslexia (Beauvois & Derouesné, 1985; Howard & Best, 1995).

With respect to surface dyslexia, Coltheart et al. (1996) succeeded in quantitatively accurate simulations of data from the relatively mild acquired surface dyslexic MP and the severe acquired surface dyslexic KT, both discussed above. They did this by impairing the functioning of the DRC model's orthographic input lexicon. Entries in the model's orthographic lexicon are frequency-sensitive. The slope of the function relating this sensitivity to word frequency can be varied; the shallower this slope, the less responsive to input the entries are, and the lower a word's frequency the greater will be the loss of responsivity of its orthographic lexical entry when this is done. Such a manipulation does not affect the response of the model's nonlexical route. This means that if the orthographic lexicon is made sufficiently sluggish, the reading of low-frequency words will become dominated by the nonlexical route, and so, if these words are irregular, they will be regularised in reading aloud. Accuracy of reading for regular words and nonwords will remain normal. In the simulations reported by Coltheart et al. (1996), a mild reduction in the sensitivity of the orthographic lexicon produced an irregular-word error rate similar to that shown by MP (with intact reading of regular words and nonwords by the model); a larger reduction in the sensitivity of the orthographic lexicon produced an irregular-word error rate similar to that shown by KT (with intact reading of regular words and nonwords by the model).

As noted above, if with the DRC model the operation of the nonlexical route is greatly slowed, nonword reading accuracy will be 0% with word reading accuracy 100%; if the operation of the orthographic lexicon is abolished rather than merely made sluggish, irregular word reading accuracy will be 0% with nonword reading accuracy 100%. The second question raised above was: Since such extreme dissociations do not seem to occur in cases of acquired dyslexia even though they occur in the model with sufficiently severe model lesioning, is this a difficulty for the model?

Whether one route of the dual-route reading system can be completely abolished by brain damage without any effect at all on the other route is not a question to do with the model, but a question to do with how the various components of the model are represented in the brain. For example, if the vasculature of the brain is such that no stroke can completely eliminate one route's operation whilst leaving the other route completely intact, that neuroanatomical fact is not relevant to the psychological reality of the model. So nonoccurrence of complete dissociations between irregular-word reading and nonword reading does not represent any difficulty for the model.

THE SUMMATION HYPOTHESIS

Not every dual-route modeller has accepted the idea that the reading system includes a lexical but not a semantic pathway from print to speech. The existence of patients with impaired semantics but intact reading of irregular words is what has motivated the inference that reading can be lexical but not semantic. If the semantic system is badly enough damaged that it cannot support correct word reading, and since the nonlexical route cannot support correct reading of irregular words, intact reading of irregular words in the presence of severe semantic damage would seem to require that word reading can be lexical yet not semantic. However, this reasoning has been challenged by Hillis and Caramazza (1991), who propose that there is just a semantic reading route and a nonlexical reading route, and offer the summation hypothesis as a way of reconciling intact reading of irregular words in the presence of semantic damage with the claim that there is no lexical nonsemantic reading route.

Suppose the word *yacht* is presented to a patient whose semantic damage is such that only a rather generic semantic representation of the word can be accessed —"aquatic vessel," for example. Reading via semantics would thus be inaccurate (semantic errors or superordinate responses would occur). But although the word *yacht* cannot be read correctly via an intact nonlexical route, its first letter can be correctly translated into the phoneme/j/by that route. So a patient with this kind of semantic damage would have imperfect information from each route about the stimulus word: The word refers to an aquatic vessel and its first phoneme is/j/. Summing these two partial representations is enough to identify the correct reading response "yacht"; here there is no need to invoke a lexical nonsemantic reading route.

Early cases of preserved irregular word reading with semantic impairment, such as WLP (Schwartz et al., 1980), might well be interpretable in this way, but it seems doubtful that recent and more rigorously studied patients will be. For example, Blazely et al. (2005) reported two cases of semantic dementia, PC and EM. Both had severe semantic impairments. PC was surface dyslexic. EM's reading aloud of irregular words was intact. Suppose there is no lexical nonsemantic route for reading aloud. PC's surface dyslexia must therefore arise because on some occasions the semantic route provides no support for reading and the reading response is entirely mediated by the nonlexical route. EM's semantic impairment was, if anything, worse than PC's. So she should also sometimes be unable to use any information from the semantic system to support reading—that is, she should have been at least as surface dyslexic as PC. But she was not: Her irregular reading was normal. It is not easy to see how to reconcile this result with the view that there is no lexical nonsemantic reading route. The view that there is a lexical nonsemantic reading route has no difficulty explaining the data from both patients: Both have an impaired lexical semantic reading route, but EM has an intact lexical nonsemantic reading route whereas this route is also impaired in PC. So these results favour the idea that the reading system includes a lexical nonsemantic reading route.

A NEWLY EMERGING ISSUE: DO SINGLE PATIENTS REALLY COUNT?

As discussed above, the triangle model account of acquired surface dyslexia requires either that all patients with semantic impairments will have this form of acquired dyslexia or the adoption of a position on premorbid reading systems that is circular and makes the account unfalsifiable. There are patients with semantic impairments whose irregular word reading is normal. Plaut et al. (1996) took these patient data as requiring a response, and the response they made was to adopt the position that makes the account unfalsifiable. Fushimi, Komori, Ikeda, Patterson, Ijuin, and Tanabe (2003), however, seem to be suggesting that these patient data can be overlooked.

> It is important to note that, because this version of the triangle model assumes a causal relationship between semantic dementia and surface dyslexia, its adequacy is challenged by any observations of semantically impaired patients whose reading does not reveal a surface dyslexic pattern ... It is nevertheless worth noting that a very small number of observations of a dissociation between substantial semantic impairment and preserved exception-word reading has to be evaluated in the context of by-now hundreds of observations of the association between semantic impairment and surface dyslexia in many different languages; and furthermore that—although traditional neuropsychological reasoning gives more weight to one reported dissociation than to 100 observations of an association—this is not a universally accepted view (Fushimi et al., 2003, p. 1655).

This discussion by Fushimi and colleagues is about the exceptions to the rule that semantic impairment is always accompanied by surface dyslexia, exceptions that are problematic for the triangle model account of surface dyslexia. Exceptions to the rule that phonological dyslexia is always accompanied by phonological impairments are, as noted above, also problematic for the triangle model account of phonological dyslexia; and Patterson (2000, pp. 59–60) has proposed making the same defensive move here: "before concluding that phonological alexia can be interpreted as a specific reading impairment, we might like to see more than one or two documented cases without an accompanying non-reading phonological deficit."

One is tempted to ask here: How many more than one or two?

More importantly, it is not at all easy to see why, if a theory predicts that a certain dissociation simply cannot occur, a single properly documented discovery of this dissociation in just one patient is not sufficient to refute that theory; and neither Fushimi et al. (2003) nor Patterson (2000) say anything about why this isn't so. However, in support of this position both refer the reader to Goldberg, who does indeed seem to be espousing such a position: "strong dissociations ... must be approached with a degree of wariness, pending the demonstration of their high prevalence" (Goldberg, 1995, p. 195). Given that this strikes at the very heart of orthodox cognitive neuropsychology, it is important to consider what arguments Goldberg makes in support of this position.

Goldberg makes just two brief points re this issue. His first point is that "strong dissociations ... are rare and may very well represent statistical aberrations" (Goldberg, 1995, p. 193). If a certain dissociation has been observed in only one patient and with only a single test, this point would apply; but if the key pattern of intact and impaired performance in a single patient has been documented with multiple tests of the relevant abilities, as is characteristic in cognitive neuropsychology, the point about statistical aberrations simply does not arise. For example, Caccappolo-van Vliet et al. (2000b) documented the poor nonword reading performance of their phonological dyslexic patient RG on nine different lists of nonwords and documented her good phonological skills on eight different phonological tasks. The conclusion that she was impaired at nonword reading and intact at phonology could not be based on a statistical aberration here.

Goldberg's second point was that the inference from a strong dissociation to a cognitive architecture makes "the tacit assumption of the invariant (across individuals) nature of cognitive architecture" and claims that this assumption "is probably wrong" (Goldberg, 1995, p. 194). If this assumption is wrong, then cognitive psychology (and cognitive neuropsychology) has no subject matter, since that aim of these disciplines is to discover the architectures of cognitive systems. Fortunately for these disciplines, however, Goldberg provides no arguments or evidence which show that this assumption is wrong.

Hence there is nothing in the paper by Goldberg that could be used to defend the practice of ignoring dissociations that refute one's theory just on the ground that any such dissociations have been reported in relatively few patients. Indeed, it is hard to see how this practice could ever be defended. I therefore suggest that the five reported patients with semantic impairment but normal reading, and the six reported patients whose phonological dyslexia cannot be attributed to phonological impairment, provide evidence that is inconsistent with the triangle model account of reading, despite the fact that the majority of patients with semantic impairment are surface dyslexic and the fact that the majority of patients with phonological dyslexia exhibit phonological impairments. In contrast, all of this evidence is consistent with the DRC model of reading.

As Fushimi et al. (2003, p. 1655) correctly noted, "traditional neuropsychological reasoning gives more weight to one reported dissociation than to 100 observations of an association." Although outside the domain of triangle modelling of acquired dyslexia, no cognitive neuropsychologist has gone so far as to suggest that theorists may ignore dissociations if they are infrequent enough, a less extreme position that still accords associations some privileged status sometimes appears to be adopted in cognitive-neuropsychological work. For example, Grodzinsky (1990, 1995, 2000) has proposed a neurolinguistic account of Broca's aphasia that is aimed at explaining why a particular *set* of symptoms co-occur in that disorder. On this theory, all of the symptoms of Broca's aphasia stem from an inability to co-index syntactic traces. One task in which such co-indexation is obligatory for successful performance is the understanding of semantically reversible passive sentences. So it follows from the theory that no Broca's aphasic will be above chance on understanding such sentences. However, above-chance performance on such sentences *has* been reported in Broca's aphasia (for a review covering such cases, see Berndt, Mitchum, & Haendiges, 1996). Grodzinsky, Pinango, Serif, and Drai (1999) have responded to this by arguing that in any group of patients in whom comprehension of such sentences is impossible, performance of some patients will be statistically significantly better than chance just by chance, and so individual patients who are performing better than chance can be ignored and the theory can survive. But this ignores two crucial points. The first is that if in a group study there are patients scoring significantly above chance and this is a chance phenomenon, then there should in the group be just as many patients scoring significantly below chance. That is not what is observed. The second point is one already mentioned above: In single case studies of Broca's aphasia, key abilities such as comprehension of reversible passive sentences is typically assessed by multiple tests with large numbers of items, making the likelihood of above-chance performance occurring by chance vanishingly small. Hence it does not seem that Grodzinsky and colleagues have successfully defended their practice of privileging association data over dissociation data.

Somewhat similar specific emphasis on association data is evident in work on computational modelling of aphasic speech production by Dell and colleagues (see, e.g., Dell, Schwartz, Martin, Saffran, & Gagnon, 1997). This model contains a semantic level, a lexical level, and a phonological level. Impaired access from semantics to phonology is modelled by a uniform change in the connections between adjacent levels of the model and a uniform change in the decay rate of units in all three levels of the model. This is the "globality assumption": Brain damage affects the system globally. This approach emphasises associations because it neglects the dissociations between symptoms that are caused just by impaired decay rate at one level of the model. It also neglects the possibility of dissociations of symptoms due to impaired connections just between semantic and lexical levels from symptoms due to impaired connections just between lexical and phonological levels. Ruml and Caramazza (2000) showed that dissociations of these kinds are observed in studies of aphasic speech production. More recent work with the Dell model (Foygel & Dell, 2000) has moved away from the emphasis on association of symptoms by allowing model lesioning in which just the connections from semantic to lexical level, or just the connections from lexical to phonological level, can be lesioned in the model. This form of simulation of damage, with its dissociation-oriented local-damage assumption, behaved similarly in many cases to simulations based on the association-oriented global-damage assumption, but when the results of the two approaches differed they typically favoured the local-damage approach "and we attribute this to its ability to effectively dissociate the two levels of processing" (Foygel & Dell, 2000, p. 213).

CONCLUSIONS

The cognitive neuropsychology of reading consists of using data collected from people with disorders of reading to develop, refine, and adjudicate between models of reading. During the 21 years that have elapsed since the birth of the journal *Cognitive Neuropsychology*, this has been one of the most active areas of research in the field of cognitive neuropsychology, and it continues to be so. In recent years this work has taken the form of *computational* cognitive neuropsychology—lesioning of computer programs that instantiate computational models of reading to see whether it is possible to cause particular models to yield patterns of impaired reading that match the patterns of impaired reading seen in reported cases of reading disorder. Proponents of both the triangle framework approach to the modelling of reading and the DRC model have taken seriously the task of attempting to simulate acquired surface dyslexia and acquired phonological dyslexia. I have argued that studies of both forms of acquired dyslexia have yielded data that are inconsistent with models developed within the triangle model framework but have not yielded any data that are inconsistent with the DRC model. Results from cognitive-neuropsychological work on reading thus currently favour the DRC model over its competitors.

PrEview proof published online 18 October 2005

REFERENCES

Beauvois, M.-F., & Derouesné, J. (1985). Phonological alexia: Three dissociations. *Journal of Neurology, Neurosurgery and Psychiatry*, *42*, 1115–1124.

Behrmann, M., & Bub, D. (1992). Surface dyslexia and dysgraphia: Dual routes, single lexicon. *Cognitive Neuropsychology*, *9*, 209–251.

Berndt, R. S., Haendiges, A. N., Mitchum, C. C., & Wayland, S. C. (1996). An investigation of nonlexical reading. *Cognitive Neuropsychology*, *13*, 763–801.

Berndt, R. S., Mitchum, C. C., & Haendiges, A. N. (1996). Comprehension of reversible sentences in "agrammatism": A meta-analysis. *Cognition*, *58*, 289–308.

Besner, D., Twilley, L., McCann, R. S., & Seergobin, K. (1990). On the association between connectionism and data: Are a few words necessary? *Psychological Review*, *97*, 432–446.

Bisiacchi, P. S., Cipolotti, L., & Denes, G. (1989). Impairment in processing meaningless verbal material in several modalities: The relationship between short-term memory and phonological skills. *Quarterly Journal of Experimental Psychology*, *41A*, 293–319.

Blazely, A., Coltheart, M., & Casey, B. (2005). Semantic impairment with and without surface dyslexia: Implications for models of reading. *Cognitive Neuropsychology*, *22*, 695–717.

Caccappolo-van Vliet, E., Miozzo, M., & Stern, Y. (2004a). Phonological dyslexia: A test case for reading models. *Psychological Science*, *15*, 583–590.

Caccappolo-van Vliet, E., Miozzo, M., & Stern, Y. (2004b). Phonological dyslexia without phonological impairment? *Cognitive Neuropsychology*, *21*, 820–839.

Cipolotti, L., & Warrington, E. K. (1995). Semantic memory and reading abilities: A case report. *Journal of the International Neuropsychological Society*, *1*, 104–110.

Coltheart, M. (1985). Cognitive neuropsychology and the study of reading. In M. I. Posner & O. S. M. Marin (Eds.), *Attention and performance, Vol. XI*. Hillsdale, NJ: Lawrence Erlbaum Associates Inc.

Coltheart, M. (Ed.). (1996). *Phonological dyslexia*. Hove, UK: Psychology Press.

Coltheart, M. (2005). Modelling reading: The dual-route approach. In M. J. Snowling & C. Hulme (Eds.), *The science of reading*. Oxford: Blackwell Publishing.

Coltheart, M., Langdon, R., & Haller, M. (1996). Computational cognitive neuropsychology. In B. Dodd, L. Worrall, & R. Campbell (Eds.), *Evaluating theories of language: Evidence from disordered communication*. London: Whurr Publishers.

Coltheart, M., Rastle, K., Perry, C., Langdon, R., & Ziegler, J. (2001). DRC: A dual route cascade model of visual word recognition and reading aloud. *Psychological Review, 108*, 204–256.

Dell, G. S., Schwartz, M. F., Martin, N., Saffran, E. M., & Gagnon, D. A. (1997). Lexical access in aphasic and nonaphasic speakers. *Psychological Review, 104*, 801–838.

Derouesné, J., & Beauvois, M.-F. (1985). The "phonemic" stage in the non-lexical reading process: Evidence from a case of phonological alexia. In K. E. Patterson, J. C. Marshall, & M. Coltheart (Eds.), *Surface dyslexia: Neuropsychological and cognitive studies of phonological reading.* Hove, UK: Lawrence Erlbaum Associates Ltd.

Ellis, A. W., & Young, A. W. (1988). *Human cognitive neuropsychology.* Hove, UK: Lawrence Erlbaum Associates Ltd.

Forster, K. I., & Chambers, S. M. (1973). Lexical access and naming time. *Journal of Verbal Learning and Verbal Behavior, 12*, 627–635.

Foygel, D., & Dell, G. S. (2000). Models of impaired lexical access in speech production. *Journal of Memory and Language, 43*, 182–216.

Fushimi, T., Komori, K., Ikeda, M., Patterson, K., Ijuin, M., & Tanabe, H. (2003). Surface dyslexia in a Japanese patient with semantic dementia: Evidence for similarity-based orthography-to-phonology translation. *Neuropsychologia, 41*, 1644–1658.

Gerhand, S. (2001). Routes to reading: A report of a non-semantic reader with equivalent performance on regular and irregular words. *Neuropsychologia, 39*, 193–208.

Goldberg, E. (1995). The rise and fall of modular orthodoxy. *Journal of Clinical and Experimental Neuropsychology, 17*, 193–208.

Grodzinsky, Y. (1990). *Theoretical perspectives on language deficits.* Cambridge, MA: MIT Press.

Grodzinsky, Y. (1995). A restrictive theory of trace deletion in agrammatism. *Brain and Language, 51*, 26–51.

Grodzinsky, Y. (2000). The neurology of syntax: Language use without Broca's area. *Behavioral and Brain Sciences, 23*, 1–71.

Grodzinsky, Y., Pinango, M. M., Zurif, E., & Drai, D. (1999). The critical role of group studies in neuropsychology: The regular nature of comprehension in Broca's aphasia. *Brain and Language, 67*, 134–147.

Harm, M. W., & Seidenberg, M. S. (1999). Phonology, reading acquisition and dyslexia: Insights from connectionist models. *Psychological Review, 106*, 491–528.

Harm, M. W., & Seidenberg, M. S. (2001). Are there orthographic impairments in phonological dyslexia? *Cognitive Neuropsychology, 18*, 71–92.

Hillis, A. E., & Caramazza, A. (1991). Mechanisms for accessing lexical representation for output: Evidence for a category-specific semantic deficit. *Brain and Language, 40*, 497–539.

Hinton, G. E., McClelland, J. L., & Rumelhart, D. E. (1986). Distributed representations. In J. L. McClelland, D. E. Rumelhart, & the PDP Research Group (Eds.), *Parallel distributed processing: Explorations in the microstructure of cognition: Vol 1. Foundations* (pp. 77–109). Cambridge, MA: MIT Press.

Howard, D., & Best, W. M. (1996). Developmental phonological dyslexia: Real word reading can be completely normal. *Cognitive Neuropsychology, 13*, 887–934.

Jacobs, A. M., & Grainger, J. (1994). Models of visual word recognition: Sampling the state of the art. *Journal of Experimental Psychology: Human Perception and Performance, 20*, 1311–1334.

Lambon Ralph, M. A., Ellis, A. W., & Franklin, S. (1995). Semantic loss without surface dyslexia. *Neurocase, 1*, 363–369.

Marshall, J. C., & Newcombe, F. (1973). Patterns of paralexia: A psycholinguistic approach. *Journal of Psycholinguistic Research, 2*, 175–199.

McCarthy, R., & Warrington, E. K. (1986). Phonological reading: Phenomena and paradoxes. *Cortex, 22*, 359–380.

McClelland, J. L., & Rumelhart, D. E. (1986). Amnesia and distributed memory. In J. L. McClelland, D. E. Rumelhart, & the PDP Research Group. *Parallel distributed processing: Explorations in the microstructure of cognition: Vol 2. Psychological and biological models* (pp. 503–528). Cambridge, MA: MIT Press.

Newcombe, F., & Marshall, J. C. (1980). Transcoding and lexical stabilization in deep dyslexia. In M. Coltheart, K. E. Patterson, & J. C. Marshall (Eds.), *Deep dyslexia* (pp. 176–188). London: Routledge & Kegan Paul.

Patterson, K. (1990). Alexia and neural nets. *Japanese Journal of Neuropsychology, 6*, 90–99.

Patterson, K. (2000). Phonological alexia: The case of the singing detective. In E. Funnell (Ed.), *Case studies in the neuropsychology of reading* (pp. 57–83). Hove, UK: Lawrence Erlbaum Associates Ltd.

Plaut, D. C. (1999). A connectionist approach to word reading and acquired dyslexia: Extension to sequential processing. *Cognitive Science, 23*, 543–568.

Plaut, D. C., McClelland, J. L., Seidenberg, M. S., & Patterson, K. (1996). Understanding normal and impaired word reading: Computational principles in quasi-regular domains. *Psychological Review, 103*, 56–115.

Rumelhart, D. E., & McClelland, J. L. (1986). PDP models and general issues in cognitive science. In J. L. McClelland, D. E. Rumelhart, & the PDP Research Group (Eds.), *Parallel distributed processing: Explorations in the microstructure of cognition: Vol 1. Foundations.* (pp. 110–146). Cambridge, MA: MIT Press.

Ruml, W., & Caramazza, A. (2000). An evaluation of a computational model of lexical access: Comments on Dell et al. (1997). *Psychological Review, 107*, 609–634.

Schwartz, M. F., Saffran, E. M., & Marin, O. S. M. (1980). Fractionating the reading process in dementia: Evidence for word-specific print-to-sound associations. In M. Coltheart, K. Patterson, & J. C. Marshall (Eds.), *Deep dyslexia* (pp. 259–269). London: Routledge & Kegan Paul.

Seidenberg, M. S., & McClelland, J. L. (1989). A distributed, developmental model of word recognition and naming. *Psychological Review, 96*, 523–568.

Shallice, T. (1988). *From neuropsychology to mental structure.* Cambridge: Cambridge University Press.

COGNITIVE NEUROPSYCHOLOGY, 2006, 23 (1), 110–134

Spelling and dysgraphia

Gabriele Miceli
Università Cattolica and Fondazione Santa Lucia IRCCS, Rome, Italy, and Harvard University, Cambridge, USA

Rita Capasso
Università Cattolica and Fondazione Santa Lucia IRCCS, Rome, Italy

Early cognitive models of spelling assumed that orthographic word representations are linear, ordered sequences of abstract letter identities (graphemes), activated only by word meaning information, and in some cases proposed that activating phonological information is a necessary stage of the spelling process. Over the past 20 years, studies on dysgraphia have shown that orthographic representations are autonomous from phonological representations and, just like the latter, are directly activated from semantics. The selection of an orthographic form for output relies on the convergence of activation from lexical-semantic information and from sublexical phoneme–grapheme conversion procedures. In addition, it is increasingly clear that orthographic representations are multidimensional objects that separately represent the graphosyllabic structure (or perhaps the nucleus/non-nucleus positions) of the target, and the identity, the CV status, and the quantity (doubling) of each grapheme. In spelling, the structure of orthographic knowledge and the mechanisms involved in processing serial order interact in complex ways and constrain performance accuracy. Further research is needed to clarify some critical issues: We need to specify in greater detail the mechanisms involved in the interaction between meaning and sublexical information; we must consider the possibility that orthographic representations have texture, in addition to structure; we must provide explicit hypotheses on the mechanisms that process orthographic knowledge; and we must gain a better understanding of the interaction between structure and serial order.

INTRODUCTION

Reports of acquired disorders of spelling appeared very early in the neuropsychological literature (Grashey, 1885; Head, 1926; Lichtheim, 1885; Wernicke, 1906; etc.). However, for a long time interest in spelling was very limited. To mention the obvious, individuals learn to speak before they learn to write, and even the most prolific writer engages in speaking much more often than he or she does in spelling. Furthermore, the ability to read and write is tightly linked to the educational level of the individual, and until not so many years ago, proficient spelling abilities were achieved only by a limited number of individuals in the population, and were rather infrequently called upon in everyday life. Thus, the finding that an infrequently developed, rarely

Correspondence should be addressed to Gabriele Miceli, MD, Neurologia, Università Cattolica, Largo A. Gemelli 8, 00168 Roma, Italia, (Email: g.miceli@mclink.it).

Preparation of this manuscript was supported in part by grants from MIUR (FIRB) and from the Università Cattolica, and from NIH/NICDC grant R01 DC 045442-01.

DOI:10.1080/02643290500202730

practised, and often unnecessary skill could be impaired as a consequence of brain damage did not raise particular attention. Consequently, research on dysgraphia did not focus on the qualitative features of the disorder as much as on the general issue of the relationships between the disorders of spelling and those of other cognitive and linguistic functions. This taxonomic approach is clearly evident from the classifications of the disorder in use until the late 1970s, which distinguished agraphia with aphasia, agraphia with alexia, pure agraphia, apractic agraphia, and spatial dysgraphia (e.g., Benson, 1979).

The 1980s marked a turning point for neuropsychological research. Models based on the assumptions of the information processing approach (e.g., Morton's logogen model, 1969) began to influence psycholinguistic research on language in normal subjects. In addition, seminal studies began to analyse aphasic performance in the light of these models, and showed that investigations in the cognitive neuropsychology of language could provide powerful tools for the development of detailed theories of normal cognition and for a better understanding of mind–brain relationhsips, as well as for the diagnosis and treatment of cognitively impaired individuals.

Initially, research on dyslexia benefited the most from the new approach (e.g., Marshall & Newcombe, 1973; Shallice & Warrington, 1975). In a very short time, many cases of "phonological alexia," that is, the selective inability to read aloud novel words (e.g., Dérouesné & Beauvois, 1979; Shallice & Warrington, 1980), and of the complementary disorder "surface dyslexia," that is, the selective inability to read aloud familiar words (e.g., Patterson, Marshall, & Coltheart, 1985), were put on record. These contrasting patterns were attributed to the selective impairment of sublexical grapheme–phoneme conversion procedures (required to read alound unfamiliar strings), and to the selective impairment of lexical-semantic mechanisms (necessary to read aloud familiar words), respectively. More complex patterns of performance, such as those reported for the so-called "deep dyslexic" subjects, also received sophisticated interpretations in this context (e.g., Coltheart, Patterson, & Marshall, 1980).

Soon after, the cognitive approach extended to research on dysgraphia. Its rapid development in this area led to the abandonment of the purely taxonomic approach, and to ask questions regarding the mechanisms and representations involved in spelling familiar and novel words. The first obvious question addressed by these studies was whether or not it was possible to support a functional architecture of the spelling system parallel to that suggested for reading by cognitive neuropsychological investigations of dyslexia. The answer was clearly in the affirmative.

A critical assumption in the domain of reading, motivated by the contrasting patterns of performance described in subjects with "phonological" and "surface" dyslexia, had been that lexical-semantic mechanisms and grapheme/phoneme conversion procedures are autonomous. Throughout the 1980s, empirical support for an analogous distinction in spelling was provided by analyses of dysgraphia. Subjects with disproportionate difficulty spelling novel words as compared to familiar words ("phonological agraphia"; e.g., Assal, Buttet, & Jolivet, 1981; Baxter & Warrington, 1985; Bub & Kertesz, 1982a; Roeltgen, Sevush, & Heilman, 1983; Shallice, 1981) and subjects with the reverse pattern, consisting of more accurate performance on novel words than on familiar words ("surface dysgraphia"; e.g., Baxter & Warrington, 1987; Beauvois & Dérouesné, 1981; Hatfield & Patterson, 1983; Roeltgen & Heilman, 1984) were described. These contrasting patterns of performance were attributed, just as the twin syndromes described in the domain of reading, to a disproportionate impairment of sublexical (in this case, phoneme–grapheme) conversion procedures, and of lexical-semantic mechanisms, respectively. The independence of lexical-semantic mechanisms, involved in retrieving stored forms from a long-term memory system, from sublexical conversion procedures, that assemble plausible orthographic strings by serially converting a phonological input, also easily accounted for the qualitative features of the errors typically found in "surface dysgraphia."

When the target orthographic form is unavailable due to damage in the lexical-semantic system, the subject is forced to spell via sublexical conversion procedures. Since these are by definition insensitive to lexical constraints, phonologically plausible errors will occur on words with "opaque" orthography (e.g., in English, "yacht" → *yot*, "subtle" → *suttel*; in French, "cyprés," cypress → *sipré*; in Italian, "cuore," heart → *quore*). In these errors, segments of the target word are mapped onto a graphemic option that, although permissible, is inappropriate for that specific word (for example,/ɔt/ → *ot* is appropriate for *knot*, but not for *yacht*). It has also been shown that selection of a sound/print mapping from sublexical conversion procedures is sensitive to the frequency of occurrence of that mapping in the language and to the orthographic context in which it appears (Goodman & Caramazza, 1986a).

Neuropsychological studies also provided evidence on other critical aspects of the spelling process. Some analyses shed light on the functional architecture of the components of the lexical-semantic system involved in spelling, demonstrating the role played by conceptual categories, grammatical class, and morphological structure information (e.g., Badecker, Hillis, & Caramazza, 1990; Caramazza & Hillis, 1991). Other investigations focused on later stages of processing, namely those involved in the oral or written spelling of an orthographic string. They have shown that a brain lesion may at times result in selective damage to oral spelling or to written spelling (e.g., Friedman & Alexander, 1989; Goodman & Caramazza, 1986b; Hodges & Marshall, 1992; Kinsbourne & Rosenfield, 1974; Kinsbourne & Warrington, 1965), indicating that the processes converting graphemes into allographs and graphonomotor routines (for writing) are independent from those that convert graphemes into letter names (for oral spelling). Finally, even though the relevant empirical evidence was not yet available, it was commonly assumed (e.g., Ellis, 1982; Wing & Baddeley, 1980) that, since both familar and novel words correspond to multi-graphemic strings but the mechanisms that convert them into motor output or letter names only process one grapheme at a time, spelling requires an orthographic working memory system—a graphemic buffer.

By the end of the 1980s, then, the overall structure of the spelling system was largely laid out (see schematic representation in Figure 1). In order to spell a familiar word, the corresponding stored form is retrieved in the orthographic lexicon from semantic input. It is subsequently placed in an orthographic working memory, while grapheme–letter name and grapheme–allograph conversion procedures serially convert it into the format appropriate for oral spelling and for written spelling, respectively. When a pseudoword must be spelled, an independent mechanism is activated, which serially transforms the

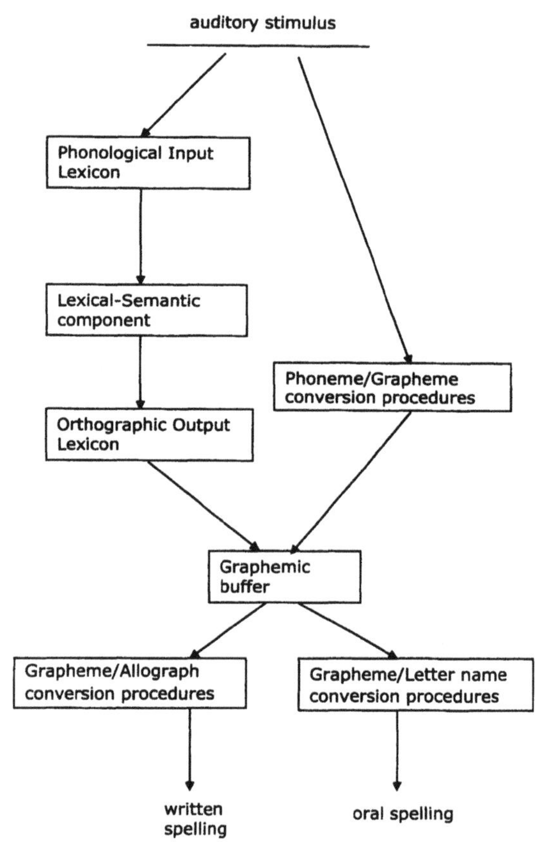

Figure 1. *Schematic representation of the spelling system.*

phonological input into a graphemic string for output, based on contextual constraints and on the frequency of mapping options. Importantly, even though for the most part the functional architecture of the system had been assumed by analogy to reading, it soon found the necessary support in studies of dysgraphia.

Since the mid 1980s, research in this area has addressed increasingly more detailed questions, relevant to a deeper understanding of the computations taking place at various stages of spelling, and of the representations and processes involved. This paper reviews the progress made on some of these issues. It focuses on three topics: the sources of input to lexical orthographic representations, the formal properties of orthographic knowledge, and the mechanisms involved in processing orthographic representations during spelling tasks. The first topic is discussed in the first section of this paper, the other two in the second section.

THE MECHANISMS INVOLVED IN THE ACTIVATION OF ORTHOGRAPHIC FORMS

The interaction between lexical-semantic and sublexical conversion procedures in spelling

Until the late 1980s, research focused on demonstrating the separability of lexical-semantic and nonlexical mechanisms involved in spelling. As a result of these analyses, it was assumed that only information retrieved within the lexical-semantic system can serve as input to an orthographic form when that form has to be spelled. This view was to change substantially in the early 1990s, when evidence from brain-damaged subjects began to suggest that the lexical-semantic and sublexical mechanisms may interact.

This possibility had been initially raised by the performance of normal subjects on pseudoword writing-to-dictation tasks (Barry & Seymour, 1988; Campbell, 1983), in which the spelling of a pseudoword appeared at times to be "primed" by a word presented previously. For example, /slOd/was more likely to be spelled as *sload* when preceded by *broad*, but as *slod* when preceded by *nod*. If the two sets of mechanisms were completely independent, the selection of graphemic options by phoneme–grapheme conversion procedures when spelling a pseudoword should be unaffected by the orthographic string retrieved from the lexicon when spelling a just-presented word that contains the same phoneme.

Neuropsychological studies of dysgraphic subjects provided evidence consistent with the interaction between lexical-semantic and sublexical routines, by demonstrating that aphasic performance in spelling is constrained by the status of *both* lexical-semantic mechanisms and sublexical conversion procedures.

A first relevant pattern is observed in subjects with semantic damage and spared sublexical conversion procedures (e.g., Hillis & Caramazza, 1991; Patterson & Hodges, 1992). Subjects of this type produce phonologically plausible errors in spelling irregular words. However, these errors do not occur at random—to the contrary, they are constrained by comprehension. Thus, case JJ (Hillis & Caramazza, 1991) spelled correctly 38/40 (95%) words that he fully comprehended, produced phonologically plausible errors to 22/22 (100%) words that he completely failed to understand; and spelled correctly 13/18 (72.2%) words he comprehended partially. Correct spelling of fully comprehended words, and phonologically plausible errors in responses to uncomprehended words, can be accommodated without assuming any interaction: In the first case, intact meaning activates the correct orthographic string; in the second case, lack of any meaning information forces exclusive reliance on sublexical conversion procedures that, in an opaque language such as English, results in phonologically plausible errors. To account for correct spelling of partially comprehended words, it was hypothesised that an impoverished semantic representation activates several meaning-related orthographic entries in the lexicon, and that under these circumstances lexical selection is eventually constrained by converging information from semantics and from sublexical phoneme–grapheme conversion procedures. For example, the auditorily presented stimulus "tiger" might activate incomplete semantic information (let's say [wild animal, feline, has fur, carnivorous], but not [lives in Asia, fur is black-and-yellow striped, etc.]), which will raise the activation level of orthographic forms sharing several of those conceptual features (e.g., *lion, panther, tiger, leopard*, etc.). If semantic information were the only input to orthographic lexical representations, any of these strings would be selected for output. In dictation tasks, however, sublexical conversion procedures work in parallel to lexical-semantic mechanisms. They probabilistically assemble a plausible orthographic string corresponding to the auditory stimulus (in our example, *tiger, tyger, tygur, tigur* would all be plausible mappings for/taigə/), which constrains the selection of the lexical entry orthographically most similar to the target and/or blocks the selection of semantically similar but orthographically dissimilar strings.

The second relevant pattern of performance is observed in cases of "deep dysgraphia". These subjects, who produce semantic errors in spelling words to dictation as a consequence of damage to either the semantic (e.g., Bub & Kertesz, 1982b; Miceli, Benvegnù, Capasso, & Caramazza, 1997a; Nolan & Caramazza, 1983) or the orthographic lexical component (Beaton, Guest, & Ved, 1998; Caramazza & Hillis, 1991; Rapp, Benzing, & Caramazza, 1997), also invariably present with very severe damage to sublexical phoneme–grapheme conversion procedures, as shown by their inability to spell pseudowords. In these cases, lack of any input from sublexical conversion mechanisms prevents blocking orthographic forms that are semantically related, but orthographically unrelated to the target. Thus, in a situation similar to the example sketched in the previous paragraph, any of the activated orthographic forms (*lion, panther, tiger, leopard*, etc.) may be selected for output.

Observations from subject RCM (Hillis, Rapp, & Caramazza, 1999) are also consistent with an interaction between lexical-semantic and sublexical mechanisms. One week after a stroke, this subject suffered from damage to orthographic lexical forms and from severe impairment of phoneme–grapheme conversion procedures. At this stage, performance on pseudowords was severely impaired (only 42% of the phonological segments included in a list of 25 pseudowords were spelled correctly), and she produced semantic errors to 56% of the words dictated by the examiner. When RCM was tested again, 21 days post onset, phoneme–grapheme conversion rules had recovered to some extent (this time, she correctly mapped 67% of the phonemes included in 34 pseudowords), and semantic paragraphias correspondingly dropped to 11%. In this case, at an early stage, loss of phoneme–grapheme conversion rules prevented the possibility to constrain the selection of the correct orthographic form and/or to block the selection of semantically related forms, and subsequent partial recovery of the same rules reinstated this possibility, at least to some extent.

To sum up, neuropsychological observations suggest that lexical-semantic and sublexical mechanisms, even though they are autonomous and thus exposed to dissociated impairment, as shown by cases of "phonological" and "surface" dysgraphia, do interact in spelling, and that sublexical phoneme–grapheme conversion procedures may influence the process of lexical selection.

The role of phonological lexical information in the selection of orthographic forms

Autonomy and interaction of orthographic and phonological forms

Before dealing with the question of how the lexical-semantic/sublexical interaction may actually take place, an additional question must be addressed, which also concerns the sources of input to orthographic forms. It regards the possibility that, in addition to the semantic component, the phonological lexical component may also activate orthographic forms.

The view that retrieving a phonological lexical form is critical for the activation of its corresponding orthographic lexical form was proposed very early in dysgraphia research. The fact that speaking skills develop before writing skills influenced the notion, implicit in early papers (Grashey, 1885; Head, 1926; Lichtheim, 1885; Wernicke, 1906), that spelling is *only* possible via phonological mediation. When spelling a word, an individual cannot activate an orthographic form directly from semantics. To the contrary, word meaning must activate the phonological lexical form, which in turn activates the corresponding orthographic form (Figure 2, top). There are two readings of this hypothesis.

The "sublexical" phonological mediation hypothesis was stated most explicitly by Luria (1966): The phonological form activated from semantics is converted, phoneme by phoneme, into the corresponding orthographic segments, until the complete orthographic representation is realised. This possibility is unlikely on logical

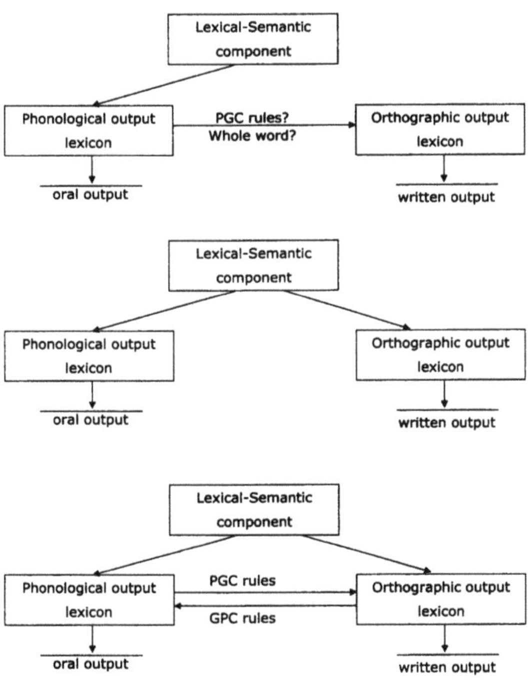

Figure 2. *Possible relationships between phonological and orthographic lexical forms. Top: The obligatory phonological mediation hypothesis. Centre: The complete autonomy hypothesis. Bottom: The autonomy and interaction hypothesis. PGC: phoneme-grapheme conversion; GPC: grapheme/phoneme conversion.*

grounds alone. Except for fully regular languages, it would inevitably yield phonologically plausible errors; and it would not be able to account for ideographic writing (e.g., Japanese kanji). Sublexical phonological mediation has been disconfirmed also by empirical observations. For example, subject PR (Shallice, 1981) could not convert a phonological string into an orthographic string, as documented by extremely poor performance in writing pseudowords to dictation (9/50 correct, or 18%), and yet spelled correctly most words (overall, 568/622 correct, or 91.3%; he was also 100% correct on high-frequency, high-concreteness stimuli).

Alternatively, phonological mediation might be "lexical": The phonological form retrieved from semantics directly activates the corresponding orthographic representation. Recent research unequivocally disproves this hypothesis as well. If lexical phonological mediation were indeed necessary, attempts at naming a picture should result either in the correct response or in the same incorrect word, both in speaking and in spelling. For example, if the picture of a tiger activates the correct phonological entry, then the orthographic entry *tiger* should also be activated. If the same picture incorrectly activates "lion," *lion* should be activated in the orthographic lexicon as well. In other words, the same stimulus should never activate different entries in the phonological and in the orthographic lexical output component (say, the correct response in one modality and a semantically related word in the other, or the correct response in one modality and no response in the other, or two distinct semantically related words; and so on).

Three types of neuropsychological evidence contradict this prediction. In the first place, some aphasic subjects make semantic errors only in one output modality. For example, RGB and HW (Hillis & Caramazza, 1991) produced semantic errors only in speaking, and SJD (Hillis & Caramazza, 1991) and WML (Hillis et al., 1999) only in writing. Second, some subjects with semantic damage, or with damage to both phonological and orthographic forms, produced different, semantically related words in response to the same target, in the context of multiple-naming tasks, which require two or more consecutive responses in both output modalities (Alario, Schiller, Domoto-Reilly, & Caramazza, 2003; Beaton et al., 1997; Miceli et al., 1997a; Rapp et al., 1997). For example, when presented with the picture of a trumpet and asked to say, then write, the corresponding name, subject WMA (Miceli et al., 1997a) said "piano" and spelled *trumpet.* Third, some subjects with lexical damage present with selective grammatical-class-by-modality effects. A selective deficit for verbs may be restricted to speech in some subjects and to spelling in others (Caramazza & Hillis, 1991); a selective deficit for nouns may be restricted to speech (Hillis & Caramazza, 1995) and a selective deficit for verbs to spelling (Rapp & Caramazza, 1998); or, the same subject may have difficulty with verbs in speaking and with nouns in spelling (Rapp & Caramazza, 2002). These observations show the autonomy of orthographic from phonological forms, and indicate that orthographic forms are directly accessed from semantics (Figure 2, centre).

Phonological and orthographic forms cannot be *completely* autonomous, however. Neuropsychological evidence suggests that they interact. In fact, most subjects with semantic damage or with lexical damage obtain comparable performance in oral and written naming (e.g., Lambon Ralph, Sage, & Roberts, 2000), and do not present with dissociated responses in multiple (spoken-then-written or written-then-spoken) naming tasks (e.g., Miceli & Capasso, 1997), even though they produce several semantically incorrect responses in both modalities. How is it possible to reconcile the apparently contrasting observations that suggest both the autonomy and the interaction of phonological and orthographic lexical representations?

The possible role of sublexical conversion procedures in the interaction between orthographic and phonological forms

All the subjects who produced lexically discordant responses in spoken and written tasks, due to either semantic or lexical damage, also presented with a severe disruption of sublexical conversion procedures—they were virtually unable to read aloud pseudowords or to write them to dictation. In addition, there was a correspondence between the damaged set of sublexical mechanisms and the tasks that yielded semantic errors. Thus, subjects who produced discordant responses both in the spoken-then-written and in the written-then-spoken naming tasks presented with severe damage to both phoneme–grapheme and grapheme–phoneme mappings (e.g., Beaton et al., 1997; Miceli et al., 1997a; Rapp et al., 1997); whereas both sets of conversion procedures were spared (or at least functional) in the subjects who never produced lexically discordant responses in similar tasks (e.g., Miceli & Capasso, 1997). And, subjects who presented with dissociated performance in spoken and written responses also suffered from asymmetric damage to sublexical conversion procedures. Thus, ECA (Miceli, Capasso, & Caramazza, 1999) produced semantically discordant responses in the spoken-then-written naming task (lion: "tiger" → *panther*), and suffered from severe damage to phoneme–grapheme conversion procedures; however, he never produced incorrect sequences of the same type in the written-then-spoken naming task (piano: *trumpet* → "trumpet"), and showed spared sublexical grapheme–phoneme mapping abilities. Furthermore, in subjects RGB and HW (Caramazza & Hillis, 1990), who only produced semantic errors in spoken tasks, severe sublexical damage was restricted to grapheme–phoneme mappings, whereas in subjects SJD (Caramazza & Hillis, 1991) and WML (Hillis et al., 1999), who only produced semantic errors in spelling tasks, sublexical damage was severe only for phoneme–grapheme conversion procedures.

Three facts help understand the relationships between phonological and orthographic lexical representations, and between these and sublexical conversion mechanisms. First, damage to semantics or to output lexical representations systematically co-occurs with severe damage to sublexical procedures in subjects who produce discrepant responses in double naming tasks. Second, damage to semantics or to output lexical representations is systematically associated with spared sublexical conversion procedures in subjects who do not produce such responses. And third, in the presence of damage to semantic or to lexical representations, the status of a specific set of sublexical conversion rules (e.g., phoneme–grapheme conversion rules) allows the prediction of the asymmetric occurrence of discordant responses in double naming tasks. These three facts converge toward the same conclusion: Phonological and orthographic lexical representations are autonomous, but they may interact, via sublexical mechanisms (Figure 2, bottom). The "ecological" value of the interaction is obvious. When a word has to be spelled, such as in a picture naming task, an orthographic representation is selected for output on the basis of direct input from semantic information, but also to some extent on the basis of the concurrent activation (also from semantics) of the corresponding phonological representation, which may be sublexically converted into an orthographic string by phoneme–grapheme conversion procedures, and then used to further constrain target selection. This will ensure that as much information as possible, from all available sources, converges on the linguistic component involved in production. A reverse interaction may take place in spoken naming. It is quite possible that this complex flow of information is unnecessary in the cognitively unimpaired subject, who relies on intact semantic and lexical systems; however, the behaviour of neurologically impaired subjects clearly suggests that these mechanisms may be recruited to compensate for functional damage in the spelling system.

The interaction between lexical and sublexical conversion procedures: Some open questions

The observations discussed in the previous sections force us to reject some long-held views on the spelling system, such as the complete independence of orthographic lexical forms from sublexical phoneme–grapheme conversion rules, and the hypothesis that phonological mediation is necessary for spelling. They are consistent with the notion that phonological and orthographic lexical forms are autonomous, and that lexical-semantic mechanisms and sublexical conversion procedures are represented in independent cognitive and neural substrates. However, they are also consistent with the idea that sublexical mechanisms may serve, at least under some circumstances, as a "backup" procedure for lexical access. More particularly, these mechanisms may interact with lexical information in constraining the selection of an orthographic form.

Although the proposal of an interaction between lexical-semantic and sublexical conversion mechanisms allows us to account for a variety of patterns of performance observed in subjects with damaged spelling (and reading), the locus and the procedural aspects of the interaction are still open to discussion. Three alternative proposals have been put forward concerning the possible mechanisms underlying the interaction between lexical-semantic and sublexical mechanisms in spelling.

Hillis and Caramazza (1991) and Patterson and Hodges (1992) (Figure 3, top) hypothesised that semantic information and the letter string probabilistically assembled by sublexical conversion procedures converge at the output lexical level. On this view, the output lexical system would be "permeable" to sublexical information, and there would be a direct interaction between the string retrieved in the lexicon and the string (or strings) assembled by sublexical conversion procedures. The latter would boost the level of activation of the lexical entry that shares the largest number of orthographic features with the assembled string (and, perhaps, reduce the activation level of the competing strings).

Houghton and Zorzi (2003) proposed a two-route computational model of spelling that includes a lexical component and a sublexical conversion component. In this framework, the interaction takes place at the segmental level (Figure 3, centre). When a word is dictated, a lexeme entry (activated from semantic information) activates segments at the grapheme layer level. At the same time, phoneme–grapheme conversion procedures convert the stimulus into an orthographic string, which also activates elements in the grapheme layer. On this view, then,

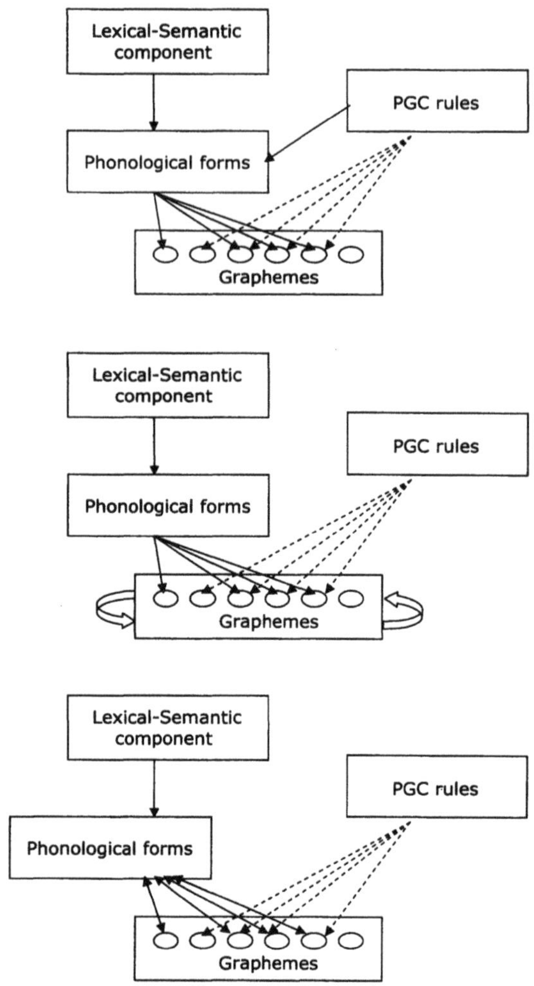

Figure 3. *Possible mechanisms of interaction between the lexical-semantic and the sublexical component, schematically representing the positions held by Hillis and Caramazza (1991), Houghton and Zorzi (2003), and Rapp et al. (2002). Note that in Houghton and Zorzi (2003) the semantic component is not modelled. It is shown here only to maintain symmetry with the other proposals.*

orthographic segments are eventually selected as the result of combined activation in the grapheme layer from semantic-lexical and sublexical input. A further hypothesis, related to both this and the previous proposal (Newcombe & Marshall, 1980), assumes interaction both at the level of the orthographic lexicon and at the level of the graphemic buffer, in the context of intrinsic instability in the semantic system.

A different hypothesis has been put forth by Rapp, Epstein, and Tainturier (2002), in the context of a two-route, connectionist-type architecture (Figure 3, bottom). This model draws on a similar proposal for speech production (Rapp & Goldrick, 2000). It assumes that word production requires a meaning layer, a lexeme layer, and a segment layer. Information flows unidirectionally from the meaning to the lexeme layer, but bidirectionally between the lexeme and the segmental layer. In order to simulate spelling, a phoneme–grapheme conversion layer is added. When a word is dictated, information retrieved in the meaning layer activates a lexeme-level representation, which in turn activates segments in the grapheme layer. At the same time, the input string is processed by phoneme–grapheme conversion rules, which also feed information to the grapheme layer. Information feeds back from the grapheme to the lexeme layer, and the interaction between these two layers eventually leads to encoding the output string. Under normal conditions, lexical information suffices to select the correct target. When the lexeme layer is damaged, at times the target node will be activated only partially. In this case, the highest activation in the grapheme layer may be reached by segments activated in the lexicon, interspersed with segments activated by sublexical conversion procedures. This hypothesis might account for the pattern of performance observed in LAT (Rapp et al., 2002), who suffered from acquired dysgraphia due to lexical damage, and produced several phonologically plausible errors containing low-frequency mappings, like "bouquet" → *bouket*, where *et* is a very low-probability graphemic mapping for/ɛi/. Errors of this type should be uncommon, since selection of phoneme–grapheme correspondences relies on their probability of occurrence in the language (Goodman & Caramazza, 1986a). According to Rapp et al., an error like "bouquet" → *bouket* results from the interaction of lexical information (which would provide the low-probability mapping *et*) and sublexical information (which would provide *k* as a plausible mapping for/k/). Evidence consistent with the hypothesis has been reported in other studies (Folk, Rapp, & Goldrick, 2002; McCloskey, Macaruso, & Rapp, in press).

A clear-cut choice among the different proposals cannot be made as yet. The idea of a direct interaction between sublexical information and lexical entries is interesting, but it is insufficiently developed. The notion of an interaction at the buffer level is interesting and reasonable, but it remains to be seen whether or not it can accommodate the actual results observed in brain-damaged subjects. The proposal by Rapp et al. calls for a radical change in the functional architecture of the spelling system but it, too, needs further testing. In particular, the role of the lexeme layer/grapheme layer interaction remains underspecified. Albeit in a different (phonological output) context, such interaction did not seem to play a significant role in accounting for aphasic performance on picture naming tasks. Removing the interaction from the model proposed by Rapp and Goldrick (2000) had no effect on the model's ability to accommodate the patterns of performance observed in a group of 50 aphasic subjects (Ruml, Caramazza, Capasso, & Miceli, 2005).

Whatever their answers, the questions on the nature of the interaction between lexical and sublexical procedures, and on its role in constraining lexical selection, are critical topics for future research on dysgraphia. However, regardless of how graphemic strings are encoded, available neuropsychological observations from dysgraphia allow tentative solutions to another crucial theoretical issue: the nature of the information represented in the orthographic string.

THE STRUCTURE AND PROCESSING OF ORTHOGRAPHIC REPRESENTATIONS

The linear hypothesis

The earliest observations relevant for understanding the properties of orthographic representations were provided by studies of slips-of-the-pen in normal subjects (e.g., Ellis, 1979, 1982; Hotopf, 1980; Wing & Baddeley, 1980). Many misspellings in these corpora resulted in the substitution (will → *wull*), omission (answered → *aswered*), insertion (several → *severeal*) or transposition (piece → *peice*) of letters. The fact that these errors could be attributed to loss of information on the identity of the segments comprising the string (substitutions, insertions, omissions of letters) or on their position within the string (transpositions of letters) prompted the hypothesis that an orthographic representation is a linear sequence of graphemes, each defined by its identity and position in the sequence. For example, *tavoli*, tables and *stella*, star would be represented as <t>$_1$ <a>$_2$ <v>$_3$ <o>$_4$ <l>$_5$ <i>$_6$ and <s>$_1$ <t>$_2$ <e>$_3$ <l>$_4$ <l>$_5$ <a>$_6$, respectively.

These assumptions on the nature of orthographic representations were integrated with those on the functional architecture of the system, resulting in the notion that spelling consists of activating a linear string of ordered graphemes (retrieved in the lexicon or assembled by phoneme–grapheme conversion procedures), and serially converting it into a sequence of allographs (for writing) or letter names (for oral spelling). This view of how orthographic forms are activated and produced logically requires the additional assumption of a graphemic buffer. The reason for this assumption is straightforward. The information retrieved from the lexicon or assembled by sublexical conversion procedures corresponds to a string of graphemes, whereas the mechanisms that transform it into letter shapes or letter names can only operate on one segment at a time. Hence, the orthographic string must remain active in a working memory component (the buffer), while later stages of processing are completed.

When the graphemic buffer is damaged, the subject's performance will reflect its role in spelling. Since the buffer is shared by all the tasks that require the activation of an orthographic string for output, spelling accuracy will not be influenced by task type (e.g., narrative writing vs. picture naming vs. spelling to dictation), or by response modality (oral spelling vs. written spelling). In addition, since the buffer is a post-lexical stage of processing, performance will be unaffected by lexical-semantic dimensions (e.g., grammatical category, abstractness/concreteness, morphological structure, frequency of usage), or by the lexical status of the string (familiar as opposed to novel word). By contrast, error rate will be influenced by stimulus length, since reduced buffer capacity (either because fewer graphemes than normal are maintained active, or because the normal number of graphemes remains active for a pathologically short time) will result in longer words being spelled less accurately than short words. In addition, misspellings will reflect loss of segmental information, and will be exclusively constrained by the properties of the orthographic representation in the buffer. Since *ex hypothesi* the latter is an ordered string of graphemes, errors will result in letter substitutions (colore, color → *conore*), insertions (usanza, habit → *unsanza*), omissions (grembo, womb → *gembo*) or transpositions (two letters exchanging positions, as in denaro, money → *derano*; or a letter shifting to an incorrect position, as in centro, centre → *crento*).

The performance of a number of dysgraphic subjects (Table 1) fully bears out the predictions based on the buffer's role in spelling putatively linear graphemic representations. In all these cases, performance is significantly affected only by stimulus length, and errors result from damage to segmental information (Table 2).

Limitations of the linear hypothesis

Detailed analyses of the errors made by the subjects listed in Table 1 support the postulated architecture, and in particular the assumption of a graphemic buffer. However, they also force a profound revision of the linear hypothesis. The view that orthographic representations only code the identity of each grapheme and its position in the string makes three unambiguous predictions: spelling performance should be unaffected by orthographic structure (other than conditional frequencies); all two-consonant clusters should behave similarly; and spelling accuracy should be unaffected by the CV (consonant/vowel) status of a grapheme. Each of these predictions was disconfirmed by neuropsychological analyses.

Table 1. *Case studies meeting the criteria for the diagnosis of graphemic buffer damage*

Case initials	*Authors*
FV	(Miceli, Silveri, & Caramazza, 1985)
SE	(Posteraro, Zinelli, & Mazzucchi, 1988)
LB	(Caramazza, Miceli, Villa, & Romani, 1987; Caramazza, & Miceli, 1990)
CF, CW	(Cubelli, 1991)
Case 1	(Piccirilli, Petrillo, & Poli, 1992)
HE	(McCloskey, Badecker, Goodman-Schulman, & Aliminosa, 1994)
JH	(Kay & Hanley, 1994)
SFI	(Miceli, Benvegnù, Capasso, & Caramazza, 1995)
AS	(Jónsdottír, Shallice, & Wise, 1996)
PM	(Cantagallo & Bonazzi, 1996)
AZO	(Miceli, Benvegnù, Capasso, & Caramazza, 1997a)
BA	(Ward & Romani, 1998)
LM	(Blanken, Schäfer, Tucha, & Lange, 1999)
Case 2	(Kokubo, Suzuki, Yamadori, & Satou, 2001)
PB, TH	(Schiller, Greenhall, Shelton, & Caramazza, 2001)
LiB	(Cotelli, Aboutalebi, Cappa, & Zorzi, 2003)
BWM, FM	(Tainturier & Rapp, 2004)
GSI	(Miceli, Benvegnù, Capasso, & Caramazza, 2004)

The effect of orthographic structure on spelling accuracy

If all that matters in orthographic representations is grapheme identity and position, performance on words of the same length should be identical,

Table 2. *Distribution of error types in some subjects with graphemic buffer deficits*

Case initials	*N of errors in corpus*	*% subst*	*% insert*	*% omiss*	*% transp*
LB	1339	43	6	27	28
CW	409	83	3	6	7
HE	674	31	19	30	20
JH	442	57	25	15	3
AS	418	29	21	28	22
BA	269	51	13	28	8
LM	183	5	1	17	77
LiB	112	96	3	1	0
BWN	720	30	5	52	13
FM	109	58	7	22	13
GSI	546	9	0	78	13

regardless of their orthographic structure. For example, performance on six-letter stimuli like *tavolo, albero, onesto, attimo, anello, stadio, guasto, scorso, nostro, strano, frutta, frusta* (table, tree, honest, instant, ring, stadium, out of order, previous, our, strange, fruit, whip) should be comparable. The qualitative analysis of the errors produced by case LB in response to six-letter words failed to meet this expectation (Caramazza & Miceli, 1990). This subject spelled correctly 1300/1777 (73%) trisyllabic items with simple-CV structure, i.e., consisting of regularly alternating consonant/vowel sequences (e.g., *tavolo*), but only 921/1523 (51.8%) words with complex-CV structure. More in detail, he spelled corerctly 244/441 (55%) trisyllabic words and 437/762 (57%) bisyllabic words with complex-CV structure, (i.e., words containing consonant or vowel clusters, like *albero, onesto, stadio, guasto, scorso, nostro, strano*), 41/55 (74.5%) trisyllabic words and 199/265 (75%) bisyllabic words with complex-CV structure and a geminate consonant (e.g., *attimo, anello, frutta*). Furthermore, simple-CV and complex-CV stimuli yielded very different error types. Incorrect responses to simple-CV stimuli resulted essentially only in letter substitutions and transpositions (81% and 18% of total errors, respectively), with less than 1% of incorrect responses resulting in letter omissions or insertions; whereas errors to complex-CV strings yielded many substitutions and transpositions (31% and 22% of total errors, respectively), but also substantial numbers of insertions (10%) and omissions (37%).

On the hypothesis that performance on a letter is unaffected by its orthographic context, any segment of a six-letter string should be equally likely to be omitted. Contrary to this prediction, LB almost always (311/313 cases, or 99.8%) omitted consonants or vowels in clusters, but not singletons. For example, in a word like *stadio*, omissions involved letters *s, t, i, o*, but not letters *a* or *d*.

Results consistent with an effect of orthographic structure on spelling accuracy were observed also in subjects BA (Ward & Romani, 1998), TH and PB (Schiller, Greenhall, Shelton, & Caramazza, 2001), but not in cases JH (Kay & Hanley, 1994) and AS (Jónsdottír, Shallice, & Wise, 1996). Thus, even though the effect was not demonstrated in all the subjects reported on so far, there is sufficient evidence on this issue to cast serious doubts on the linear hypothesis.

Geminates are special

As stated in the previous section, LB spelled correctly 74.5% trisyllabic and 75% bisyllabic six-letter words with a geminate consonant (e.g., *anello*, ring; *frutta*, fruit), as opposed to 55% trisyllabic and 57% bisyllabic six-letter words with one or more consonant clusters, but without a geminate consonant (e.g., *onesto*, honest; *frusta*, whip). He made far fewer errors in spelling geminate consonants than two-consonant clusters in the same position (Figure 4). A qualitative analysis of his incorrect responses to words containing geminate consonants is also revealing. Table 3 shows error types that occurred and that did not occur (or occurred only very rarely) in his writing corpus. The errors that LB produced frequently on geminates can be construed as the result of damage to both geminate consonants, to a "doubling" feature, or to both; whereas he essentially never made errors involving only one of the geminate consonants.

The "special" behaviour of geminates has been documented in several dysgraphic subjects

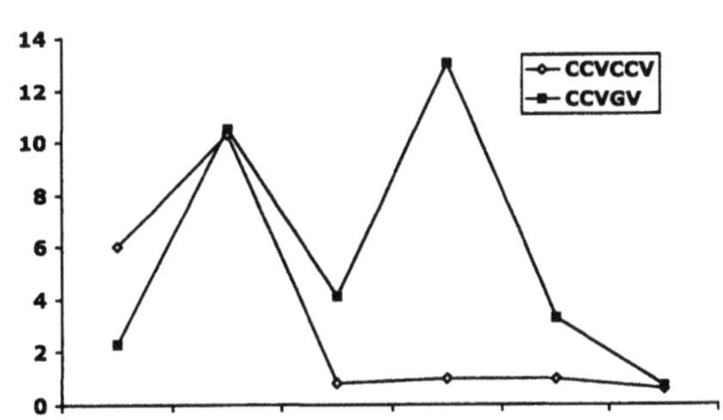

Figure 4. *Evidence for geminate consonants as spelling units. Case LB: comparison of response accuracy on each grapheme when spelling 6-letter words with similar orthographic structure, but with a letter cluster (CCVGV, as in* frusta, *whip) or a geminate consonant (CCVCCV, as in* frutta, *fruit) in the same position.*

Table 3. *Types of errors that occurred and that did not occur in the error corpus collected from subject LB*

	Example
a. Errors that occurred in LB's corpus	
Substitution of both geminate consonants	troppo, too much → *trocco*
Duplication of the geminate feature	cavallo, horse → *cavvallo*
Deletion of the geminate feature	passare, to pass → *pasare*
Exchange involving the geminate consonant	cellula, cell → *leccula*
Shift of the geminate feature	avviso, notice → *avisso*
Exchange of the geminate feature and consonant	sorella, sister → *sollera*
b. Errors that did not occur in LB's corpus	
Substitution of one geminate consonant (there were 10 such errors out of 4400 opportunities, 0.2%)	grammo, gram → *granmo*
*Transposition (shift or exchange) involving one geminate consonant	anello, ring → *alenlo*
*Insertion of a letter between the two geminate consonants	frutta, fruit → *frutita*

(e.g., Jónsdottír et al., 1996; Miceli, Benvegnu, Capasso, & Caramazza, 1995; Schiller et al., 2001; Tainturier & Caramazza, 1996; Venneri, Cubelli, & Caffarra, 1994). Thus, subject RT (Venneri et al., 1994) produced perseverative errors only on geminate consonants (e.g., "modello" → *modelllo*). In case FM (Tainturier & Caramazza, 1996), incorrect responses to stimuli containing a geminate consonant frequently preserved a geminate consonant (albeit incorrect), in the context of otherwise unrelated grapheme strings (e.g., umbrella → *umallow* and, in another instance, → *emmucpe*). A selective disorder spelling geminates was reported in subject SFI (Miceli et al., 1995), who deleted a consonant in 52/197 (26.4%) geminate clusters, but in only 12/405 (3%) two-consonant clusters in similar within-word position. Evidence on geminates suggests that they behave as spelling units, and is impossible to reconcile with the linear hypothesis.

The role of CV structure

The clearest evidence against the linear hypothesis is provided by neuropsychological observations showing that the CV status of a grapheme is a crucial feature of orthographic knowledge.

In the first place, there is the fact that letters involved in substitution errors tend to preserve the CV status of the target. In a linear representation, when a vowel or a consonant is substituted, the relationships between the target letter and the incorrect letter should be random — a vowel should be substituted sometimes by another vowel (tesori, treasures → *tesari*) and sometimes by a consonant (tesori → *tespri*), and the same should be true of consonants (stadio, stadium → *spadio*; stadio → *siadio*). This prediction is disconfirmed by a number of reports (Table 4) showing that substituted letters overwhelmingly preserve the CV status of the target (Caramazza & Miceli, 1990; Cotelli, Aboutalebi, Zorzi, & Cappa, 2003; Cubelli, 1991; Jónsdottír et al., 1996; Kay & Hanley, 1994; McCloskey, Badecker, Goodman-Shulman, & Aliminosa, 1994; Miceli, Benvegnù, Capasso, & Caramazza, 2004; Schiller et al., 2001; Ward & Romani, 1998).

The second type of evidence results from the observation that some dysgraphic subjects present with a selective inability to spell consonants or vowels. The linear hypothesis does not distinguish

Table 4. *Preservation of the CV status of the target letter in substitution errors*

Subject's initials	*N of substitutions in the error corpus*	*% of errors preserving CV status*
LB (Caramazza & Miceli, 1990)	520	99.8
CW (Cubelli, 1991)	340	99.0
JH (Kay & Hanley, 1994)	253	93.0
HE (McCloskey et al., 1994)	207	95.0
AS (Jonsdottir et al., 1996)	121	85.0
BA (Ward & Romani, 1998)	138	88.0
TH (Schiller et al., 2001)	291	86.0
LiB (Cotelli et al., 2003)	283	100.0
GSI (Miceli et al., 2004)	46	100.0

between consonant and vowel graphemes, and consequently predicts that the two grapheme types should be comparably affected by spelling errors. Even though comparable involvement of consonants and vowels was observed in several cases (e.g., Caramazza, Miceli, Villa, & Romani, 1987; Jónsdottír et al., 1996; McCloskey et al., 1994; Schiller et al., 2001), a selective impairment for vowels was observed in three cases, and a selective impairment for consonants in three others. Subject CF (Cubelli, 1991), who had a transient spelling disorder, wrote correctly all the consonants but omitted all the vowels of seven words he was asked to write (first and last name, birthplace, some object names). For example, he spelled the name of the city he lived in as B L GN (Bologna). A similar dissociation was observed in CW (Cubelli, 1991) and LiB (Cotelli et al., 2003), whose errors involved vowels in 87% and 85% of the cases, respectively. The reverse picture, that is, greater involvement of consonants than of vowels, was observed in JH (Kay & Hanley, 1994) and TI (Cubelli, Paolieri, & Borghi, 2003), whose substitution errors involved consonants in 79% and 85% of the cases, respectively. In a recent case, a complete dissociation between consonants and vowels was documented (Miceli et al., 2004). Subject GSI made errors almost exclusively on consonants (98.2% of total errors), across a variety of tasks. For example, he misspelled *discendere*, to step down, as *dis_e__e_e* in writing to dictation; *albero*, tree, as "*a, l, b, e*, I cannot see the consonant, *o*" in oral spelling to dictation; and, *pizza* as *pi__a* in picture naming. Selective damage for consonants in this subject cannot be attributed to the fact that consonants are less frequent than vowels, since the error rate on a consonant did not correlate with its frequency in Italian. In addition, the vowel *u* was always spelled correctly, whereas consonants of comparable frequency were often misspelled (up to 47.3% errors on *d*).

Taken together, the preservation of CV status in letter substitutions and the cases with selective impairment of consonants or vowels strongly suggest that CV status of graphemes is an essential part of orthographic representations, and that consonants and vowels are categorically separate grapheme types, represented by distinct cognitive and neural processes.

The multidimensional hypothesis

The observations that graphosyllabic structure affects spelling performance, that geminate consonants behave as spelling units, and that the CV status of a grapheme constrains the dysgraphic subject's errors support the proposal (Caramazza & Miceli, 1990) that, far from being a linear string of ordered graphemes, orthographic representations are multidimensional objects (Figure 5). Some aspects of this proposal must be further analysed. For example, it remains unclear whether orthographic representations encode graphosyllabic information (Caramazza & Miceli, 1990), or simply the nucleus/non nucleus status of a letter (Caramazza & Miceli, 1989; McCloskey et al., 1994). On the whole, however, the available evidence supports the separate representation of grapheme identity, CV status and quantity in orthographic strings.

There is also evidence that, even though the structural properties of orthographic representations are similar to those assumed for phonological representations, their content is specifically

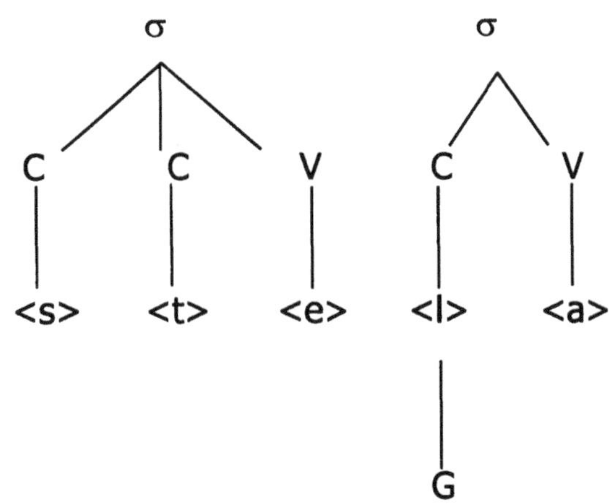

Figure 5. *The multidimensional hypothesis of orthographic representations, as it emerges from the analysis of subjects with "graphemic buffer damage" (modified from Caramazza & Miceli, 1990).*

orthographic (for a different view, see Jónsdottír et al., 1996). In fact, the spelling errors produced by LB and by other subjects with similar writing disorders frequently violate the sonority principle, a crucial principle in phonology—for example, LB produced *stzazi* in response to "strazi." In addition, segmental damage at the orthographic and at the phonological level is dissociable. Thus, subjects LB (Caramazza et al., 1987), GSI (Miceli et al., 2004), and LiB (Cotelli et al., 2003) suffer from segmental orthographic damage in the presence of spared phonology; whereas subject AS (Caramazza, Chialant, Capasso, & Miceli, 2000) presents with a marked phonological disorder but only minor difficulty in spelling.

Recent neuropsychological studies suggest that the multidimensional hypothesis sketched above might have to be further elaborated. Subjects FM and BWN (Tainturier & Rapp, 2004) behaved differently when spelling digraphs, that is, two-letter clusters corresponding to one phoneme (e.g., *ck* in *bucket*), than when spelling two-letter clusters (e.g., *sk* in *basket*). In particular, they were less likely to spell incorrectly only one letter of a digraph, or to "break" a digraph by producing both letters, but in distant position within the response. A related pattern of performance was observed in German-speaking case LM (Blanken, Schäfer, Tucha, & Lange, 1999), whose errors frequently resulted in adjacent letter exchanges involving consonants in digraphs (e.g., "die", the → *dei*). The status of digraphs is not yet fully established. For example, the behaviour described for LM, FM, and BWN was not observed in LB (Caramazza & Miceli, 1990), who spelled similarly *sc* when it corresponded to a digraph (as in *scelta*, choice,/ʃelta/) or to a consonant cluster (as in *sconto*, discount,/skonto/). However, even though a clear-cut decision as to whether or not digraphs are spelling units is premature, the performance observed in LM, FM, and BWN justifies the proposal that orthographic representations are unevenly textured, the amount of information being greater at some junctures of the grapheme identity tier (e.g., those corresponding to digraphs) than at other junctures of the same tier. Digraphs behave as spelling units also in a connectionist model of spelling (Houghton & Zorzi, 2003).

The processing of serial order in orthographic representations

Any comprehensive theory of a cognitive ability should specify, in addition to the nature of the representations involved in that ability, the mechanisms that process them. Analyses of serial position effects in the error corpora collected from subjects with impairments at the graphemic buffer stage have stimulated speculations on this issue.

Most subjects listed in Table 1 show one of two error distributions. In some cases (Blanken et al., 1999; Cantagallo & Bonazzi, 1996; Caramazza et al., 1987; Jónsdottír et al., 1996; McCloskey et al., 1994; Piccirilli, Petrillo, & Poli, 1992; Tainturier & Rapp, 2004), errors follow a reverse-U distribution (Figure 6, top). Typically, the beginning and the end of the written string are relatively spared, and errors peak at central positions. This distribution was shown to be sensitive to neglect phenomena—it may be skewed to the left in subjects with right brain damage and left-sided neglect, and to the right in subjects with left brain damage and right-sided neglect (Hillis & Caramazza, 1989). The other error distribution is characterised by sparing of initial letters and monotonic increase of errors from initial to final positions within the string (Figure 6, bottom). (This distribution has been documented by Katz, 1991; Kokubo, Suzuki, Yamadori, & Satou, 2001; Miceli et al., 2004; Schiller et al., 2001; Ward & Romani, 1998.) There are also rare cases who do not fit into either pattern, and show an undulating or flat error distribution (Cotelli et al., 2003; Miceli et al., 1985; Posteraro, Zinelli, & Mazzuchi, 1989).

Different accounts have been offered for the two most frequent error distributions. Since the bow-shaped distribution recorded for dysgraphic subjects is similar to that observed in free recall tasks, it was initially attributed to a failure to recall the elements of the to-be-written string.

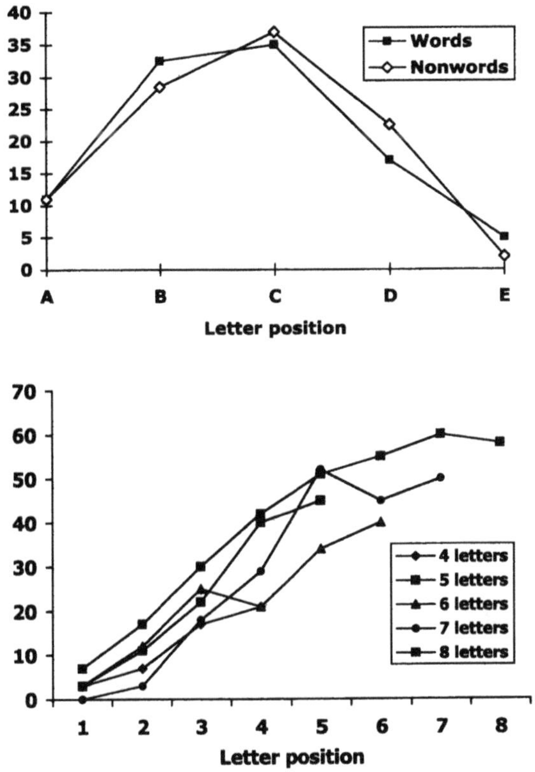

Figure 6. *Serial order effects. The error position effects most frequently observed in subjects with graphemic buffer damage. Top: the bow-shaped (or reverse-U shaped) distribution, as observed in case LB (Caramazza & Miceli, 1990); bottom: the monotonic error increase, as observed in case PB (Schiller et al., 2001).*

However, this similarity is probably only a coincidence, as spelling is indeed a recall task, but one of the serial type—and the error distribution in serial recall tasks is not bow-shaped (Caramazza et al., 1987; Wing & Baddeley, 1980). More recently, this distribution has been attributed to the fact that the string selected for output reaches only a lower-than-normal level of activation (Schiller et al., 2001). Even under normal conditions, graphemes toward the centre of a string are more vulnerable to errors, perhaps due to the number of items competing for production to their left and their right (larger for these graphemes than for graphemes in initial or final positions) (Glasspool, 2000). The observation that slips-of-the-pen in normal subjects have a bow-shaped distribution (Wing & Baddeley, 1980) is in line with this possibility. In the event of brain damage, the orthographic string may not reach optimal activation, which will make letters in medial positions even more likely to be misspelled. Consistent with this proposal, the bow-shaped distribution of spelling errors in graphemic buffer cases is similar to that observed in normal controls, but usually has a steeper slope.

The monotonic error increase is "easier" to account for by assuming a graphemic buffer deficit in the strict sense (Schiller et al., 2001). Due to reduced buffer capacity, orthographic information decays very rapidly. Consequently, string-initial elements are spared, as they are the first to be processed, whereas graphemes in the following positions are increasingly less likely to be spelled correctly, in proportion to their distance from the beginning of the string. This is because they are processed later, and are likely to be no longer active when their turn to be produced arrives.

The meaning of the "flat" error distribution is unclear. In at least one case, however, an intriguing explanation is possible. Subject LiB (Cotelli et al., 2003) produced comparable numbers of errors across all positions in a string (see Figure 10, bottom right). Since he presented with a selective impairment for vowels in both spelling and reading tasks, it is not unreasonable to assume that he suffered from a categorical impairment of orthographic vowels. Such an impairment would affect all vowels to the same extent, regardless of their position within a string.

The interaction between orthographic structure and serial order

So far, we have discussed structure and serial order separately, as if these were two independent dimensions of the spelling process. However, empirical evidence clearly shows that they are related.

An example of the interaction between structure and serial order is provided by case LB (Caramazza & Miceli, 1990). This subject's performance on six-letter words like *nastro* (ribbon) and *strani* (strange), which contain a CCC cluster in different within-string positions, yielded an interesting contrast (Figure 7). In both cases, errors peaked at the third consonant in the cluster—except that this letter was in position 3 in one type of stimulus, and in position 5 in the other. Thus, performance on these words was constrained by an interaction between structure (presence of a CCC cluster) and serial order (position of the cluster within the string).

The performance of subject GSI (Miceli et al., 2004) also clearly demonstrates that structure and serial order interact. This subject produced errors essentially only on consonants, and presented with a monotonically increasing pattern of errors—the first consonant in a string was invariably produced correctly, and error probability on the following consonants steadily increased with the distance from a word's beginning; at the same time, vowel spelling accuracy was excellent throughout the string (Figure 8). This pattern rules out accounts based purely on structure or purely on serial order. Poor performance cannot result from selective damage to a graphemic type (consonants), as in this case all consonants should have been misspelled, regardless of their position within the string (this was probably the case for the errors on vowels observed by Cotelli et al., 2003). It cannot be attributed to a serial

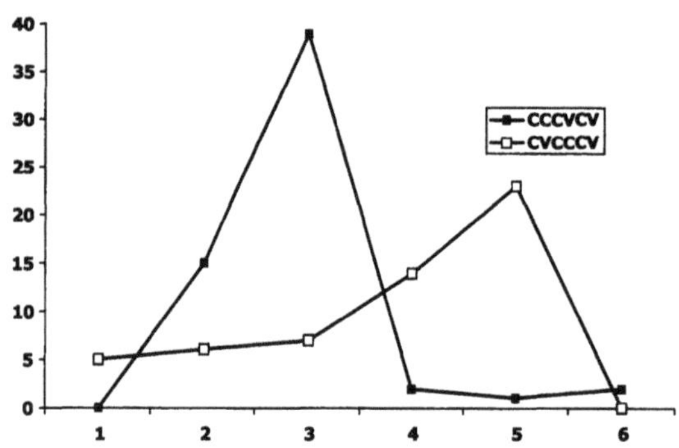

Figure 7. *The interaction of structure and serial position. Case LB: Comparison of response accuracy when spelling words with a string-initial CCC cluster (CCCVCV, as in* strano, *strange) and with a string-medial CCC cluster (CVCCCV, as in nastri, ribbons). In both cases errors peak at the third consonant in the cluster, but this corresponds to position 3 in* strano, *and to position 5 in* nastri.

order effect, either, as in this case error rate should have increased steadily across letter positions, regardless of whether the to-be-spelled letter was a consonant or a vowel (this was probably the case in the subjects studied by Schiller et al., 2001). For GSI, the critical constraint on performance accuracy is provided by an interaction of structure (letter type) and serial order (the distance of a letter from the beginning of the word).

Open questions on the representation and processing of orthographic knowledge

A fact that clearly emerges from the data discussed in this section is the variability of the performance observed in subjects who meet the criteria for the diagnosis of "graphemic buffer damage." This variability concerns the distribution of error types, the position of errors in the string, and the degree to which the various elements of orthographic representations are damaged. The occurrence rate of letter substitutions, insertions, omissions, and transpositions shows a wide range, as the various error types occur with roughly comparable frequency in some subjects (e.g., HE and AS), whereas in others one error type is substantially more frequent than the others—substitutions in CF and LiB, omissions in GSI, and transpositions in LM (see Table 2). Errors show sometimes a reverse-U distribution, and in other cases a monotonic increase (see Figure 6). In some cases damage affects consonants and vowels to the same extent, whereas in others one grapheme type is disproportionately affected (see earlier). Thus, across-subject variability seems to affect dimensions related to the structural properties of orthographic representations (e.g., the CV dimension), to the mechanisms involved in their processing (e.g., the error position effects), or to a complex interaction between the two (see Figures 7 and 8). This leads to the unavoidable conclusion that the term "graphemic buffer deficit" is no more than a convenient label for a pattern of behaviour shown by a group of subjects (see Table 1), and does not reflect a homogeneous cognitive deficit. The pattern documented in subjects with damage to the "graphemic

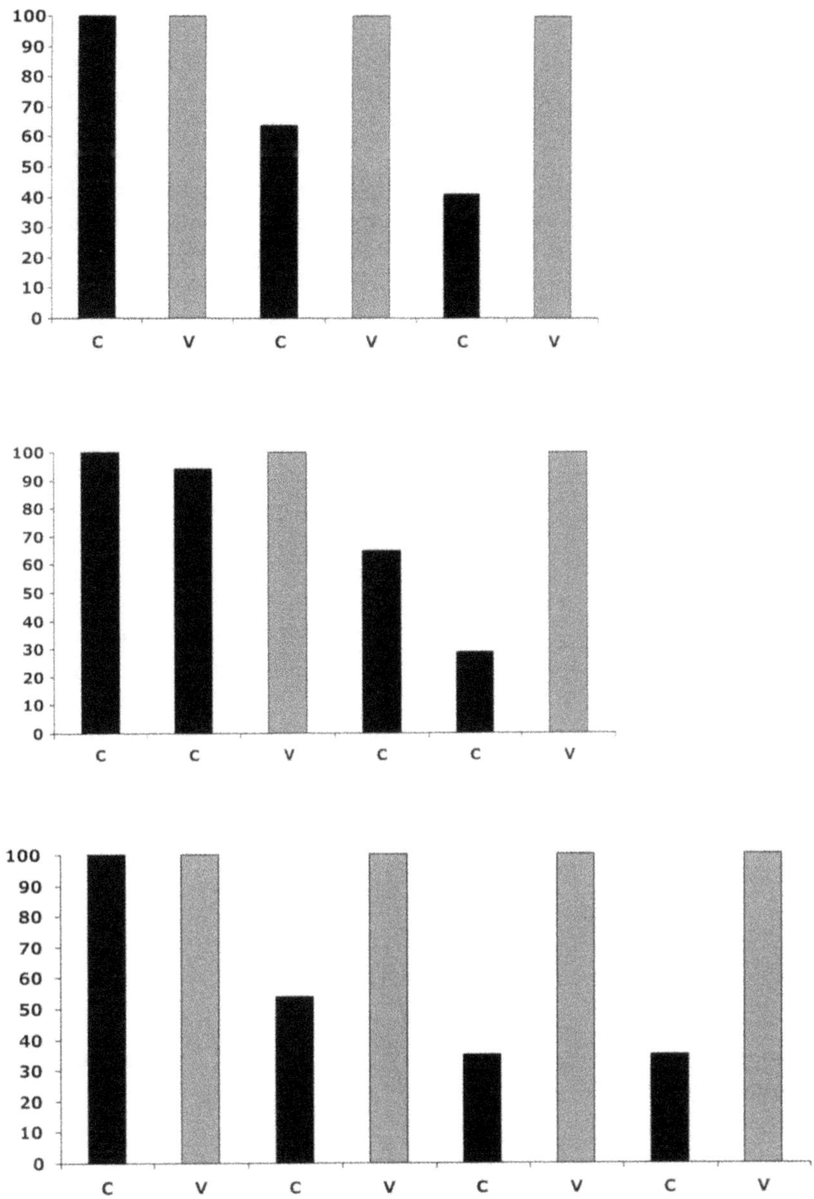

Figure 8. *The interaction of structure and serial order. Case GSI: performance in spelling words like* tavolo *(table),* tromba *(trumpet), and* recenti *(recent). In this subject, spelling performance is constrained by grapheme type (errors only affect consonants) and by letter position (error rate on consonants increases with distance from the beginning of the word).*

buffer" stage may result from a variety of cognitive impairments, affecting any of several stages of spelling in which an orthographic form is encoded and stored while format-specific representations are obtained.

No current model of spelling can account for all the observations discussed thus far, and a fully explicit theory of the representations and processes involved in encoding and producing a string of graphemes is not yet available. Cognitive neuropsychological investigations have been very successful in prompting increasingly complex hypotheses on the nature of orthographic representations, but have been less concerned with providing computationally detailed theories of how orthographic knowledge is processed during spelling. The proposals put forth on the basis of the serial position effects documented in dysgraphic subjects and briefly discussed earlier are reasonable, but not very explicit.

Largely out of dissatisfaction with the lack of procedural detail of cognitive theories, alternative approaches to the study of spelling were developed, beginning in the mid-1990s. These models start from radically different views about the representation and processing of orthographic information. They reject the notion of a graphemic buffer, and try to account for the observed patterns of performance in the context of connectionist architectures based on competitive queueing (CQ); these are particularly well-suited to deal with serial behaviour. A typical model of this type (Figure 9) includes three layers: control nodes, letter nodes, and a competitive filter (Shallice, Glasspool, & Houghton, 1995). Each word is specified by a series of letter nodes, each of which has weighted connections to two control nodes. The first, or *initiate* (*I*) node, has strongest connections to the initial letter, and progressively weaker connections to the letters following the first; the second, or *end* (*E*) node, has strongest connections to the final letter of the string, and increasingly weak connections to the preceding letters, in proportion to their distance from the word's end. The *I* node is most active at the beginning of the word, and the *E* node at the end. Control nodes activate the letters corresponding to the target word in the second layer, where various graphemes may compete for production at each node (note that letters are not associated with information on their CV status). The competitive filter (i.e., the units in the third layer) selects the grapheme most active at each letter node; this is specified for output. After a grapheme is selected for production, it is inhibited, and selection of the following letter takes place. The inhibition of a

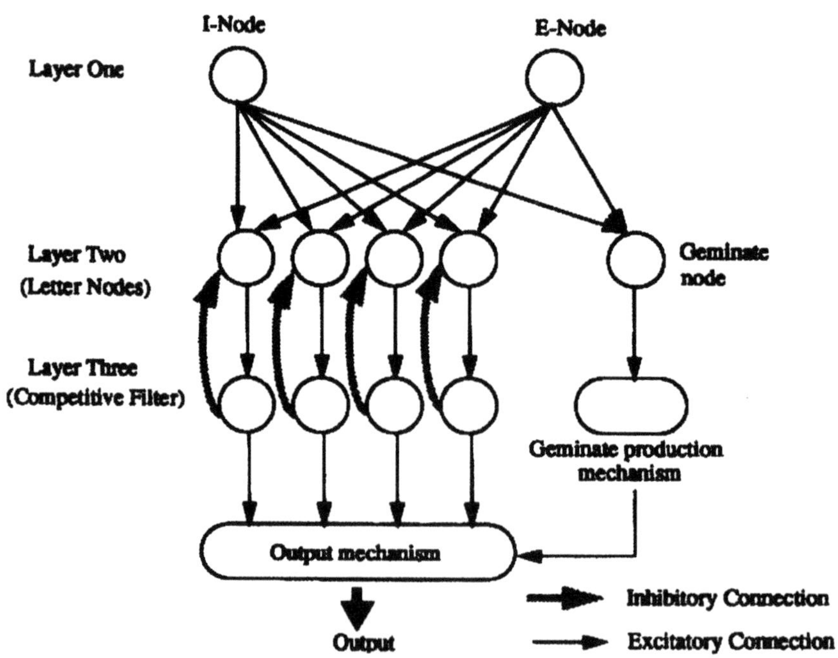

Figure 9. *Schematic representation of a CQ model of spelling (from Shallice et al., 1995).*

just-selected grapheme ensures that, after reaching threshold, a letter returns to a low level of activation and is not selected again inappropriately. The model also includes a *geminate* node involved in the production of doubled letters (e.g., the double <l> in <stella>, star). In a model of this type, serial order is encoded by the time frame of activation in the *I* and *E* nodes and by the strength of the connections between the *I* and *E* nodes and the segments in the letter layer.

When damaged, this model is able to reproduce the error position effects observed in dysgraphic subjects. Adding noise to the competitive filter results in a bow-shaped error distribution, similar to that observed in dysgraphic subjects (Blanken et al., 1999; Cantagallo & Bonazzi, 1996; Caramazza et al., 1987; Jónsdottír et al., 1996; McCloskey et al., 1994; Piccirilli et al., 1992; Tainturier & Rapp, 2004), whereas reducing the level of activation of the *E* node results in a monotonically increasing error rate, similar to that observed in other cases of dysgraphia (Katz, 1991; Kokubo et al., 2001; Miceli et al., 2004; Schiller et al., 2001; Ward & Romani, 1998).

In addition to many virtues, this model, as well as others of the same type, has several drawbacks. A major problem concerns the processing of repeated letters such as *t* in *start*. Under normal conditions, the first *t* is activated mostly by the *I* node. After being produced, the letter *t* is inhibited, so as not to be activated again. However, since the second *t* is controlled mostly by the *E* node, the letter *t* can be activated again, in due time. Suppose, however, that the *E* node is damaged, as might be the case in a subject with a monotonic error increase. In this case, both *t*'s are controlled by the *I* node. Thus, after the first *t* is spelled, the letter *t* has a very low level of activation and it is unlikely to be produced again at the end of the string. In other words, a dysgraphic subject with a monotonically increasing error rate should spell repeated letters with particular difficulty. This prediction was not borne out in the corpus of errors collected from BA (Ward & Romani, 1998), PB, and TH (Schiller et al., 2001).

Another major problem of models of this type is the difficulty in accounting for the CV status effects. These models try to simulate dysgraphic behaviour by assuming no symbolic information (or as little symbolic information as possible). Coherent with this assumption, the original CQ model did not distinguish between vowels and consonants. Consequently, it cannot accommodate the neuropsychological data. In later versions of the model (Glasspool & Houghton, 1997), some distinction between consonants and vowels was introduced, and selective damage to consonants or to vowels was simulated by lowering the level of activation of the segments marked as vowels, or of those marked as consonants in the target string. An example of the CV status effects yielded by this model is shown in Figure 10, top left (adapted from Cotelli et al., 2003). The comparison with the error distribution documented for GSI (Figure 10, top right) clearly shows the distance between the error pattern predicted by the model when lesioned and that documented in our subject. In the simulation, the graphemes of the damaged type (consonants) are produced less accurately at each position than those of the spared type; however, all letters in the string are spelled incorrectly to some extent. This contrasts with consistently accurate spelling of all the vowels and of the first consonant in a string documented for GSI. A similar but complementary function is predicted in the event of selective damage to vowels (Figure 10, bottom left), but also in this case the lesioned model fails to accommodate the error distribution documented in a dysgraphic subject with disproportionate damage for vowels (Cotelli et al., 2003). In subject LiB, errors affected vowels in 87% of the cases, but were not sensitive to the within-string position of the vowel (Figure 10, bottom right).[1] In other words, since they reject an explicit, "categorical" distinction between consonants and vowels,

[1] The error position effect is different in GSI and LiB because functional damage affects the interaction between consonants and serial order in the first object, and vowels per se in the second.

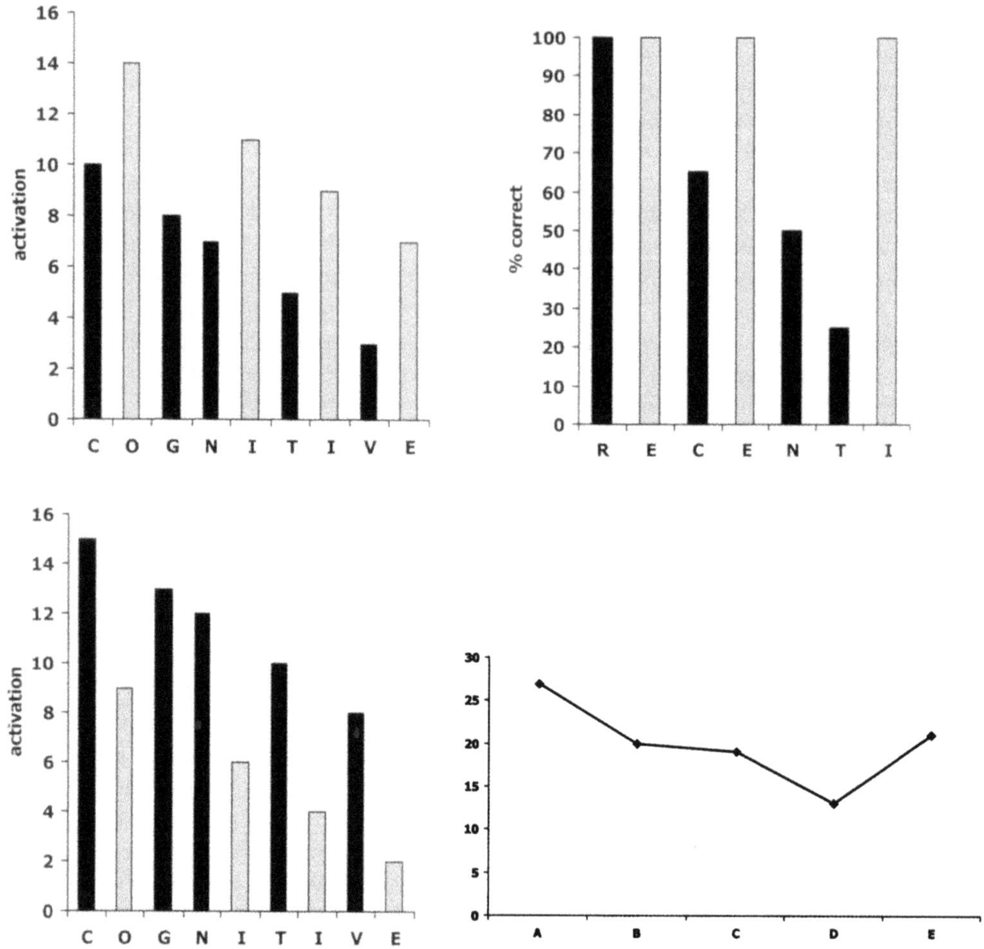

Figure 10. *Error distributions predicted by CQ models of spelling (Houghton & Glasspool, 1998; figures redrawn from Cotelli et al., 2003) and observed in dysgraphic subjects. Top left: distribution obtained after selectively "lesioning" consonants in a CQ model (performance on the word* cognitive *is presented as an example). Top right: distribution observed in GSI (Miceli et al., 2004), who suffered from selective damage to consonants (performance on words like* recenti, *recent). Bottom left: performance obtained after selectively "lesioning" vowels in a CQ model (performance on the word* cognitive*). Bottom right: error distribution observed in LiB (Cotelli et al., 2003), who suffered from selective impairment in spelling vowels (performance collapsed across five abstract letter positions).*

CQ models when lesioned are also unable to reproduce the contrasting patterns of performance resulting from damage to a grapheme type (as might be the case in Cotelli et al., 2003), to the mechanisms that process serial order information in the presence of comparable damage to both grapheme types (as might be the case in most subjects in Table 1), or to the interaction between CV status and serial order (as might be the case in Miceli et al., 2004). In an attempt to overcome these limitations, the latest version of the model includes a fourth layer, and assumes an explicit distinction between consonants and vowels (Glasspool & Houghton, 2005). It remains to be seen whether this new version will accommodate the data reported in this paragraph.

Investigations in the cognitive neuropsychology of spelling have provided detailed hypotheses on the formal properties of orthographic representations. They have also documented intriguing interactions between structure and serial order in spelling, but on the whole they have been less detailed on the procedural aspects of the spelling process. CQ models, on the other hand, are more explicit from the procedural viewpoint, but their failure to include enough symbolic structure in orthographic representation prevents them from accounting for the structural effects and for the interactions between structure and serial order that unequivocally emerge from neuropsychological investigations of dysgraphia. Whether or not these models will at some point be able to accommodate the whole range of phenomena documented in the neuropsychological literature is unclear. However, it is interesting to note that, in order to do so, they are progressively including just the symbolic features included in the original symbolic proposal: CV status, identity, and quantity information.

CONCLUSIONS

Cognitive models adopted in the 1980s assumed that spelling a word involves the selection of an orthographic form exclusively on the basis of information originating within the lexical-semantic system. It was debated whether orthographic forms were activated directly from meaning representations, or whether they could only be activated by an intermediate phonological representation; however, there was a general consensus on the view that they are ordered, linear strings of abstract letter identities (graphemes). Neuropsychological research conducted over the past 20 years has fruitfully dealt with many of these issues. It is now firmly established that orthographic lexical forms are autonomous from phonological lexical forms, and that they are directly accessed from semantics in the course of spelling. There is also evidence that, contrary to the views put forth in the 1980s, the selection of orthographic lexical representations may be affected, at least under some circumstances, by sublexical information. In particular, orthographic information assembled by sublexical phoneme–grapheme conversion procedures might serve to constrain lexical selection. Finally, analyses of dysgraphic subjects demonstrate that orthographic knowledge consists of a multidimensional representation of specifically orthographic content, specifying suprasegmental (graphosyllabic structure or nucleus/non-nucleus positions) information, as well as information about the identity, CV status, and quantity of each segment in the string. The solutions proposed for these issues allow us to ask other, more detailed questions. The most pressing issues for future research concern the cognitive locus and the procedural aspects of the interaction between lexical-semantic and sublexical information in lexical selection; the mechanisms that encode the target string; the formal properties of orthographic representations and the ways in which the encoded string is processed during spelling. Regardless of how these questions are ultimately answered, the fact that most current theories of spelling have been developed on the basis of neuropsychological investigations shows that the theory-driven analysis of subjects with brain damage is a powerful tool for understanding cognitive and linguistic performance. The integration of neuropsychological studies of spelling with similar studies in related areas (e.g., reading), with simulation studies and with neuroimaging research in normal and neurological subjects, holds the promise of further advancing our knowledge on these topics.

PrEview proof published online 18 October 2005

REFERENCES

Alario, F. X., Schiller, N. O., Domoto-Reilly, K., & Caramazza, A. (2003). The role of phonological and orthographic information in lexical selection. *Brain and Language*, *84*, 372–398.

Assal, G., Buttet, J., & Jolivet, R. (1981). Dissociations in aphasia: A case report. *Brain and Language*, *13*, 223–240.

Badecker, W., Hillis, A. E., & Caramazza, A. (1990). Lexical morphology and its role in the writing process: Evidence from a case of acquired dysgraphia. *Cognition*, *35*, 205–244.

Barry, C., & Seymour, P. H. K. (1988). Lexical priming and sound-to-spelling contingency effects in nonword spelling. *Quarterly Journal of Experimental Psychology, 40A*, 5–40.

Baxter, D. M., & Warrington, E. K. (1985). Category-specific phonological dysgraphia. *Neuropsychologia, 23*, 653–666.

Baxter, D. M., & Warrington, E. K. (1987). Transcoding sound to spelling: Single or multiple sound unit correspondence? *Cortex, 23*, 11–28.

Beaton, A., Guest, J., & Ved, R. (1997). Semantic errors of naming, reading, writing, and drawing following left-hemisphere infarction. *Cognitive Neuropsychology, 14*, 459–478.

Beauvois, M.-F., & Dérouesné, F. (1981). Lexical or orthographic agraphia. *Brain, 104*, 21–49.

Benson, F. (1979). Agraphia. In K. M. Heilman & E. Valenstein (Eds.), *Clinical neuropsychology* (1st ed.). Oxford: Oxford University Press.

Blanken, G., Schäfer, C., Tucha, O., & Lange, K. W. (1999). Serial processing in graphemic encoding: Evidence from letter exchange errors in a multilingual patient. *Journal of Neurolinguistics, 12*, 13–39.

Bub, D., & Kertesz, A. (1982a). Evidence for lexicographic processing in a patient with preserved writing over single word naming. *Brain, 105*, 697–717.

Bub, D., & Kertesz, A. (1982b). Deep agraphia. *Brain and Language, 17*, 146–165.

Campbell, R. (1983). Writing nonwords to dictation. *Brain and Language, 19*, 153–178.

Cantagallo, A., & Bonazzi, S. (1996). Acquired dysgraphia with selective damage to the graphemic buffer: A single case report. *Italian Journal of Neurological Science, 17*, 249–254.

Caramazza, A., Chialant, D., Capasso, R., & Miceli, G. (2000). Separable processing of consonants and vowels. *Nature, 403*, 428–430.

Caramazza, A., & Hillis, A. (1990). Where do semantic errors come from? *Cortex, 26*, 95–122.

Caramazza, A., & Hillis, A. (1991). Lexical organization of nouns and verbs in the brain. *Nature, 349*, 788–790.

Caramazza, A., & Miceli, G. (1989). Orthographic structure, the graphemic buffer and the spelling process. In C. von Euler (Ed.), *Brain and reading. Macmillan/Wenner-Gren International Symposium Series* (pp. 257–268). Basingstoke, UK: Macmillan.

Caramazza, A., & Miceli, G. (1990). The structure of graphemic representations. *Cognition, 37*, 243–297.

Caramazza, A., Miceli, G., & Villa, G. (1986). The role of the (output) phonological buffer in reading, writing and repetition. *Cognitive Neuropsychology, 3*, 37–76.

Caramazza, A., Miceli, G., Villa, G., & Romani, C. (1987). The role of the graphemic buffer in spelling: Evidence from a case of acquired dysgraphia. *Cognition, 26*, 59–85.

Coltheart, M., Patterson, K. E., & Marshall, J. C. (Eds.). (1980). *Deep dyslexia*. London: Routledge & Kegan Paul.

Cotelli, M., Aboutalebi, J., Zorzi, M., & Cappa, S. (2003). Vowels in the buffer: A case study of acquired dysgraphia with selective vowel substitutions. *Cognitive Neuropsychology, 20*, 99–114.

Cubelli, R. (1991). A selective deficit for writing vowels in acquired dysgraphia. *Nature, 353*, 258–260.

Cubelli, R., Paolieri, D., & Borghi, E. (2003). *Un caso di disgrafia selettivas per le consonanti: Si tratta di un deficit ortografico?* Talk presented at the meeting of the Società Italiana di Neuropsicologia. Bologna, November 21, 2003.

Dérouesné, J., & Beauvois, M.-F. (1979). Phonological processing in reading: Data from alexia. *Journal of Neurology, Neurosurgery and Psychiatry, 42*, 1125–1132.

Ellis, A. W. (1979). Slips of the pen. *Visible Language, 13*, 265–282.

Ellis, A. W. (1982). Spelling and writing (and reading and speaking). In A. W. Ellis (Ed.), *Normality and pathology in cognitive functions* (pp. 113–145). New York: Academic Press.

Folk, J., Rapp, B., & Goldrick, M. (2002). Interaction of lexical and sublexical information in spelling: What's the point? *Cognitive Neuropsychology, 19*, 653–671.

Friedman, R. B., & Alexander, M. P. (1989). Written spelling agraphia. *Brain and Language, 36*, 503–517.

Glasspool, D. W. (2000). Serial order in behaviour: Evidence from performance slips. In G. Houghton (Ed.), *Connectionist modeling in psychology*. Hove, UK: Psychology Press.

Glasspool, D., & Houghton, G. (1997). The representation of categorial constraints in connectionist models of serial order. In J. Bullinaria, D. Glasspool, & G. Houghton (Eds.), *Connectionist representation* (pp. 269–282). London: Springer-Verlag.

Glasspool, D. W., & Houghton, G. (2005). Serial order and consonant-vowel structure in a graphemic output buffer model. *Brain and Language* (e-published February 25, 2005).

Goodman, R. A., & Caramazza, A. (1986a). Aspects of the spelling process: Evidence from a case of acquired

dysgraphia. *Language and Cognitive Processes, 1,* 263–296.
Goodman, R. A., & Caramazza, A. (1986b). Dissociation of spelling errors in written and oral spelling: The role of allographic conversion in writing. *Cognitive Neuropsychology, 3,* 179–206.
Grashey, H. (1885). Über Aphasie und ihre Beziehungen zur Wahrnehmung. *Archiv für Psychiatrie und Nervnekrankheiten, 16,* 654–688.
Hanna, R. R., Hanna, J. S., Hodges, R. E., & Rudorf, E. H. (1966). *Phoneme–grapheme correspondences as cues to spelling improvement.* Washington, DC: US Department of Health, Education, and Welfare, Office of Education, US Government Printing Office.
Hatfield, F. M., & Patterson, K. E. (1983). Phonological spelling. *Quarterly Journal of Experimental Psychology, 35A,* 451–458.
Head, H. (1926). *Aphasia and kindred disorders of speech.* London: Cambridge University Press.
Hillis, A. E., & Caramazza, A. (1989). The graphemic buffer and mechanisms of unilateral spatial neglect. *Brain and Language, 36,* 208–235.
Hillis, A. E., & Caramazza, A. (1991). Mechanisms for accessing lexical representations for output: Evidence from a category-specific semantic deficit. *Brain and Language, 40,* 106–144.
Hillis, A. E., & Caramazza, A. (1995). Representation of grammatical categories in the brain. *Journal of Cognitive Neuroscience, 7,* 396–407.
Hillis, A. E., Rapp, B., & Caramazza, A. (1999). When a rose is a rose in speech but a tulip in writing. *Cortex, 35,* 337–356.
Hodges, J. R., & Marshall, J. C. (1992). Discrepant oral and written spelling after left hemisphere tumour. *Cortex, 28,* 643–656.
Hotopf, W. H. N. (1980). Slips of the pen. In U. Frith (Ed.), *Cognitive processes in spelling* (pp. 287–307). New York: Academic Press.
Houghton, G., & Zorzi, M. (2003). Normal and impaired spelling in a connectionist dual-route architecture. *Cognitive Neuropsychology, 20,* 115–162.
Jónsdottír, M. K., Shallice, T., & Wise, R. (1996). Phonological mediation and the graphemic buffer disorder in spelling: Cross-language differences? *Cognition, 59,* 169–197.
Katz, R. B. (1991). Limited retention of information in the graphemic buffer. *Cortex, 27,* 111–119.
Kay, J., & Hanley, J. (1994). Peripheral disorder of spelling: The role of the graphemic buffer. In G. D. A. Brown & N. C. Ellis (Eds.), *Handbook of spelling: Theory, process and intervention* (pp. 295–315). New York: John Wiley.
Kinsbourne, M., & Rosenfield, D. B. (1974). Agraphia selective for written spelling. *Brain and Language, 1,* 215–225.
Kinsbourne, M., & Warrington, E. K. (1965). A case showing selectively impaired oral spelling. *Journal of Neurology, Neurosurgery, and Psychiatry, 28,* 563–566.
Kokubo, K., Suzuki, K., Yamadori, A., & Satou, K. (2001). Pure kana agraphia as a manifestation of graphemic buffer impairment. *Cortex, 37,* 187–195.
Lambon Ralph, M. A., Sage, K., & Roberts, J. (2000). Classical anomia: A neuropsychological perspective on speech production. *Neuropsychologia, 38,* 186–202.
Lichtheim, L. (1885). Über Aphasie. *Deutsches Archiv für klinische Medizin, 36,* 204–268.
Luria, A. R. (1970). *Traumatic aphasia.* The Hague: Mouton de Gruyter.
Marshall, J. C., & Newcombe, F. (1973). Patterns of paralexia. *Journal of Psycholinguistic Research, 2,* 175–199.
McCloskey, M., Badecker, W., Goodman-Shulman, R., & Aliminosa, D. (1994). The structure of graphemic representations in spelling: Evidence from a case of acquired dysgraphia. *Cognitive Neuropsychology, 11,* 341–392.
McCloskey, M., Macaruso, P., & Rapp, B. (in press). Grapheme-to-lexeme feedback in the cognitive spelling system: Evidence from a dysgraphic patient. *Cognitive Neuropsychology.*
Miceli, G., Benvegnù, B., Capasso, R., & Caramazza, A. (1995). Selective deficit in processing double letters. *Cortex, 31,* 161–71.
Miceli, G., Benvegnù, B., Capasso, R., & Caramazza, A. (1997a). The independence of phonological and orthographic lexical forms: Evidence from aphasia. *Cognitive Neuropsychology, 14,* 35–70.
Miceli, G., Benvegnù, B., Capasso, R., & Caramazza, A. (2004). The categorical distinction of consonant and vowel representations: Evidence from dysgraphia. *Neurocase, 10,* 109–121.
Miceli, G., & Capasso, R. (1997). Semantic errors as evidence for the independence and the interaction of orthographic and phonological word forms. *Language and Cognitive Processes, 12,* 733–764.
Miceli, G., Capasso, R., & Caramazza, A. (1994). The interaction of lexical and sublexical processes in reading, writing, and repetition. *Neuropsychologia, 32,* 317–333.

Miceli, G., Capasso, R., & Caramazza, A. (1999). Sublexical conversion procedures and the interaction of phonological and orthographic lexical forms. *Cognitive Neuropsychology*, *16*, 557–572.

Miceli, G., Capasso, R., Ivella, A., & Caramazza, A. (1997b). Acquired dysgraphia in a professional stenographer. *Cortex*, *33*, 355–367.

Miceli, G., Silveri, M. C., & Caramazza, A. (1985). Cognitive analysis of a case of pure dysgraphia. *Brain and Language*, *25*, 187–212.

Morton, J. (1969). The interaction of information in word recognition. *Psychological Review*, *76*, 165–178.

Newcombe, F., & Marshall, J. C. M. (1980). Transcoding and lexical stabilization. In M. Coltheart, K. E. Patterson, & J. C. Marshall (Eds.), *Deep dyslexia* (pp. 176–188). London: Routledge & Kegan Paul.

Nolan, K. A., & Caramazza, A. (1983). An analysis of writing in a case of deep dyslexia. *Brain and Language*, *20*, 305–328.

Patterson, K., & Hodges, J. R. (1992). Deterioration of word meaning: Implications for reading. *Neuropsychologia*, *30*, 1025–1040.

Patterson, K. E., Marshall, J. C., & Coltheart, M. (1985). *Surface dyslexia*. Hove, UK: Lawrence Erlbaum Associates Ltd.

Piccirilli, M., Petrillo, S., & Poli, R. (1992). Dysgraphia and selective impairment of the graphemic buffer. *Italian Journal of Neurological Science*, *3*, 113–117.

Posteraro, L., Zinelli, P., & Mazzucchi, A. (1988). Selective impairment of the graphemic buffer in acquired dysgraphia: A case study. *Brain and Language*, *35*, 274–286.

Rapp, B. C., Benzing, L., & Caramazza, A. (1997). The autonomy of lexical orthography. *Cognitive Neuropsychology*, *14*, 71–104.

Rapp, B., & Caramazza, A. (1997). The modality-specific organization of grammatical categories: Evidence from impaired spoken and written sentence production. *Brain and Language*, *56*, 248–286.

Rapp, B., & Caramazza, A. (1998). A case of selective difficulty in writing verbs. *Neurocase*, *4*, 127–140.

Rapp, B., & Caramazza, A. (2002). Selective difficulties with spoken nouns and written verbs: A single case study. *Journal of Neurolinguistics*, *15*, 373–402.

Rapp, B., Epstein, C., & Tainturier, M.-J. (2002). The integration of information across lexical and sublexical processes in spelling. *Cognitive Neuropsychology*, *19*, 1–30.

Rapp, B., & Goldrick, M. (2000). Discreteness and interactivity in spoken word production. *Psychological Review*, *107*, 460–499.

Roeltgen, D. P., Gonzalez Rothi, L., & Heilman, K. M. (1986). Linguistic semantic agraphia: A dissociation of the lexical spelling system from semantics. *Brain and Language*, *27*, 257–280.

Roeltgen, D. P., & Heilman, K. M. (1984). Lexical agraphia. Further support for the two-system hypothesis of linguistic agraphia. *Brain*, *107*, 811–827.

Roeltgen, D. P., Sevush, S., & Heilman, K. M. (1983). Phonological agraphia: writing by the lexical-semantic route. *Neurology*, *33*, 755–765.

Ruml, W., Caramazza, A., Capasso, R., & Miceli, G. (2005). Interactivity and continuity in normal and aphasic language production. *Cognitive Neuropsychology*, *22*, 131–168.

Schiller, N. O., Greenhall, J. A., Shelton, J. R., & Caramazza, A. (2001). Serial order effects in spelling errors: Evidence from two dysgraphic patients. *Neurocase*, *7*, 1–14.

Shallice, T. (1981). Phonological agraphia and the lexical route in writing. *Brain*, *104*, 413–429.

Shallice, T., Glasspool, D. W., & Houghton, G. (1995). Can neuropsychological evidence inform connectionist modeling? Analyses of spelling. *Language and Cognitive Processes*, *10*, 192–225.

Shallice, T., & Warrington, E. K. (1975). Word recognition in a phonemic dyslexic patient. *Quarterly Journal of Experimental Psychology*, *27*, 187–199.

Shallice, T., & Warrington, E. K. (1980). Single and multiple component central dyslexic syndromes. In M. Coltheart, K. E. Patterson, & J. C. Marshall (Eds.), *Deep dyslexia* (pp. 119–145). London: Routledge & Kegan Paul.

Tainturier, M.-J., & Caramazza, A. (1996). The status of double letters in graphemic representations. *Journal of Memory and Language*, *35*, 53–73.

Tainturier, M.-J., & Rapp, B. C. (2004). Complex graphemes as functional spelling units: evidence from acquired dysgraphia. *Neurocase*, *10*, 122–131.

Venneri, A., Cubelli, R., & Caffarra, P. (1993). Perseverative dysgraphia: A selective disorder in writing double letters. *Neuropsychologia*, *32*, 923–931.

Ward, J., & Romani, C. (1998). Serial position effects and lexical activation in spelling: Evidence from a single case study. *Neurocase*, *4*, 189–206.

Wernicke, C. (1906). Nervenheilkunde. Die neueren Arbeiten über Aphasie. *Fortschritte der Medizin*, *4*, 463–482.

Wing, A. M., & Baddeley, A. D. (1980). Spelling errors in handwriting: A corpus and a distributional analysis. In U. Frith (Ed.), *Cognitive processes in spelling* (pp. 251–286). New York: Academic Press.

COGNITIVE NEUROPSYCHOLOGY, 2006, 23 (1), 135–155

Memory systems: The case of phonological short-term memory. A festschrift for *Cognitive Neuropsychology*

Giuseppe Vallar
University of Milano-Bicocca, Milano, Italy

Developments in the understanding of the neurological and functional architectures of phonological short-term memory that took place in the 1984–2004 time period are reviewed. Phonological short-term memory is discussed as a case that illustrates a number of issues shared by other research domains in cognitive neuropsychology, with particular reference to memory systems. *Modularity:* Phonological short-term memory includes two main components, an input storage system (the *phonological short-term store*), and an output *rehearsal process.* These components are functionally connected with other verbal processes and systems, such as visual-verbal short-term memory, phonological recoding (grapheme-to-phoneme conversion), verbal long-term episodic memory, and the lexical systems. *Neurological specificity:* The store and the rehearsal components of phonological memory are implemented in discrete parts of the left cerebral hemisphere. Both the traditional anatomo-clinical correlation studies in brain-damaged patients, and neuroimaging activation experiments in neurologically unimpaired subjects, have contributed to elucidate the neural basis of the system. *Converging operations:* A distinctive feature of these advances has been the close interaction and cross-fertilisation among the related domains of cognitive experimental psychology and cognitive neuropsychology, based on observations in adult and child populations, in brain-damaged patients, and in subjects with developmental and genetically based cognitive disorders. Data from these different research areas have converged to specify the structure of phonological short-term memory (the distinction between storage and rehearsal) and its relationships with other memory systems (the long-term learning of novel phonological lexical entries).

1984 KNOWLEDGE

The 1984 (when *Cognitive Neuropsychology* was founded) to 2004 time period covers important advances in understanding of the neurological and functional architecture of memory systems. As often happens in the development of scientific knowledge, however, two main principles, which informed subsequent scientific progress in cognitive neuropsychology in depth, had been already laid down.

The functional modularity of memory systems

It had long been known that memory was not a monolithic unitary system, including instead at least two main components: *short-term memory and long-term memory* (James, 1895; Locke, 1700/1975). It was, however, in the late 1950s and early 1960s that the short-term vs. long-term memory distinction was supported by investigations in neurologically unimpaired individuals

Correspondence should be addressed to Giuseppe Vallar, MD, Department of Psychology, University of Milano-Bicocca, Piazza dell'Ateneo Nuovo 1, 20126 Milano, Italy (Email: giuseppe.vallar@unimib.it).

DOI:10.1080/02643290542000012

(Baddeley & Hitch, 1974; Glanzer, 1972). However, the two-component model of memory was not uncontroversial at that time (Melton, 1963), and unitary views are sometimes resurrected (Ruchkin, Grafman, Cameron, & Berndt, 2003, and commentaries therein).

In the 1970s, the process of fractionation of memory systems proceeded both within long (the "episodic" vs. "semantic" memory distinction: Tulving, 1972) and short-term memory (Baddeley, 1976; Crowder, 1976). In the following years it became clear that both short- and long-term memory are multicomponent systems. A current taxonomy is shown in Figure 1. In this article, the case of phonological short-term memory will be considered, together with its relationships with learning and retention, namely with long-term memory systems.

In the early 1970s, the standard view of short-term memory was in terms of a unitary system that received modality-specific inputs (e.g., auditory, visual) from different sensory registers. This "amodal" (i.e., not specific for any sensory modality) system is effectively illustrated by the information processing flow charts of Waugh and Norman (1965) and, most of all, by the "modal" (in the statistical sense of more frequent) model of Atkinson and Shiffrin (1971), shown in Figure 2.

At the time when these models were published, it was already clear that they were to be conceived as highly schematic flow chart representations of memory, with each store including a number of subsystems. Within short-term memory, the fractionation process had produced models such as that shown in Figure 3 in the early 1960s. This distinguishes, in the verbal domain, three main short-term memory components: a *visual store*, an *auditory store*, and a *rehearsal process*.

The neuroanatomical modularity of memory systems

In the late 1960s, the definite demonstration of the validity of the constructs of short- and

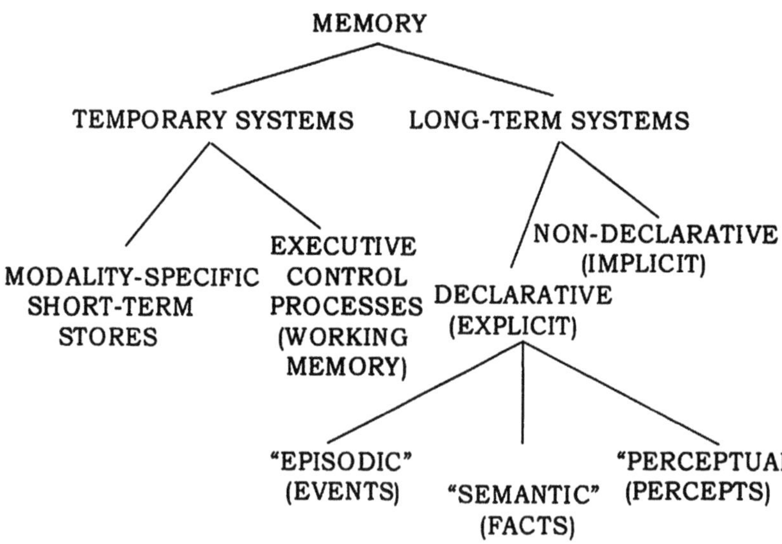

Figure 1. *A schematic functional taxonomy of memory systems, based on Squire (1992) and Baddeley (2002). Modality-specific short-term memory includes phonological (auditory-verbal), visual, and spatial storage and rehearsal systems, and a number of comparatively less investigated components based on other sensory modalities (e.g., auditory nonverbal). Non-declarative memory includes further subsystems, not shown here. The functional connections among these memory systems are not shown (modified from Vallar, Kopelman, Markowitsch, Miceli, & Papagno, 2005).*

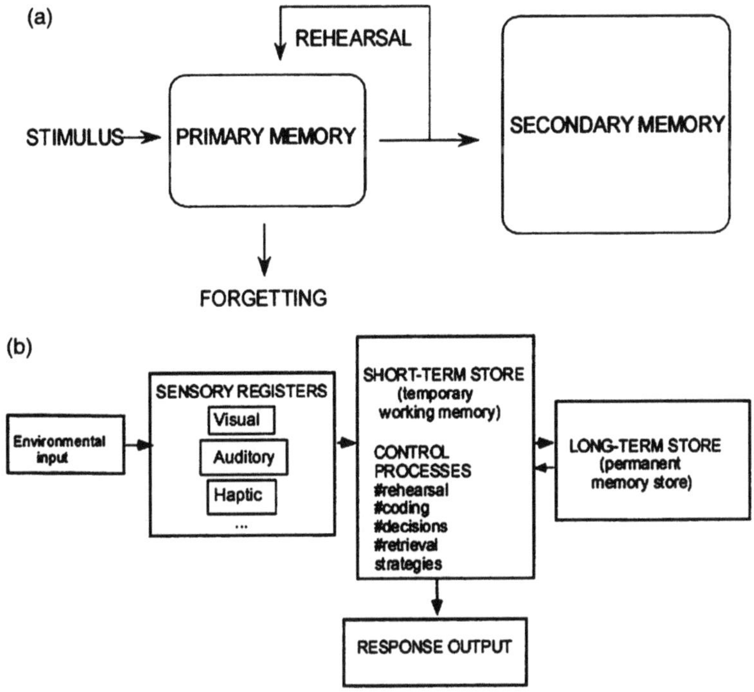

Figure 2. *Two-component models of memory. (a) Waugh and Norman (1965). Stimuli enter a very limited primary memory store; rehearsal may transfer items to a larger and more stable secondary memory store; unrehearsed items tend to be quickly forgotten. (b) Atkinson and Shiffrin (1971). This, the most influential model of memory in the early 1970s, explicitly illustrates the "supramodal"- or "amodal" nature of the short-term store, to which a number of relevant cognitive functions are assigned, hence the term working memory. Both (a) and (b) models have a serial architecture.*

long-term memory as discrete systems came from the study of brain-damaged patients suffering from amnesia, interpreted in terms of a deficit of the long-term component of memory. Atkinson and Shiffrin (1968, pp. 96–97) wrote:

> Workers of a traditional bent have argued against dichotomizing memory. However, we feel there is much evidence indicating the parsimony and usefulness of such a division. The argument is often given that one memory is somehow "simpler" than two; but quite the opposite is usually the case... Any single-process system making a fair attempt to explain the mass of data currently available must, of necessity, be sufficiently complex that the term *single process* becomes a misnomer. ...Still, some justification of our decision[1] would seem indicated at this point. For this reason we turn to what is perhaps the single most convincing demonstration of a dichotomy in the memory system: the effects of hippocampal lesions reported by [Brenda] Milner ....

The argument put forward by Atkinson and Shiffrin is relevant in two respects. First, neuropsychological findings are taken as relevant evidence to adjudicate between two competing functional accounts of the organisation of memory (single vs. double component models). Second, note is taken of the localisation of the responsible lesion, which involves the medial temporal lobes bilaterally (Milner, 1966). The neuroanatomical part of the argument was not developed by Atkinson and Shiffrin (1971), but became relevant

[1] i.e., to fractionate memory into short- and long-term stores.

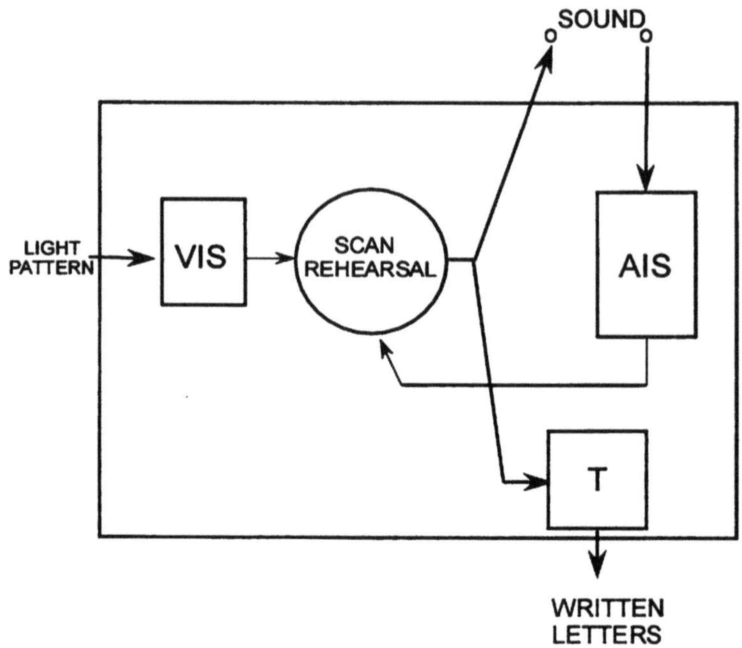

Figure 3. *Model 2 of short-term memory by Sperling (1967). A distinction is drawn between a visual (VIS) and an auditory information (AIS) storage. The process of scan-rehearsal has the dual function of (1) conveying visual traces to the auditory store, through a visual-to-auditory conversion, and (2) refreshing the traces stored in the AIS, preventing their decay.*

a few years later, when deficits of the other component of the system (short-term memory) were described. Luria, Sokolov, and Klimkowski (1967) described two patients with damage to the left temporal lobe (B and K), and a selective impairment of repetition of series of words spoken by the experimenter (auditory presentation), with repetition of visual-verbal stimuli being comparatively preserved. Luria et al. (1967) interpreted these findings as a modality-specific disturbance of auditory-verbal memory traces, and distinguished this *partial* deficit from the *general memory disorder* of amnesia described by Milner (1966). Luria's (1971) distinction was not only functional, but also neuroanatomical: Damage to the medial temporal lobe (hippocampal region) and parts of the diencephalon brought about amnesia, while deficits of immediate repetition were associated with damage to the convexity of the cerebral hemispheres, particularly the left temporal lobe.

An information processing approach to short-term memory and its neuropsychological impairments

In the late 1960s and early 1970s, Warrington and Shallice (Shallice & Butterworth, 1977; Shallice & Warrington, 1970, 1974; Warrington & Shallice, 1969, 1972) investigated extensively the memory performance of three left-brain-damaged patients (WH, and—most of all—KF, and JB), who, as Luria et al.'s (1967) patients, showed a disproportionate deficit of immediate repetition of auditory-verbal material. In their experiments, Warrington and Shallice made use of the whole repertoire of paradigms suitable to assess short-term memory performance, in the context of the short- vs long-term memory distinction shown in Figures 1 and 2, and, within short-term memory, with reference to Sperling's (1967) model, shown in Figure 3. A number of experiments were performed using paradigms such as immediate memory span (Conrad & Hull, 1964), short-term forgetting (Peterson & Peterson, 1959), and free recall (Glanzer, 1972), with results providing evidence that converged to suggest a deficit of the AIS component of Figure 3. At variance with the interpretation of Luria and his co-workers (Luria, 1971; Luria et al., 1967), the account of Warrington and Shallice was explicitly framed theoretically within the information-processing view of cognitive processes depicted in Figure 2.

In addition to an information processing based account of the impairments of verbal short-term memory, Warrington and her co-workers (Warrington, Logue, & Pratt, 1971) localised more precisely the anatomical basis of the system in the supramarginal gyrus of the inferior parietal lobule of the left cerebral hemisphere.

Neuropsychological fractionations: The case of auditory-verbal short-term memory

System fractionation has been one hallmark of the development of research in cognitive neuropsychology. Fractionation refers to a development of knowledge in a particular scientific domain whereby a given system, previously considered a single monolith, divides (fractionates) into two or more subcomponents. Fractionation is not confined to cognitive neuropsychology but has occurred in related domains, including for example neuroanatomy (the number of cerebral cortical areas) and diseases of the peripheral nervous systems (Vallar, 1994). More generally, the increase of known components is a main feature of the advances of knowledge in subatomic particles (Close, Marten, & Sutton, 1987), in genetics (McKusick, 1998) (see also Online Mendelian Inheritance in Man: http://www.ncbi.nlm.nih.gov/omim/), and in classification of organisms (taxonomy) (Margulis & Schwartz, 1998).

In the particular case of auditory-verbal short-term memory, fractionation distinguished between an input, relatively passive, storage component (the *phonological short-term store*) and a process (*articulatory rehearsal*) concerned with refreshing the phonological memory trace held in the store, and conveying written material to it. This latter function involves grapheme-to-phoneme conversion or phonological recoding. The basic functional relationships between the store and the rehearsal components of phonological memory are shown in Sperling's 1967 model (Figure 3). Data from neurologically unimpaired subjects provided support for the store–rehearsal distinction (Baddeley, Lewis, & Vallar, 1984; Salamé & Baddeley, 1982). The neuropsychological evidence was not complete, however. The pattern of impairment of patients with a disproportionate deficit of auditory-verbal span (Luria et al., 1967; Warrington & Shallice, 1969) was interpreted as a selective deficit of the auditory input store component on the basis of two sets of converging empirical data. (1) the auditory-visual dissociation both in span, and in other tasks assessing immediate memory (short-term forgetting, e.g., Basso, Spinnler, Vallar, & Zanobio, 1982; the recency effect in free recall, Vallar & Papagno, 1986), with the patients' performance being most often higher with visual input. (2) The evidence that speech output was entirely preserved in some of these patients (Shallice & Butterworth, 1977, patient JB; Vallar & Baddeley, 1984a, patient PV). This second argument was based on the assumption that articulatory rehearsal was a process making use of systems primarily concerned with the programming and realisation of speech output (Shallice & Vallar, 1990). The neuropsychological impairment of the process of rehearsal had not been described in the mid-1980s, however. Review chapters about the neuropsychological impairments of auditory-verbal (phonological) short-term memory describe and discuss the functional deficit of the input phonological short-term store (Shallice & Vallar, 1990; Vallar & Papagno, 1995), not of the process of rehearsal.

The 1984-vintage neuropsychological evidence was even more ambiguous when a number of specific short-term effects are considered. It has long been known that sequences of acoustically or phonologically dissimilar items are recalled better than lists of similar items, both with auditory and with visual presentation (Baddeley, 1966; Conrad, 1964; Conrad & Hull, 1964; Wickelgren, 1965). This *phonological similarity effect* is abolished by the concurrent task of *articulatory suppression* (the continuous uttering of an irrelevant speech sound, such as *the, the, the*...), which interferes with speech output processes and, plausibly therefore, with articulatory rehearsal. Suppression also reduces memory span performance, disrupting the rehearsal process. Suppression abolishes the effect of phonological similarity only when input is visual, however, but not with auditory presentation (Baddeley et al., 1984; Levy, 1971). This dissociation is consistent with a distinction between a nonarticulatory phonological input store, and an articulatory process concerned, among other activities, with conveying written material to the input store.

The store–rehearsal distinction was also corroborated by the finding that the *word length effect* (the serial recall advantage in immediate span of *short* words, in terms of number of syllables or time to pronounce, over *long* words; Baddeley et al., 1984; Baddeley, Thomson, & Buchanan, 1975; Levy, 1971) is abolished by articulatory suppression both with visual and articulatory input. This suggests that the effect of word length may be taken as an index of the activity of the process of rehearsal.

Patients with a selective impairment of auditory-verbal span, interpreted, as noted earlier, as a deficit of the phonological store, exhibit however a pattern of effects comparable to that shown by neurologically unimpaired individuals when engaged in articulatory suppression. Although span is disproportionately low, in most patients the phonological similarity effect is still present with auditory presentation, suggesting some residual phonological storage capacity. However, the effect is consistently absent, across patients, when input is visual. Furthermore, the effect of word length is absent with either input modality. Finally, suppression does not further impair visual verbal span (patient PV: Vallar & Baddeley, 1984a). This pattern, replicated in many published patients with a defective auditory-verbal span (meta-analysis in Vallar & Papagno, 2002), suggests that the process of rehearsal is either primarily damaged, or unimpaired but not utilised by these patients. This latter interpretation (lack of use due to a strategic choice: There may be no advantage in rehearsing items held in a damaged store) attempted to accommodate evidence for the absence of rehearsal in immediate retention on the one hand, and the preserved speech production of some patients on the other. Furthermore, the anatomical localisation of the patients' lesions in the inferior-posterior parietal region of the left hemisphere (meta-analysis in Vallar & Papagno, 2002) was consistent with the hypothesis of a damaged auditory-verbal input store, and a preserved rehearsal process.

1984–2004

Functional neuroimaging: Phonological short-term memory

In the late 1980s, the availability of Positron Emission Tomography, coupled with the utilisation of the subtraction method (whereby the cerebral regional pattern of activation, or deactivation, associated with the execution of a given task A, is subtracted from the activation pattern associated with task B), gave novel impetus to the investigation of the cerebral correlates of mental activity. The relevant cognitive issue here is that task B should ideally differ from A only for the factor of interest (review in Frackowiak, Friston, Frith, Dolan, & Mazziotta, 1997).

Activation studies in neurologically unimpaired subjects provided definite neural evidence for the articulatory rehearsal-phonological short-term store model of phonological short-term memory. In accord with the neuropsychological evidence reviewed earlier, the supramarginal gyrus of the left inferior parietal lobule is a main correlate of the phonological short-term store, while the premotor regions (Broca's area BA 44 and BA 6) in the left hemisphere are associated with the process of rehearsal (Paulesu, Frith, &

Frackowiak, 1993; Paulesu et al., 1996). The view that a premotor frontal-posterior parietal left-hemisphere based neural network constitutes the neural basis of the rehearsal process-phonological short-term store memory system was subsequently confirmed by a number of activation experiments (review and meta-analyses in Smith & Jonides, 1999).

In the original activation studies that provided evidence for the store-rehearsal two-component architecture, two theoretically driven tasks were used (Paulesu et al., 1993, 1996). Storage was assessed by a span task, which brought about activation of the whole network, in line with the behavioural evidence that articulatory suppression affects immediate retention (Baddeley et al., 1984). The articulatory rehearsal process was localised instead by a rhyming task. This was used because, in neurologically unimpaired subjects, suppression has minor, but significant, detrimental effects on some phonological judgments, involving segmentation and deletion, such as deciding whether or not two words rhyme, or share the same stress pattern or initial sound (Besner, 1987; Burani, Vallar, & Bottini, 1991).

Neuropsychological impairments of the rehearsal process

The pattern of impairment associated with a deficit of the process of articulatory rehearsal has been described in one left-brain-damaged patient with subcortical damage to the premotor region and the anterior part of the insula. Patient TO (Vallar, Di Betta, & Silveri, 1997) was a nonfluent aphasic, who exhibited in part the pattern of short-term memory performance shown by patients with a defective phonological short-term store: reduced auditory-verbal span, absence of the word length effect and of the phonological similarity effect when the input was visual. The patient's performance, however, improved when a nonverbal response (recognition by pointing) was used. The span performance of patients with a deficit of the phonological short-term store does not improve when a nonarticulatory response modality is used, and this is taken as evidence that output or production factors cannot explain the memory deficit (review in Shallice & Vallar, 1990) Furthermore, the patient showed a comparatively preserved recency effect, while a defective recency in free recall of auditory-verbal lists is a hallmark of the impairment of the phonological short-term store (Vallar & Papagno, 1986; Warrington et al., 1971). Finally, patient TO was unable to perform judgments on written words, involving phonological articulatory codes (Burani et al., 1991).

By contrast, a control patient, LA, was a fluent aphasic with a lesion involving the left inferior parietal lobule and the superior and middle temporal gyri. Patient LA exhibited the pattern of impairment associated with damage to the phonological short-term store. The patient had a disproportionately reduced auditory-verbal span that did not improve with a nonarticulatory response modality. The patient also showed no effects of phonological similarity and word length, suggesting very severe damage of the phonological store and an impairment (or lack of use) of the process of rehearsal. Patient LA also showed no recency effect in auditory free recall. Crucially, LA's ability to perform phonological judgments on visually presented words was preserved (see also Vallar & Baddeley, 1984b; Vallar et al., 1997). Accordingly, the output phonological articulatory components that, in the unimpaired system, also support the phonological store, were undamaged in patient LA, though not utilised for the purpose of immediate retention due to total damage of the phonological store.

The contrasting patterns of short-term memory impairments found in patients TO and LA illustrate also the relationships between *classical* syndromes, such as the aphasias of the Wernicke-Lichtheim taxonomy (Lichtheim, 1885; Wernicke, 1874/1966–1968) and the *modern* syndromes, based on information processing models. The deficit of patient LA would be classified as a fluent aphasia (Benson, 1979): conduction aphasia, with a disproportionate impairment of repetition, with little if any phonemic paraphasias in speech production

(Luria, 1977; Shallice & Warrington, 1977). The deficit of patient TO would be classified as a non-fluent mild Broca's aphasia (hypo-fluent slightly agrammatic speech, with preserved articulation and prosody and some phonological errors).

The interpretation of the patients' repetition impairments in the light of the information processing framework of the phonological short-term store-rehearsal system illustrates a syndrome-based approach, based on cognitive models. This was very controversial in the late 1980s (Caramazza, 1986; Ellis, 1987; Patterson, Marshall, & Coltheart, 1985). In the early 1990s, Shallice and Vallar (1990) argued that the selective impairment of auditory-verbal short-term memory (i.e., of the phonological short-term store, shown in Figure 4) was a successful example of functional syndrome, based on cognitive models of memory. Successive meta-analyses of the patients' impairment confirmed this conclusion, showing that the similarities across patients extend to effects such as phonological similarity and word length, which provide insight into the kind of representations involved (Vallar & Papagno, 2002).

Premotor vs. motor components of the rehearsal process

Both the activation data in neurologically unimpaired individuals and the neuropsychological findings in brain-damaged patients concur to

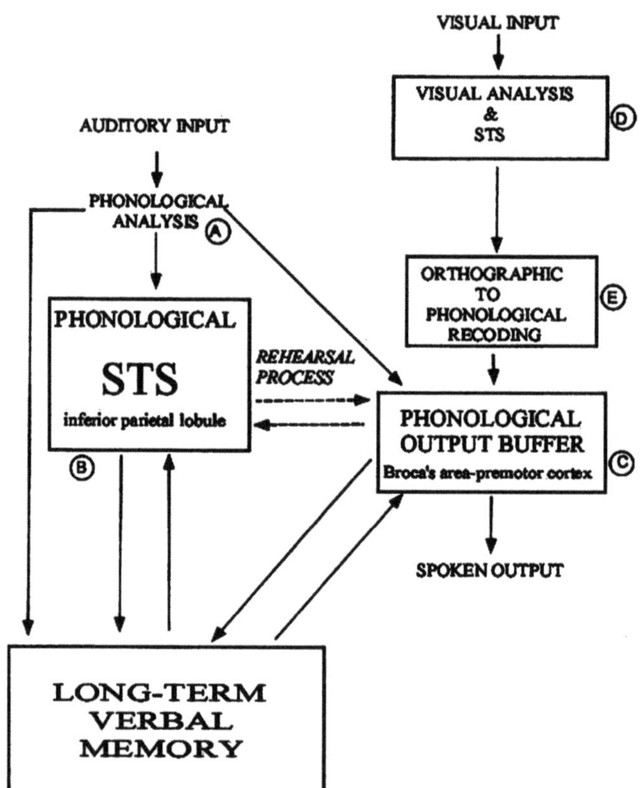

Figure 4. *An anatomo-functional model of phonological short-term memory, based on Vallar et al. (1997), and Baddeley, Gathercole, and Papagno (1998).*

suggest a role of the premotor cortex of the left cerebral hemisphere in the process of articulatory rehearsal. Since Vygotsky's (1934/1962) seminal work, *inner speech* has been also associated with the more peripheral components of the articulatory system, including the motor cortex, the descending motor efferent pathways in the brainstem, the cranial nerves, and the muscles involved in articulation. Also in Sperling's (1967) model, shown in Figure 3, articulatory rehearsal may have a peripheral component involving sound generation.

Children with damage presumably involving brainstem neural structures, who are congenitally unable to produce articulated speech (*anarthria*) (for a terminological discussion, with reference to the classic aphasic syndromes, see Lecours, Lhermitte, & Bryans, 1983), show a preserved immediate memory, with the presence of the effects of phonological similarity and item length, as well as a preserved ability to make phonological judgments (Bishop & Robson, 1989b).

Adult patients with neurological damage bringing about anarthria show a preserved auditory-verbal span, both when the lesion involves the motor regions of the cerebral hemispheres and when the descending motor pathways in the brainstem are affected (Table 1). All patients show the effect of phonological similarity with auditory presentation (Table 2). This finding, together with the preserved auditory digit span, suggests a preserved phonological short-term store. This conclusion is also consistent with the localisation of the lesion in these patients, which largely spares the parietal lobe, involving either the frontal cortex or the brainstem. The normal span suggests that the process of rehearsal may be preserved, since in neurologically unimpaired subjects articulatory suppression has mildly disruptive effects on span performance (Baddeley et al., 1984; Paulesu et al., 1993, for related neural evidence). In line with this conclusion, patients with brainstem damage (GF, Vallar & Cappa, 1987; CM, Cubelli & Nichelli, 1992) show the effect of phonological similarity with visual input, suggesting that visual material gains access to the phonological short-term store. In patients GF and CM the absence of the effect of word length may be attributed to the complex response modalities that were used, due to the presence of severe motor deficits (Cubelli & Nichelli, 1992; Vallar & Cappa, 1987). Patients with cortical damage (MDC, Vallar & Cappa, 1987; FC, Cubelli & Nichelli, 1992; XX, Silveri, Cappa, & Salvigni, 2003) make use of the rehearsal process when presentation is auditory, namely to refresh material held in the phonological store: This is indexed by the presence of the word length effect with auditory input. Visual material, however, does

Table 1. *Auditory-verbal digit span in patients with acquired anarthria*

Patient	*Aetiology*	*Lesion site (assessment)*	*Digit span*
A. Cortico-subcortical damage			
MDC[a]	CVA	L-T anterior, central atrophy, R-FT operculum (CT)	7
FC[b]	CVA	L-F precentral, L/R subcortical (MRI)	6
XX[c]	DD	L-R F cortical atrophy A (MRI)	6.2
B. Brainstem damage			
GF[a]	CVA (?)	Pontine(?) (CT)	4.6
CM[b]	CVA	Pontine (MRI)	5
C. Diffuse damage			
EC[d]	CVA	L(?)	6
GB[e]	HI	?	7

CVA: cerebrovascular attack; DD: degenerative disease; HI: head injury; L/R: left, right; F: frontal; T: temporal; CT: Computerized Tomography; MRI: Magnetic Resonance Imaging.
[a]Vallar & Cappa, 1987; [b]Cubelli & Nichelli, 1992; [c]Silveri et al., 2003; [d]Nebes, 1975; [e]Baddeley & Wilson, 1985.

Table 2. *Phonological similarity and word length effects in patients with acquired anarthria*

	Phonological similarity effect		*Word length effect*	
Patient	*Auditory*	*Visual*	*Auditory*	*Visual*
A. Cortico-subcortical damage				
MDC	+	−	+	−
FC	+	−	+	−
XX	+	−	na	na
B. Brainstem damage				
GF	+	+	−	na
CM	+	+	−	−
C. Diffuse or nondefinite damage				
EC	na	+	na	na
GB	na	+	+	na

not gain access to the process of rehearsal, as suggested by the absence of the similarity and length effects with visual input. Performance on rhyme judgments are in line with these conclusions. Cortical damage disrupts rhyme judgments (patients MDC, Vallar & Cappa, 1987; and XX, Silveri et al., 2003), but brainstem damage does not GF, Vallar & Cappa, 1987). In the two patients (EC, Nebes, 1975; and GB. Baddeley & Wilson, 1985) where the responsible lesion was not definitely localised, the available data suggest a preservation of the phonological short-term store-rehearsal system.

To summarise, relatively peripheral damage (brainstem) to the motor efferent pathways bringing about anarthria does not disrupt the process of rehearsal. These data are consistent with developmental neuropsychological evidence (Bishop & Robson, 1989a, 1989b). Also, the data from patients where the lesion was not adequately localised (EC and GB) suggest that the complete inability to produce articulated speech does not disrupt per se the process of rehearsal. Cortical damage, by contrast, may prevent access of visual verbal material to a preserved process of rehearsal, possibly at the level of phonological recoding. Tables 1 and 2 also suggest a note of caution, considering the limited number of patients and the fact that not all the effects of interest have been tested in all patients.

Short-term storage and long-term learning

The standard serial architecture of the relationships between short- and long-term storage shown in Figure 2 implies that deficits of short-term storage cause defective learning and retention, because temporary storage is required for long-term learning. The serial model is entirely compatible with the case of amnesia (defective learning and retention in declarative memory; see review in Vallar, 1999). Early investigations of the learning abilities of patients with defective phonological memory showed, however, that learning of both verbal (word lists, passages of prose) and nonverbal material was entirely preserved (Basso et al., 1982; Shallice & Warrington, 1970; Warrington et al., 1971). This pattern, together with the opposite impairment of amnesia, constitutes a double dissociation of deficits, which may be explained by a parallel rather than a serial organisation of memory systems: inputs may gain independent access to short- and long-term memory systems for the purpose of temporary vs. more long lasting learning (Shallice & Warrington, 1970).

In the late 1980s, Baddeley, Papagno, and Vallar (1988) confirmed that the short-term memory patient PV (Basso et al., 1982) was indeed able to learn verbal material. However, they also showed that PV was dramatically impaired when required to learn pseudowords, or new words, that did not possess a pre-existing phonological (and lexical-semantic) representation in her mental lexicon. This learning deficit, associated with the impairment of phonological short-term memory, in principle could be explained either in terms of associated disorders of two independent memory components (the phonological short-term store, and some other phonological learning device), or, more parsimoniously, by a specific role of phonological short-term storage in verbal learning (Weiskrantz, 1990). This latter view, whereby temporary storage is necessary in order to set up novel phonological representations, is fully compatible with the serial model shown in Figure 2.

In the 1990s, results from different subjects' populations supported the view that the phonological short-term store contributes to long-term learning. In children, the capacity of phonological memory is a main predictor of vocabulary acquisition (Gathercole & Baddeley, 1993). In neurologically unimpaired adult subjects, the same variables that affect immediate retention (phonological similarity, item length, articulatory suppression) influence the acquisition of new words, but have minor or null effects on word learning (Papagno, Valentine, & Baddeley, 1991; Papagno & Vallar, 1992). Subjects with developmental disorders of phonological memory show defective learning of novel words (Baddeley, 1993). By contrast, subjects with an isolated sparing of phonological short-term memory, in the context of a developmental cognitive deficit (Williams' syndrome: Barisnikov, Van der Linden, & Poncelet, 1996; Down's syndrome: Vallar & Papagno, 1993), may show a preserved learning of new words, together with an excellent acquisition of foreign languages (Vallar & Papagno, 1993) and unfamiliar highly specific words (terms about horse riding: Papagno & Vallar, 2001). Finally, neurologically unimpaired subjects who prove able to learn efficiently novel languages (polyglots) have a higher short-term memory capacity (verbal span) and show superior learning of novel words, compared with matched nonpolyglot subjects. The differences between polyglots and nonpolyglots are confined to phonological short-term memory and learning of unfamiliar letter strings, such as words of an unknown foreign language (Papagno & Vallar, 1995).

A current neurofunctional model of phonological short-term memory is shown in Figure 4. Auditory-verbal material, after early acoustic and phonological analysis (A) enters the main retention component of the system, the phonological short-term store (STS) (B), where material is coded in a phonological format. The phonological STS is an input-system, to which auditory material has a direct and automatic access. The process of rehearsal is conceived as involving a recirculation of the memory trace between the phonological STS and a phonological output system, the *phonological output buffer* or *phonological assembly system* (C), primarily concerned with the articulatory programming of speech output, with a recurring translation between input (*acoustic*) and output (*articulatory*) phonological representations. The phonological output buffer provides access for visually presented verbal material to the phonological STS, after *phonological recoding* or *grapheme-to-phoneme conversion* (E). The model also illustrates the multiple-component nature of short-term memory, showing a visual STS (D), where material is likely to be encoded in terms of shape.

The phonological STS-rehearsal system is mainly based on left hemisphere networks: (1) Broca's area (Brodmann's area 44), the premotor area (Brodmann's area 6), and the supplementary motor area are associated with the process of rehearsal; (2) the left inferior parietal lobule (supramarginal gyrus, Brodmann's area 40), at the temporo-parietal junction, with the phonological STS. These two sets of cerebral regions are likely to be connected through the arcuate fasciculus, and white matter fibre tracts in the insular region. The motor cortex may contribute to phonological recoding.

Retention in the phonological STS-rehearsal system contributes to the setting up of novel (not pre-existing) phonological representations in verbal long-term memory, as in the case of non-words or novel words, for instance of a foreign language, unknown to the subject (connections to long-term verbal memory). Episodic learning of real words that possess pre-existing stored phonological representations may not require temporary storage in phonological memory (connection from phonological analysis to long-term verbal memory).

Connections from verbal long-term memory to the phonological STS and the rehearsal process illustrate the support of long-term memory systems (e.g., stored phonological and lexical representations) to aspects of immediate retention.

CONCLUSION

This selective review of the 1984–2004 developments of knowledge concerning the functional and neural architecture of phonological short-term memory illustrates a number of issues that are shared by other aspects of cognitive neuropsychology. These include a novel interest in the *neural basis of cognition*, and the increasingly relevant role of *converging operations* (Garner, Hake, & Eriksen, 1956), from research performed in different pathological and neurologically unimpaired populations. The aim of this is to develop interpretative theoretical models fitting a wider range of empirical data, beyond the specific domain in which particular experiments were performed; in this case, neuropsychological studies in brain-damaged patients.

Neural basis of cognition

The 1984 manifesto of *Cognitive Neuropsychology* emphasised the role of behavioural investigations in brain-damaged patients, theoretically informed by information processing models of cognition, in elucidating the functional architecture of cognitive processes. In the 1984 editorial of the Journal (Editor: Max Coltheart), the role of neurological evidence for the understanding of mental processes was not a main one: "...it is likely that such separate levels of description as the neurological and the psychological are only 'loosely coupled' (1984, p. 3). The cover of the *Cognitive Neuropsychology* journal at that time was neutral in this respect: a pink hue, darker on the top. In 1998 (Vol. 15, Editor: Alfonso Caramazza) the journal changed its cover. The new cover has different hues in pink and red, but also, and more importantly, some drawings: (1) a clock face drawn by a patient with left neglect, with the hours displaced rightwards (this is likely to acknowledge the increasing relevance within cognitive neuroscience of the syndrome of spatial unilateral neglect (Bisiach & Vallar, 2000; Karnath, Milner, & Vallar, 2002; Marshall & Vallar, 2004); (2) an axial slice of the brain with two activation spots (to acknowledge the role of functional imaging methods: Frackowiak et al., 1997); (3) a schematic lateral view of the right hemisphere, alluding to electrophysiological maps of brain activity (Zani & Proverbio, 2002); (4) an information processing flow chart. The message conveyed by the cover is that cognitive neuropsychology draws on empirical data (the patient's drawing), theory (mainly, the information processing approach), and, importantly, neurological information provided by neuroimaging and electrophysiological methods.

In the 1984–1997 time period, using *PsychInfo* and *Web of Science* (from 1990) 37 papers, notes, and comments published in *Cognitive Neuropsychology*, primarily concerned with different aspects of memory processes (out a total of 328 papers, 11%), were retrieved (retrieval keywords for the title: memory/memories, learning, forgetting, store/stored, articulatory loop) (see Appendix A). Most of them provide adequate baseline neurological information (localisation of the lesion by CT Scan, MRI). The vast majority of them, at least in my reading, are primarily concerned with the relationships between the neural and the functional architectures of cognitive processes (see, however, Morris, Miotto, Feigenbaum, Bullock, & Polkey, 1997). In the comparatively shorter 1998–2004 period, using *Web of Science* and the same retrieval keywords, 38 papers have been retrieved (out of a total of 251 papers, 15%), devoted to different aspects of memory processes (see Appendix B). Again, most of them were mainly concerned with the functional organisation of memory systems and provided, when appropriate, baseline neurological information, including localisation of the lesion. In two papers, anatomical data are mentioned in the title and represent a relevant part of the arguments of the paper (Kitchener & Hodges, 1999; Mayes et al., 2001). One study used event-related potentials (Rubin, Van Petten, Glisky, & Newberg, 1999), while cerebral activation (Positron Emission Tomography, functional Magnetic Resonance Imaging) or Transcranic Magnetic Stimulation studies concerned with memory have not been published so far in the Journal. To summarise, the perusal of published papers devoted to memory shows that *Cognitive Neuropsychology* still maintains its main focus on the functional structure of cognitive systems, at least in this area of cognition. This distinctive feature may be seen positively in a scientific publishing market, where different journals "specialise" in particular issues, though being theoretically open-minded to related research areas.

Converging operations

The 1984 emphasis on models of cognitive processes, of which neuropsychological data constitute an important source of empirical evidence, has also prompted interactions between cognitive and experimental psychology on the one hand, and neuropsychology on the other.

Research in phonological short-term memory provides an illustrative example. Current knowledge is based on studies in neurologically unimpaired adult individuals (Baddeley et al., 1984; Conrad & Hull, 1964; Salamé & Baddeley, 1982), in children (Hitch, 1990), and in elderly subjects (e.g., Spinnler, Della Sala, Bandera, & Baddeley, 1988). Neuropsychological experiments are performed in adult brain-damaged patients, who provided the original evidence for a selective impairment of phonological memory (Luria et al., 1967; Warrington & Shallice, 1969), in subjects with developmental disorders (Baddeley, 1993; Barisnikov et al., 1996; Campbell & Butterworth, 1985; MacKenzie & Hulme, 1987; Vallar & Papagno, 1993), and in elderly patients with a widespread cognitive impairment, such as that brought about by Alzheimer's disease (Morris, 1986; Spinnler et al., 1988). Similarly, the neural basis of phonological memory in humans is currently investigated not only by the time-honoured anatomoclinical correlations (Vallar et al., 1997; Warrington et al., 1971), but also through a wider range of approaches, including cerebral activation studies (Smith & Jonides, 1999) and techniques, such as TMS, which brings about temporary interference with brain function (e.g., Mottaghy, Gangitano, Krause, & Pascual-Leone, 2003; Nixon, Lazarova, Hodinott-Hill, Gough, & Passingham, 2004).

To summarise, 1984–2004 investigations of phonological memory provide a case where behavioural studies of different populations of neurologically unimpaired subjects and of patients suffering from adult or developmental deficits have provided evidence that has been interpreted in a largely convergent fashion (Vallar & Papagno, 2002). Current knowledge has increased both in the functional and in the neural aspects of the phonological short-term memory system, particularly in the component part of rehearsal and the relationships between short-term storage and long-term learning. If any lesson can be learned from these 20 years, this is that converging operations may increase the probability of relevant developments.

PrEview proof published online 2 September 2005

REFERENCES

Atkinson, R. C., & Shiffrin, R. M. (1968). Human memory: A proposed system and its control processes. In K. W. Spence & J. Taylor Spence (Eds.), *The psychology of learning and motivation: Advances in research and theory, Vol. 2* (pp. 89–195). New York: Academic Press.

Atkinson, R. C., & Shiffrin, R. M. (1971). The control of short-term memory. *Scientific American, 225*, 82–90.

Baddeley, A. D. (1966). Short-term memory for word sequences as a function of acoustic, semantic and formal similarity. *Quarterly Journal of Experimental Psychology, 18*, 362–365.

Baddeley, A. D. (1976). *The psychology of memory.* New York: Basic Books.

Baddeley, A. D. (1993). Short-term phonological memory and long-term learning: A single case study. *European Journal of Cognitive Psychology, 5*, 129–148.

Baddeley, A. D. (2002). The psychology of memory. In A. D. Baddeley, M. D. Kopelman, & B. A. Wilson (Eds.), *The handbook of memory disorders* (2nd ed., pp. 3–15). Chichester, UK: Wiley.

Baddeley, A. D., Gathercole, S., & Papagno, C. (1998). The phonological loop as a language learning device. *Psychological Review, 105*, 158–173.

Baddeley, A. D., & Hitch, G. (1974). Working memory. In G. H. Bower (Ed.), *The psychology of learning and motivation: Advances in research and theory, Vol. 8* (pp. 47–89). New York: Academic Press.

Baddeley, A. D., Lewis, V., & Vallar, G. (1984). Exploring the articulatory loop. *Quarterly Journal of Experimental Psychology, 36A*, 233–252.

Baddeley, A. D., Papagno, C., & Vallar, G. (1988). When long-term learning depends on short-term storage. *Journal of Memory and Language, 27*, 586–595.

Baddeley, A. D., Thomson, N., & Buchanan, M. (1975). Word length and the structure of short-term memory. *Journal of Verbal Learning and Verbal Behavior, 14*, 575–589.

Baddeley, A. D., & Wilson, B. (1985). Phonological coding and short-term memory in patients without speech. *Journal of Memory and Language, 24*, 490–502.

Barisnikov, K., Van der Linden, M., & Poncelet, M. (1996). Acquisition of new words and phonological working memory in Williams syndrome: A case study. *Neurocase, 2*, 395–404.

Basso, A., Spinnler, H., Vallar, G., & Zanobio, M. E. (1982). Left hemisphere damage and selective impairment of auditory-verbal short-term memory. *Neuropsychologia, 20*, 263–274.

Benson, D. F. (1979). *Aphasia, alexia, and agraphia.* New York: Churchill Livingstone.

Besner, D. (1987). Phonology, lexical access in reading, and articulatory suppression: A critical review. *Quarterly Journal of Experimental Psychology, 39A*, 467–478.

Bishop, D. V. M., & Robson, J. (1989a). Accurate non-word spelling despite congenital inability to speak: Phoneme-grapheme conversion does not require subvocal articulation. *British Journal of Psychology, 80*, 1–13.

Bishop, D. V. M., & Robson, J. (1989b). Unimpaired short-term memory and rhyme judgement in congenitally speechless individuals: Implications for the notion of "articulatory coding." *Quarterly Journal of Experimental Psychology, 41A*, 123–140.

Bisiach, E., & Vallar, G. (2000). Unilateral neglect in humans. In F. Boller, J. Grafman, & G. Rizzolatti (Eds.), *Handbook of neuropsychology, Vol. 1* (2nd ed., pp. 459–502). Amsterdam: Elsevier Science.

Burani, C., Vallar, G., & Bottini, G. (1991). Articulatory coding and phonological judgements on written words and pictures: The role of the phonological output buffer. *European Journal of Cognitive Psychology, 3*, 379–398.

Campbell, R., & Butterworth, B. (1985). Phonological dyslexia and dysgraphia in a highly literate subject: A developmental case with associated deficits of phonemic processing and awareness. *Quarterly Journal of Experimental Psychology, 37A*, 435–475.

Caramazza, A. (1986). On drawing inferences about the structure of normal cognitive systems from the analysis of patterns of impaired performance: The case for single-patient studies. *Brain and Cognition, 5*, 41–66.

Close, F., Marten, M., & Sutton, C. (1987). *The particle explosion.* Oxford: Oxford University Press.

Conrad, R. (1964). Acoustic confusions in immediate memory. *British Journal of Psychology, 55*, 75–84.

Conrad, R., & Hull, A. J. (1964). Information, acoustic confusion and memory span. *British Journal of Psychology, 55*, 429–432.

Crowder, R. G. (1976). *Principles of learning and memory.* Hillsdale, NJ: Lawrence Erlbaum Associates Inc.

Cubelli, R., & Nichelli, P. (1992). Inner speech in anarthria: Neuropsychological evidence of differential effects of cerebral lesions on subvocal articulation. *Journal of Clinical and Experimental Neuropsychology, 14*, 499–517.

Editorial. (1984). *Cognitive Neuropsychology, 1*, 1–8.

Ellis, A. W. (1987). Intimations of modularity, or, the modularity of mind: Doing cognitive neuropsychology without syndromes. In M. Coltheart, G. Sartori, & R. Job (Eds.), *The cognitive neuropsychology of language* (pp. 397–408). Hove, UK: Lawrence Erlbaum Associates Ltd.

Frackowiak, R. S. J., Friston, K. J., Frith, C. D., Dolan, R. J., & Mazziotta, J. C. (Eds.) (1997). *Human brain function.* San Diego: Academic Press.

Garner, W. R., Hake, H. W., & Eriksen, C. W. (1956). Operationism and the concept of perception. *Psychological Review, 63*, 149–159.

Gathercole, S. E., & Baddeley, A. D. (1993). *Working memory and language.* Hove, UK: Lawrence Erlbaum Associates Ltd.

Glanzer, M. (1972). Storage mechanisms in recall. In G. H. Bower (Ed.), *The psychology of learning and motivation Advances in research and theory, Vol. 5*, (pp. 129–193). New York: Academic Press.

Hitch, G. J. (1990). Developmental fractionation of working memory. In G. Vallar & T. Shallice (Eds.), *Neuropsychological impairments of short-term memory* (pp. 221–246). Cambridge: Cambridge University Press.

James, W. (1895). *The principles of psychology.* New York: Holt.

Karnath, H. -O., Milner, A. D., & Vallar, G. (Eds.) (2002). *The cognitive and neural bases of spatial neglect.* Oxford: Oxford University Press.

Kitchener, E. G., & Hodges, J. R. (1999). Impaired knowledge of famous people and events with intact autobiographical memory in a case of progressive right temporal lobe degeneration: Implications for the organisation of remote memory. *Cognitive Neuropsychology, 16*, 589–607.

Lecours, A. R., Lhermitte, F., & Bryans, B. (1983). *Aphasiology*. London: Baillière Tindall.

Levy, B. A. (1971). Role of articulation in auditory and visual short-term memory. *Journal of Verbal Learning and Verbal Behavior, 10*, 123–132.

Lichtheim, L. (1885). On aphasia. *Brain, 7*, 433–484.

Locke, J. (1975). *An essay concerning human understanding*. (Original work published 1700) Oxford: Clarendon Press.

Luria, A. R. (1971). Memory disturbances in local brain lesions. *Neuropsychologia, 9*, 367–375.

Luria, A. R. (1977). *Neuropsychological studies in aphasia*. Amsterdam: Swets & Zeitlinger.

Luria, A. R., Sokolov, E. N., & Klimkowski, M. (1967). Towards a neurodynamic analysis of memory disturbances with lesions of the left temporal lobe. *Neuropsychologia, 5*, 1–11.

MacKenzie, S., & Hulme, C. (1987). Memory span development in Down's syndrome severely subnormal and normal subjects. *Cognitive Neuropsychology, 4*, 303–319.

Margulis, L., & Schwartz, K. V. (1998). *Five kingdoms: An illustrated guide to the phyla of life on earth* (3rd ed.). New York: W. H. Freeman.

Marshall, J. C., & Vallar, G. (Eds.) (2004). *Spatial neglect: A representational disorder? A festschrift for Edoardo Bisiach*. The Cortex book series. Milan: Masson.

Mayes, A. R., Isaac, C. L., Holdstock, J. S., Hunkin, N. M., Montaldi, D., Downes, J. J., MacDonald, C., Cezayirli, E., & Roberts, J. N. (2001). Memory for single items, word pairs, and temporal order of different kinds in a patient with selective hippocampal lesions. *Cognitive Neuropsychology, 18*, 97–123.

McKusick, V. A. (1998). *Mendelian inheritance in Man. A catalog of human genes and genetic disorders* (12th ed.). Baltimore, MD: Johns Hopkins University Press.

Melton, A. (1963). Implications of short-term memory for a general theory of memory. *Journal of Verbal Learning and Verbal Behavior, 2*, 1–21.

Milner, B. (1966). Amnesia following operation on the temporal lobes. In C. W. M. Whitty & O. L. Zangwill (Eds.), *Amnesia* (1st ed., pp. 109–133). London: Butterworths.

Morris, R. G. (1986). Short-term forgetting in senile dementia of the Alzheimer's type. *Cognitive Neuropsychology, 3*, 77–97.

Morris, R. G., Miotto, E. C., Feigenbaum, J. D., Bullock, P., & Polkey, C. E. (1997). Planning ability after frontal and temporal lobe lesions in humans: The effects of selection equivocation and working memory load. *Cognitive Neuropsychology, 14*, 1007–1027.

Mottaghy, F. M., Gangitano, M., Krause, B. J., & Pascual-Leone, A. (2003). Chronometry of parietal and prefrontal activations in verbal working memory revealed by transcranial magnetic stimulation. *Neuroimage, 18*, 565–575.

Nebes, R. (1975). The nature of internal speech in anarthria. *Brain and Language, 2*, 489–497.

Nixon, P., Lazarova, J., Hodinott-Hill, I., Gough, P., & Passingham, R. (2004). The inferior frontal gyrus and phonological processing: An investigation using rTMS. *Journal of Cognitive Neuroscience, 16*, 289–300.

Papagno, C., Valentine, T., & Baddeley, A. D. (1991). Phonological short-term memory and foreign language vocabulary learning. *Journal of Memory and Language, 30*, 331–347.

Papagno, C., & Vallar, G. (1992). Phonological short-term memory and the learning of novel words: The effects of phonological similarity and word length. *Quarterly Journal of Experimental Psychology, 44A*, 47–67.

Papagno, C., & Vallar, G. (1995). Verbal short-term memory and vocabulary learning in polyglots. *Quarterly Journal of Experimental Psychology, 48A*, 98–107.

Papagno, C., & Vallar, G. (2001). Understanding metaphors and idioms: A single case neuropsychological study in a subject with Down syndrome. *Journal of the International Neuropsychological Society, 7*, 516–528.

Patterson, K. E., Marshall, J. C., & Coltheart, M. (Eds.) (1985). *Surface dyslexia. Neuropsychological and cognitive studies of phonological reading*. Hove, UK: Lawrence Erlbaum Associates Ltd.

Paulesu, E., Frith, C. D., & Frackowiak, R. S. J. (1993). The neural correlates of the verbal component of working memory. *Nature, 362*, 342–345.

Paulesu, E., Frith, U., Snowling, M., Gallagher, A., Morton, J., Frackowiak, R. S. J., & Frith, C. D. (1996). Is developmental dyslexia a disconnection syndrome? Evidence from PET scanning. *Brain, 119*, 143–157.

Peterson, L. R., & Peterson, M. J. (1959). Short-term retention of individual verbal items. *Journal of Experimental Psychology, 58*, 193–198.

Rubin, S. R., Van Petten, C., Glisky, E. L., & Newberg, W. N. (1999). Memory conjunction errors in younger and older adults: Event-related potential and neuropsychological data. *Cognitive Neuropsychology, 16*, 459–488.

Ruchkin, D. S., Grafman, J., Cameron, K., & Berndt, R. S. (2003). Working memory retention systems: A state of activated long-term memory. *Behavioral and Brain Sciences, 26*, 709–769.

Salamé, P., & Baddeley, A. D. (1982). Disruption of short-term memory by unattended speech: Implications for the structure of working memory. *Journal of Verbal Learning and Verbal Behavior, 21*, 150–164.

Shallice, T., & Butterworth, B. (1977). Short-term memory impairment and spontaneous speech. *Neuropsychologia, 15*, 729–735.

Shallice, T., & Vallar, G. (1990). The impairment of auditory-verbal short-term storage. In G. Vallar & T. Shallice (Eds.), *Neuropsychological impairments of short-term memory* (pp. 11–53). Cambridge: Cambridge University Press.

Shallice, T., & Warrington, E. K. (1970). Independent functioning of verbal memory stores: A neuropsychological study. *Quarterly Journal of Experimental Psychology, 22*, 261–273.

Shallice, T., & Warrington, E. K. (1974). The dissociation between short-term retention of meaningful sounds and verbal material. *Neuropsychologia, 12*, 553–555.

Shallice, T., & Warrington, E. K. (1977). Auditory-verbal short-term memory impairment and conduction aphasia. *Brain and Language, 4*, 479–491.

Silveri, M. C., Cappa, A., & Salvigni, B. L. (2003). Speech and language in primary progressive anarthria. *Neurocase, 9*, 213–220.

Smith, E. E., & Jonides, J. (1999). Storage and executive processes in the frontal lobes. *Science, 283*, 1657–1661.

Sperling, G. (1967). Successive approximations to a model for short-term memory. *Acta Psychologica, 27*, 285–292.

Spinnler, H., Della Sala, S., Bandera, R., & Baddeley, A. D. (1988). Dementia, ageing, and the structure of human memory. *Cognitive Neuropsychology, 5*, 193–211.

Squire, L. R. (1992). Declarative and nondeclarative memory: Multiple brain systems supporting learning and memory. *Journal of Cognitive Neuroscience, 4*, 232–243.

Tulving, E. (1972). Episodic and semantic memory. In E. Tulving & W. Donaldson (Eds.), *Organization of memory* (pp. 381–403). New York: Academic Press.

Vallar, G. (1994). Left spatial hemineglect: An unmanageable explosion of dissociations? No. *Neuropsychological Rehabilitation, 4*, 209–212.

Vallar, G. (1999). Neuropsychological disorders of memory. In G. Denes & L. Pizzamiglio (Eds.), *Handbook of clinical and experimental neuropsychology* (pp. 321–368). Hove, UK: Psychology Press.

Vallar, G., & Baddeley, A. D. (1984a). Fractionation of working memory: Neuropsychological evidence for a phonological short-term store. *Journal of Verbal Learning and Verbal Behavior, 23*, 151–161.

Vallar, G., & Baddeley, A. D. (1984b). Phonological short-term store, phonological processing and sentence comprehension. *Cognitive Neuropsychology, 1*, 121–141.

Vallar, G., & Cappa, S. F. (1987). Articulation and verbal short-term memory. Evidence from anarthria. *Cognitive Neuropsychology, 4*, 55–78.

Vallar, G., Di Betta, A. M., & Silveri, M. C. (1997). The phonological short-term store-rehearsal system: Patterns of impairment and neural correlates. *Neuropsychologia, 35*, 795–812.

Vallar, G., Markowitsch, H. J., Kopelman, M., Miceli, G., & Papagno, C. (2005). The neuropsychology of human memory. *Neurocase, 11*, 151–153.

Vallar, G., & Papagno, C. (1986). Phonological short-term store and the nature of the recency effect. Evidence from neuropsychology. *Brain and Cognition, 5*, 428–442.

Vallar, G., & Papagno, C. (1993). Preserved vocabulary acquisition in Down's syndrome: The role of phonological short-term memory. *Cortex, 29*, 467–483.

Vallar, G., & Papagno, C. (1995). Neuropsychological impairments of short-term memory. In A. D. Baddeley, B. A. Wilson, & F. Watts (Eds.), *Handbook of memory disorders* (pp. 135–165). Chichester, UK: John Wiley.

Vallar, G., & Papagno, C. (2002). Neuropsychological impairments of verbal short-term memory. In A. Baddeley, B. Wilson & M. Kopelman (Eds.), *Handbook of memory disorders* (2nd ed., pp. 249–270). Chichester, UK: John Wiley.

Vygotsky, L. (1962). *Thought and language.* (Original work published 1934) Boston, MA: MIT Press.

Warrington, E. K., Logue, V., & Pratt, R. T. C. (1971). The anatomical localisation of selective impairment of auditory verbal short-term memory. *Neuropsychologia, 9*, 377–387.

Warrington, E. K., & Shallice, T. (1969). The selective impairment of auditory-verbal short-term memory. *Brain, 92*, 885–896.

Warrington, E. K., & Shallice, T. (1972). Neuropsychological evidence of visual storage in short-term memory tasks. *Quarterly Journal of Experimental Psychology, 24*, 30–40.

Waugh, N., & Norman, D. A. (1965). Primary memory. *Psychological Review, 72*, 89–1104.

Weiskrantz, L. (1990). Problems of learning and memory: One or multiple memory systems? *Philosophical Transaction of the Royal Society of London, 329B*, 99–108.

Wernicke, C. (1966–1968). *The symptom complex of aphasia.* (Original work published 1874.) Paper presented at the Boston Studies in the Philosophy of Science Proceedings of the Boston Colloquium for the Philosophy of Science.

Wickelgren, W. A. (1965). Short-term memory for phonemically similar lists. *American Journal of Psychology, 78*, 567–574.

Zani, A., & Proverbio, A. M. (Eds.). (2002). *The cognitive electrophysiology of mind and brain.* Amsterdam: Academic Press, Elsevier Science.

APPENDIX A

1984–1997 original papers published in *Cognitive Neuropsychology*, mainly concerned with *memory*.

1. Buxbaum, L. J., Schwartz, M. F., & Carew, T. G. (1997). The role of semantic memory in object use. *Cognitive Neuropsychology, 14*, 219–254.
2. Caplan, D., & Waters, G. S. (1995). Aphasic disorders of syntactic comprehension and working-memory capacity. *Cognitive Neuropsychology, 12*, 637–649.
3. Chertkow, H., Bub, D., & Caplan, D. (1992). Constraining theories of semantic memory processing—evidence from dementia. *Cognitive Neuropsychology, 9*, 327–365.
4. Cossu, G., & Marshall, J. C. (1990). Are cognitive skills a prerequisite for learning to read and write? *Cognitive Neuropsychology, 7*, 21–40.
5. Dannenbaum, S. E., Parkinson, S. R., & Inman, V. W. (1988). Short-term forgetting: Comparisons between patients with dementia of the Alzheimer type, depressed, and normal elderly. *Cognitive Neuropsychology, 5*, 213–233.
6. Forde, E. M. E., & Humphreys, G. W. (1997). A semantic locus for refractory behaviour: Implications for access-storage distinctions and the nature of semantic memory. *Cognitive Neuropsychology, 14*, 367–402.
7. Goulandris, N. K., & Snowling, M. (1991). Visual memory deficits—a plausible cause of developmental dyslexia—evidence from a single case-study. *Cognitive Neuropsychology, 8*, 127–154.
8. Hillis, A. E., Rapp, B., & Caramazza, A. (1995). Constraining claims about theories of semantic memory—More on unitary versus multiple semantics. *Cognitive Neuropsychology, 12*, 175–186.
9. Hodges, J. R., Patterson, K., & Tyler, L. K. (1994). Loss of semantic memory—implications for the modularity of mind. *Cognitive Neuropsychology, 11*, 505–542.
10. Howard, D., & Butterworth, B. (1989). Short-term memory and sentence comprehension: A reply to Vallar and Baddeley 1987. *Cognitive Neuropsychology, 6*, 455–463.
11. Kapur, N. (1994). Remembering Norman Schwarzkopf: Evidence for two distinct long-term fact learning-mechanisms. *Cognitive Neuropsychology, 11*, 661–670.
12. Knott, R., Patterson, K., & Hodges, J. R. (1997). Lexical and semantic binding effects in short-term memory: Evidence from semantic dementia. *Cognitive Neuropsychology, 14*, 1165–1218.
13. Kopelman, M. D., Ng, N., & VandenBrouke, O. (1997). Confabulation extending across episodic, personal, and general semantic memory. *Cognitive Neuropsychology, 14*, 683–712.

14. MacKenzie, S., & Hulme, C. (1987). Memory span development in Down's syndrome severely subnormal and normal subjects. *Cognitive Neuropsychology*, *4*, 303–319.
15. Marcel, A. J. (1990). What does it mean to ask whether cognitive skills are prerequisite for learning to read and write: A response. *Cognitive Neuropsychology*, *7*, 41–48.
16. Martin, N., & Saffran, E. M. (1997). Language and auditory-verbal short-term memory impairments: Evidence for common underlying processes. *Cognitive Neuropsychology*, *14*, 641–682.
17. Martin, R. C. (1995). Working-memory doesn't work—a critique of Miyake et al.'s capacity theory of aphasic comprehension deficits. *Cognitive Neuropsychology*, *12*, 623–636.
18. Martin, R. C., & Breedin, S. D. (1992). Dissociations between speech-perception and phonological short-term-memory deficits. *Cognitive Neuropsychology*, *9*, 509–534.
19. Morris, R. G. (1984). Dementia and the functioning of the articulatory loop system. *Cognitive Neuropsychology*, *1*, 143–157.
20. Morris, R. G. (1986). Short-term forgetting in senile dementia of the Alzheimer's type. *Cognitive Neuropsychology*, *3*, 77–97.
21. Morris, R. G., Miotto, E. C., Feigenbaum, J. D., Bullock, P., & Polkey, C. E. (1997). Planning ability after frontal and temporal lobe lesions in humans: The effects of selection equivocation and working memory load. *Cognitive Neuropsychology*, *14*, 1007–1027.
22. Morton, N., & Morris, R. G. (1995). Image transformation dissociated from visuospatial working-memory. *Cognitive Neuropsychology*, *12*, 767–791.
23. Mutter, S. A., Howard, D. V., Howard, J. H., & Wiggs, C. L. (1990). Performance on direct and indirect tests of memory after mild closed head-injury. *Cognitive Neuropsychology*, *7*, 329–346.
24. Nicolas, S. (1996). Experiments on implicit memory in a Korsakoff patient by Claparede (1907). *Cognitive Neuropsychology*, *13*, 1193–1199.
25. Quinlan, P. T. (1987). Theoretical notes on "Parallel models of associative memory". *Cognitive Neuropsychology*, *4*, 333–364.
26. Richardson, J. T., & Barry, C. (1985). The effects on minor closed head injury upon human memory: Further evidence on the role of mental imagery. *Cognitive Neuropsychology*, *2*, 149–168.
27. Richardson, J. T., & Snape, W. (1984). The effects of closed head injury upon human memory: An experimental analysis. *Cognitive Neuropsychology*, *1*, 217–231.
28. Snowden, J. S., Griffiths, H. L., & Neary, D. (1996). Semantic-episodic memory interactions in semantic dementia: Implications for retrograde memory function. *Cognitive Neuropsychology*, *13*, 1101–1137.
29. Spinnler, H., Della Sala, S., Bandera, R., & Baddeley, A. D. (1988). Dementia, ageing, and the structure of human memory. *Cognitive Neuropsychology*, *5*, 193–211.
30. Vallar, G., & Baddeley, A. D. (1984). Phonological short-term store, phonological processing and sentence comprehension. *Cognitive Neuropsychology*, *1*, 121–141.
31. Vallar, G., & Baddeley, A. D. (1987). Phonological short-term store and sentence processing. *Cognitive Neuropsychology*, *4*, 417–438.
32. Vallar, G., & Cappa, S. F. (1987). Articulation and verbal short-term memory. Evidence from anarthria. *Cognitive Neuropsychology*, *4*, 55–78.
33. Vallar, G., & Baddeley, A. D. (1989). Developmental disorders of verbal short-term memory and their relation to sentence comprehension: A reply to Howard and Butterworth. *Cognitive Neuropsychology*, *6*, 465–473.
34. VanderLinden, M., Bredart, S., Depoorter, N., & Coyette, F. (1996). Semantic memory and amnesia: A case study. *Cognitive Neuropsychology*, *13*, 391–413.
35. Vanderlinden, M., Coyette, F., & Seron, X. (1992). Selective impairment of the central executive component of working memory—a single case-study. *Cognitive Neuropsychology*, *9*, 301–326.

36. Waters, G., Caplan, D., & Hildebrandt, N. (1991). On the structure of verbal short-term memory and its functional role in sentence comprehension—evidence from neuropsychology. *Cognitive Neuropsychology*, *8*, 81–126.
37. Willshaw, D. (1990). Ceded memories. Review of *Sparse distributed memory* by P. Kanerva. *Cognitive Neuropsychology*, *7*, 245–246.

APPENDIX B

1998–2004 original papers published in *Cognitive Neuropsychology*, mainly concerned with *memory*.

1. Balota, D. A., Cortese, M. J., Duchek, J. M., Adams, D., Roediger, H. L., McDermott, K. B., & Yerys, B. E. (1999). Veridical and false memories in healthy older adults and in dementia of the Alzheimer's type. *Cognitive Neuropsychology*, *16*, 361–384.
2. Coccia, M., Bartolini, M., Luzzi, S., Provinciali, L., & Ralph, M. A. L. (2004). Semantic memory is an amodal, dynamic system: Evidence from the interaction of naming and object use in semantic dementia. *Cognitive Neuropsychology*, *21*, 513–527.
3. Dalla Barba, G., Nedjam, Z., & Dubois, B. (1999). Confabulation, executive functions, and source memory in Alzheimer's disease. *Cognitive Neuropsychology*, *16*, 385–398.
4. Dodhia, R. M., & Metcalfe, J. (1999). False memories and source monitoring. *Cognitive Neuropsychology*, *16*, 489–508.
5. Epstein, R., DeYoe, E. A., Press, D. Z., Rosen, A. C., & Kanwisher, N. (2001). Neuropsychological evidence for a topographical learning mechanism in parahippocampal cortex. *Cognitive Neuropsychology*, *18*, 481–508.
6. Farah, M. J., & Rabinowitz, C. (2003). Genetic and environmental influences on the organisation of semantic memory in the brain: Is "living things" an innate category? *Cognitive Neuropsychology*, *20*, 401–408.
7. Freedman, M. L., & Martin, R. C. (2001). Dissociable components of short-term memory and their relation to long-term learning. *Cognitive Neuropsychology*, *18*, 193–226.
8. Freedman, M. L., Martin, R. C., & Biegler, K. (2004). Semantic relatedness effects in conjoined noun phrase production: Implications for the role of short-term memory. *Cognitive Neuropsychology*, *21*, 245–265.
9. Funnell, E. (2001). Evidence for scripts in semantic dementia: Implications for theories of semantic memory. *Cognitive Neuropsychology*, *18*, 323–341.
10. Gagnon, S., Foster, J. K., Turcotte, J., & Jongenelis, S. (2004). Involvement of the hippocampus in implicit learning of supra-span sequences: The case of SJ. *Cognitive Neuropsychology*, *21*, 867–882.
11. Haslam, C., Cook, M. L., & McKone, E. (1998). Memory for generalities: Access to higher-level categorical relationships in amnesia. *Cognitive Neuropsychology*, *15*, 401–437.
12. Humphreys, G. W., & Rumiati, R. I. (1998). Agnosia without prosopagnosia or alexia: Evidence for stored visual memories specific to objects. *Cognitive Neuropsychology*, *15*, 243–277.
13. Kay, J., & Hanley, J. R. (2002). Preservation of memory for people in semantic memory disorder: Further category-specific semantic dissociation. *Cognitive Neuropsychology*, *19*, 113–133.
14. Keenan, J. M., & Simon, J. A. (2004). Inference deficits in women with Fragile X Syndrome: A problem in working memory. *Cognitive Neuropsychology*, *21*, 579–596.
15. Kensinger, E. A., & Schacter, D. L. (1999). When true memories suppress false memories: Effects of ageing. *Cognitive Neuropsychology*, *16*, 399–415.
16. Kitchener, E. G., & Hodges, J. R. (1999). Impaired knowledge of famous people and events with intact autobiographical memory in a case of progressive right temporal lobe degeneration: Implications for the organisation of remote memory. *Cognitive Neuropsychology*, *16*, 589–607.

17. Kopelman, M. D. (1999). Varieties of false memory. *Cognitive Neuropsychology, 16*, 197–214.
18. Kumada, T., & Humphreys, G. W. (2001). Lexical recovery from extinction: Interactions between visual form and stored knowledge modulate visual selection. *Cognitive Neuropsychology, 18*, 465–478.
19. Majerus, S., Lekeu, F., Van der Linden, M., & Salmon, E. (2001). Deep dysphasia: Further evidence on the relationship between phonological short-term memory and language processing impairments. *Cognitive Neuropsychology, 18*, 385–410.
20. Majerus, S., Van der Linden, M., Poncelet, M., & Metz-Lutz, M. N. (2004). Can phonological and semantic short-term memory be dissociated? Further evidence from Landau-Kleffner syndrome. *Cognitive Neuropsychology, 21*, 491–512.
21. Marques, J. F. (2000). The "living things" impairment and the nature of semantic memory organisation: An experimental study using PI-release and semantic cues. *Cognitive Neuropsychology, 17*, 683–707.
22. Marques, J. F. (2002). An attribute is worth more than a category: Testing different semantic memory organisation hypotheses in relation to the living/nonliving things dissociation. *Cognitive Neuropsychology, 19*, 463–478.
23. Martin, N., & Gupta, P. (2004). Exploring the relationship between word processing and verbal short-term memory: Evidence from associations and dissociations. *Cognitive Neuropsychology, 21*, 213–228.
24. Mayes, A. R., Isaac, C. L., Holdstock, J. S., Hunkin, N. M., Montaldi, D., Downes, J. J., MacDonald, C., Cazayirli, E., & Roberts, J. N. (2001). Memory for single items, word pairs, and temporal order of different kinds in a patient with selective hippocampal lesions. *Cognitive Neuropsychology, 18*, 97–123.
25. Moss, H. E., Kopelman, M. D., Cappelletti, M., Davies, P. D., & Jaldow, E. (2003). Lost for words or loss of memories? Autobiographical memory in semantic dementia. *Cognitive Neuropsychology, 20*, 703–732.
26. Parkin, A. J., Hunkin, N. M., & Squires, E. J. (1998). Unlearning John Major: The use of errorless learning in the reacquisition of proper names following herpes simplex encephalitis. *Cognitive Neuropsychology, 15*, 361–375.
27. Piolino, P., Belliard, S., Desgranges, B., Perron, M., & Eustache, F. (2003). Autobiographical memory and autoneotic consciousness in a case of semantic dementia. *Cognitive Neuropsychology, 20*, 619–639.
28. Riddoch, M. J., Humphreys, G. W., Blott, W., Hardy, E., & Smith, A. D. (2003). Visual and spatial short-term memory in integrative agnosia. *Cognitive Neuropsychology, 20*, 641–671.
29. Rubin, S. R., Van Petten, C., Glisky, E. L., & Newberg, W. N. (1999). Memory conjunction errors in younger and older adults: Event-related potential and neuropsychological data. *Cognitive Neuropsychology, 16*, 459–488.
30. Schacter, D. L. (1999). The cognitive neuropsychology of false memories: Introduction. *Cognitive Neuropsychology, 16*, 193–195.
31. Shalev, L., & Humphreys, G. W. (2002). Implicit location encoding via stored representations of familiar objects: Neuropsychological evidence. *Cognitive Neuropsychology, 19*, 721–744.
32. Swinnen, S. P., Verschueren, S. M. P., Bogaerts, H., Dounskaia, N., Lee, T. D., Stelmach, G. E., & Serrien, D. J. (1998). Age-related deficits in motor learning and differences in feedback processing during the production of a bimanual coordination pattern. *Cognitive Neuropsychology, 15*, 439–466.
33. Tippett, L. J., Miller, L. A., & Farah, M. J. (2000). Prosopamnesia: A selective impairment in face learning. *Cognitive Neuropsychology, 17*, 241–255.
34. Treves, A., & Samengo, I. (2002). Standing on the gateway to memory: Shouldn't we step in? *Cognitive Neuropsychology, 19*, 557–575.

35. Turnbull, O. H., & Laws, K. R. (2000). Loss of stored knowledge of object structure: Implications for "category-specific" deficits. *Cognitive Neuropsychology*, *17*, 365–389.
36. Westmacott, R., Freedman, M., Black, S. E., Stokes, K. A., & Moscovitch, M. (2004). Temporally graded semantic memory loss in Alzheimer's disease: Cross-sectional and longitudinal studies. *Cognitive Neuropsychology*, *21*, 353–378.
37. Westmacott, R., & Moscovitch, M. (2002). Temporally graded semantic memory loss in amnesia and semantic dementia: Further evidence for opposite gradients. *Cognitive Neuropsychology*, *19*, 135–163.
38. Whalen, J., McCloskey, M., Lindemann, M., & Bouton, G. (2002). Representing arithmetic table facts in memory: Evidence from acquired impairments. *Cognitive Neuropsychology*, *19*, 505–522.

COGNITIVE NEUROPSYCHOLOGY, 2006, 23 (1), 156–183

Features, objects, action: The cognitive neuropsychology of visual object processing, 1984–2004

Glyn W. Humphreys and M. Jane Riddoch
University of Birmingham, Birmingham, UK

We review evidence on the cognitive neuropsychology of visual object processing, from 1984–2004, dividing the work according to whether it deals with the analysis of visual features, objects, or the relations between object processing and action. Research across this period has led to (1) a more detailed analysis of disorders of feature processing and feature binding, (2) a finer-grained understanding of disorders of object recognition, how these disorders can change over time, and their relations to visual imagery, and (3) new accounts of the relations between vision and action. Cognitive neuropsychological studies have played a key part in furthering our understanding of the functional nature of object processing in the brain.

INTRODUCTION

The early 1980s saw considerable advances in our understanding of how visual processing operates in the brain. Several of these advances came through physiological studies using single cell recording or lesion techniques to investigate the basic properties of neurons in different subcortical and cortical regions (e.g., Cowey, 1979, 1985; Zeki, 1978). Other advances were linked to the development of more sophisticated computer vision systems, whose architectures could provide frameworks for understand biological vision (e.g., Marr, 1982). Cognitive neuropsychological studies prior to this time had served as the impetus for some of the theories of computer vision (e.g., Warrington & Taylor, 1973, 1978), but detailed analyses of patients in relation to these theories were yet to develop. In the subsequent 20 years of research, our analysis of neuropsychological disorders of vision has become increasingly sophisticated. We have begun to understand not only how the world is decomposed along contrasting visual dimensions, but also how the features coded through this decomposition are subsequently "bound together" so that we perceive a coherent environment. We have been able to fractionate between different recognition disorders, and to analyse the relations between disorders affecting different stimuli (objects, face, words). We have also moved beyond thinking of vision solely in terms of object recognition (semantic classification, naming), with many studies now dealing with the use of vision to guide and select actions to objects—processes that had received relatively little attention up to 1984, despite the fact that "vision for action" is fundamental for our survival. In this article we present

Correspondence should be addressed to Professor Glyn W. Humphreys, Behavioural Brain Sciences Centre, School of Psychology, University of Birmingham, Birmingham B15 2TT, UK (Email: g.w.humphreys@bham.ac.uk).
This work was supported by grants from the Medical Research Council and the Biology and Biotechnology Research Council (UK), the Stroke Association, and the Wellcome Trust.

DOI:10.1080/02643290542000030

an outline of the state of knowledge in the early 1980s, before proceeding to review some of the advances in the cognitive neuropsychology of visual object processing that have taken place subsequently. We argue that cognitive neuropsychological studies of vision have provided crucial insights into the nature of visual object processing in the brain.

THE COGNITIVE NEUROPSYCHOLOGY OF VISUAL OBJECT PROCESSING IN 1984

As is the case for much of cognitive neuropsychology, many of the basic neuropsychological disorders of visual object processing were documented in early neurological case studies conducted in the late nineteenth and early twentieth centuries. Disorders of visual object recognition, occurring alongside relatively intact sensory processing (visual agnosia), had been discussed since the time of Lissauer (1890). Disorders apparently affecting visual object naming more than recognition were noted by Freund (1889), and disorders that limited both the perception of multiple stimuli and visually guided action were reported by Balint (1909) and Holmes (1918). Holmes also noted patients who seemed to suffer processing deficits for particular visual features, such as depth, whilst cortical losses of colour vision were discussed by Quaglino (1867) and Verrey (1888). The apparent preserved processing of some types of information (such as movement) without apparent perception of other attributes of the stimulus was reported by G. Riddoch (1917), whilst disorders more specific to some types of visual stimulus than others (e.g., faces compared with objects) were highlighted by Wigan (1844) and Quaglino (1867). However, whilst numerous other case studies of patients with these and similar disorders had been documented by 1984, few studies had gone beyond either the original case descriptions or the theoretical frameworks that had been originally provided. This is perhaps best illustrated by considering the disorder of agnosia.

Lissauer (1890) first put forward a framework for understanding agnosia in terms of impairments to either of two serial and hierarchically arranged stages of visual object recognition: *apperception* and *association*. He proposed that the apperceptive stage was concerned with assembling a stable perceptual representation of the stimulus, whereas the association stage was concerned with linking the perceptual representation (formed at the apperceptive stage) to stored knowledge. Patients with a pure associative agnosia should have intact perception of a stimulus but they may fail to recognise it because the association process is disrupted. This two-part framework remained the standard view of agnosia in 1984 (e.g., Albert, Reches, & Silverberg, 1975), and, as in 1890, the distinction between apperceptive and associative disorders was still grounded on tests such as copying (e.g., see Ratcliff & Newcombe, 1982). Patients were deemed to have associative agnosia if they could copy objects that they failed to recognise (e.g., as shown by impaired object naming or impaired gesturing to the object). Defining apperceptive agnosia, however, was more controversial, as it is intrinsically difficult to rule out the contribution of sensory deficits in such cases. Efron (1968; Benson & Greenberg, 1969) devised forms of a shape-matching test (Figure 1) where the design ruled out patients basing their performance on some covarying property, such as brightness. Failure on "Efron shape-matching" has been taken as diagnostic of poor coding of shape representations. However, the problem remained in demonstrating that there was not a basic sensory loss in such cases, and indeed some theorists doubted the existence of pure agnosia isolated from either sensory loss (e.g., in apperceptive cases) or loss of stored knowledge (e.g., in associative cases) (Bender & Feldman, 1972). This last view expressed clear scepticism that

Figure 1. *Example stimuli from the Efron shape test.*

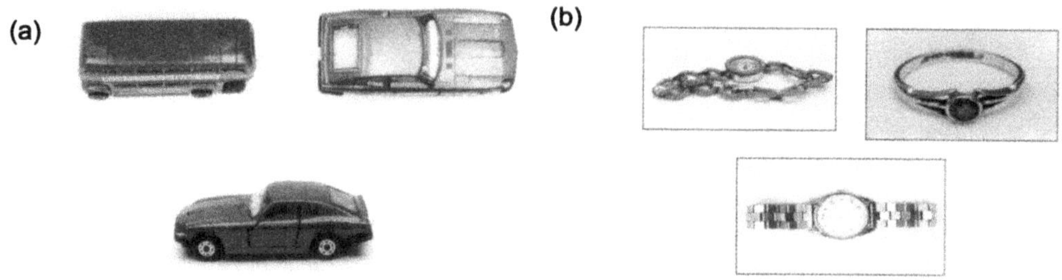

Figure 2. *(a) Example stimuli from an unusual view matching task. Participants see a target object depicted from a prototypical view (the lower picture of the car), along with two reference objects (upper pictures). One of the reference stimuli is the same object as the target, but depicted from an unusual view. The task is to match the target to the reference picture showing the same object but from a different viewpoint. (b) Example stimuli from a function match task. Here the target object (lower picture) has to be matched to the reference stimulus (upper pictures) that is the same kind of object (e.g., two watches) (following Warrington & Taylor, 1978).*

neuropsychological studies could contribute to our understanding of "intermediate" visual processes, intervening between the coding of basic sensory properties (impaired in apperceptive agnosia) and the retrieval of stored knowledge (impaired in associative agnosia).

One notable exception to this was the research conducted by Elizabeth Warrington and colleagues in the 1970s (e.g., Warrington & James, 1967; Warrington & Rabin, 1970; Warrington & Taylor, 1973, 1978; see Warrington, 1982, for a summary). This work took a localisationist approach to object processing, being concerned primarily with understanding the role of particular brain areas in vision. Warrington and colleagues reported data from groups of patients with lesions to the right posterior parietal lobe who were impaired in a variety of visual perception tasks, although the same lesions do not necessarily disrupt the ability of the patients to make discriminations based on more basic properties of stimuli, such as line orientation or size (e.g., Humphreys & Riddoch, 1984). In addition, the same investigators contrasted patients with posterior right-hemisphere lesions and patients with posterior left-hemisphere lesions on matching tasks involving: (1) different views of the same object (with one view being prototypical and the other unusual), and (2) functionally equivalent but different object exemplars, each seen from prototypical views (e.g., an armchair and a deckchair)(see Figure 2; also Warrington & Taylor, 1978). Relative to controls, the right-hemisphere patients were impaired in task (1) (so-called "unusual views matching") but not task (2) ("function matches"). The opposite result was found for the left-hemisphere lesioned patients.

These findings are informative because they indicate that the processes involved in object processing can fractionate. For example, there can be selective disturbance of processes involved in matching objects across different views without concomitant loss of more basic visual perception in right-hemisphere lesioned patients—contrary to the idea that there is a necessary sensory loss in such cases. Furthermore, the ability to match objects across views can be dissociated from the ability to access information about an object's function. In right hemisphere cases, there is intact access to functional knowledge provided the object appears in a prototypical view, but poor ability to match objects in a view-invariant manner. In left hemisphere cases, view-invariant matching can take place but there can be impaired access to knowledge about object function, irrespective of the view of the object. This last result was influential on the development of Marr's (1982) views about computer vision, since he used it to propose that high-level visual processes could encode a view-invariant representation prior to access to stored knowledge (stored knowledge being needed for function matching with visually different exemplars). According to Marr, this view-invariant representation allowed objects to be matched across different views. The dissociation between unusual views and function matching also puts some "flesh" on the skeletal framework that distinguishes apperceptive from association processes in vision. It indicates that apperception can include processes that support view invariance. The work of Warrington and colleagues also suggested that these two processes, view-invariant object coding and access to stored functional knowledge, were localised in different hemispheres.

One other combined functional and anatomical distinction, which provided a framework for understanding visual processing in 1984, was between "object" and "space" vision—between coding "what" and "where" an object was. This distinction was motivated initially from physiological work on lesion effects and single cell response properties for neurons within, respectively, the "ventral" (occipito-temporal) and "dorsal" (occipito-parietal) pathways (see Ungerleider & Mishkin, 1982). The distinction matched the lesion data for some of the well-known human clinical syndromes, including visual agnosia (linked to ventral lesions) and more apparently spatial disorders such as unilateral neglect (linked to more dorsal lesions). Since 1984 the distinction has been elaborated to provide a somewhat different view of the processes characterising ventral and more dorsal visual pathways. We discuss this in the third main section of our article, on "Action".

FROM 1984 TO 2004

As we noted in our Introduction, the early 1980s saw increasing contributions to our understanding of vision from a wide variety of sources, from neurophysiology to computer vision. It is our conjecture that this converging research has helped to advance cognitive neuropsychological studies from that time onwards. The study of what we might think of as lower-level processes, such as the coding of basic sensory attributes of visual stimuli, has been informed by neurophysiological analyses of sensory coding in cortical and subcortical structures. The study of what we will term intermediate vision (grouping and perceptual organisation of visual elements, and the achievement of view invariance) and higher-level vision (processes involved in accessing stored knowledge about stimuli) has been informed by theories coming from cognitive psychology and computer modelling. We elaborate below on the contributions that have followed from this converging approach. In addition, new knowledge has flowed from several clinical insights that had not been highlighted in the classic neurological literature, with analyses of single cases playing an important part in this process.[1] This knowledge, harnessed to the more detailed theories that characterise present-day neuropsychology, provides the framework for current thinking in this area.

To provide an overview of the advances made since 1984, we will divide our analysis of the field into three main sections, dealing with: (1) the processing of features (including: basic feature coding, feature binding, and the relations between these processes and conscious perception), (2) the processing of objects (including: intermediate processes in grouping and in view-invariant coding, higher-order processes in accessing stored knowledge, and differences in processing different visual objects), and (3) the use of vision for action (including: the computations for prehensile action and effects of object "affordances"). Where relevant, we will also attempt to link the neuropsychological studies to explicit models of normal performance, which

[1] Single case studies may have played a particularly important role in understanding disorders of early processes in object recognition because these processes are likely to be bilaterally represented, so that patients would need to have bilateral damage in order for a clinical deficit to become apparent. Given the relative rarity of isolated bilateral lesions, there are pragmatic constraints on group studies.

enable a more over-arching picture of vision to be developed.

Feature-processing

Coding basic visual features. The visual world is composed of multiple dimensions, including shape (and basic components of shape, such as edge orientations), colour, movement, depth, surface texture, and so forth. One of the contributions of physiological work in the early 1980s was to suggest that such separate dimensions may be processed in a quasi-independent manner, in different neural areas. For example, from the retina onwards, different cells may be specialised for carrying information about shape, colour, and movement, and, in areas of visually responsive cortex, there can be a subdivision of regions where cells respond selectively to shape detail (V2, V3), colour (V4), and motion (areas MT and MST; see Cowey, 1985; Zeki, 1993; amongst others). As we have noted, early neurological case studies indicated that patients could have selective deficits in processing particular visual features, such as shape (Lissauer, 1890), colour (Verrey, 1888), and depth (Holmes, 1918), but the specificity of these deficits was often unclear. From the early 1980s onwards, however, more detailed case studies have been conducted showing that some of these feature-processing deficits can be quite specific. For example, Heywood, Wilson, and Cowey (1987) demonstrated that basic brightness discrimination can be relatively spared in cerebral achromatopsia, suggesting that colour and brightness perception operate along separable pathways. Similarly, impaired colour vision can exist alongside relatively good motion vision (Humphreys & Riddoch, 1987). The opposite pattern was first documented by Zihl, Von Cramon, and Mai (1983), who reported a patient with disrupted motion vision alongside spared colour and form perception (see also Vaina, Cowey, Lemay, Bienfang, & Kikinis, 2002). Battelli, Casco, and Sartori (1997) presented a case of a patient with impaired perception of the texture of objects, along with relatively preserved processing of object shape, and Ackroyd (2003) documented a double dissociation between form and depth perception (respectively being selectively impaired or spared, in different patients). Thus our realisation of the specificity of early feature processing has increased. The data on the selective lesions in patients with problems in basic feature coding fits with neurophysiological and brain imaging studies showing specialised areas for colour motion, etc. (e.g., Lueck et al., 1989; Zeki, 1993; Zeki, Watson, Lueck, Friston, Kennard, & Frackowiak, 1991).

In addition, selective impairments affecting subtypes of feature processing have also been revealed. For example, within the motion domain, Vaina and Cowey (1996; Vaina, Cowey, & Kennedy, 1999; Vaina, Makris, Kennedy, & Cowey, 1998) have noted selective dissociations between the ability to process so-called first- and second-order motion—first-order motion being dependent on luminance differences between signals and second-order motion being based on contrast differences rather than absolute luminance levels. These investigators have also shown that patients can be impaired at extracting shape from motion, despite having both intact motion perception and intact perception of static forms (Cowey & Vaina, 2000). In addition, the ability to perceive biological motion (e.g., from a moving human light-point figure) can dissociate from the ability to perceive local motion and even form from motion. Patients with poor motion perception may nevertheless perceive biological motion (McLeod, Dittrich, Driver, Perrett, & Zihl, 1996), whilst patients with intact local motion perception may fail to discriminate biological motion (Battelli, Cavanagh, & Thornton, 2003).

Within the colour domain, Troscianko et al. (1996) have contrasted different forms of cerebral achromatopsia. They suggest that some patients respond using residual information about surface colour within a specialised channel (using "parvocellular" pathways in the cortex), whilst others respond using boundary contrast information provided by a "magnocellular" pathway, not specialised for colour perception. These patients may be distinguished by examining the effects of static and flickering luminance noise "masking" on their ability to match colours. Troscianko et al. contrasted one achromatopsic patient who was affected by static luminance noise but not rapid flicker, with another who showed the opposite pattern. The first patient was presumed to be using a residual parvocellular pathway (affected by static noise but not rapid flicker), the second to be using a magnocellular pathway (affected by rapid flicker but not luminance noise). These studies demonstrate that the "grain of analysis" for feature processing disorders has increased, providing new information about the independence of feature analysers (e.g., for first- and second-order motion) and the different types of information being computed (e.g., colour boundaries vs. surface colours). We deal with contrasting disorders of shape coding in the following section, "Objects".

Binding visual features. If visual features are processed in a quasi-independent manner, then a major question for vision researchers to understand is how the features are combined to generate a perception of a coherent world—where colours are tied to particular shapes moving in particular directions. Some insights into this binding process come from patients who seem to be impaired at linking together information from different feature dimensions. Friedman-Hill, Robertson, and Treisman (1995) reported data from a patient with Balint's syndrome (following Balint, 1909) who had sustained bilateral damage to his posterior parietal cortices. Balint's syndrome is comprised of two major clinical symptoms—simultanagnosia (an apparent ability only to see a single object at a time) and optic ataxia (poor reaching to visual targets, relative to targets defined via other modalities, such as proprioception). Friedman-Hill et al. noted that their patient made many "illusory conjunctions" in which he interchanged shapes and colours when presented with multiple letters. Illusory conjunctions have been reported in normal observers when the conditions prevent the stimuli from being fully attended (e.g., with briefly presented displays and participants dividing their attention across the stimuli; see Prinzmetal, 1981; Treisman & Schmidt, 1982). The interesting aspect of Friedman-Hill et al.'s patient is that these errors occurred even when the stimuli were presented for prolonged exposures. Similar data from another patient with Balint's syndrome have been reported by Humphreys, Cinel, Wolfe, Olson, and Klempen (2000), and, for stimuli presented in contralesional space, increased illusory conjunctions have been documented after unilateral parietal damage (Cohen & Rafal, 1991) and after unilateral pulvinar lesions (Ward, Danziger, Owen, & Rafal, 2002).

There have been at least two major accounts proposed for the way that visual elements become bound after independent feature analysis (see Robertson, 2003, for a recent review). One proposal, supported by physiological data on neuronal firing patterns, is that binding depends on the temporal synchrony of neuronal firing between features that belong to the same object (e.g., Singer & Gray, 1995). The temporal signal provided by synchronised activity in neurons responsive to different features indicates that the features "belong together." The second proposal, expressed in Treisman's Feature Integration Theory (FIT; e.g., Treisman, 1998) is that features are bound by spatial attention. By attending to the location where the stimuli fall, the activation levels for these features will be increased and those for unattended features suppressed (cf. Moran & Desimone, 1985). This solves the "binding problem" because it prevents bindings between the "enhanced" features of the attended object and the "suppressed" features of unattended objects. According to FIT, the bilateral parietal damage suffered by patients with Balint's syndrome means that they have problems in focusing spatial attention on one target when there are multiple items present. As a consequence, features from different objects may become combined inappropriately. Unilateral parietal and pulvinar damage may disrupt attention in much the same way, but only to contralesional stimuli. According to the argument for temporal synchrony, parietal (and pulvinar) damage may disrupt feedback signals to early visual cortex; such signals normally help to produce synchronous firing patterns for features at a common location and/or belonging to one object. A somewhat different account of both the binding process and of linked neuropsychological disorders is that binding is dependent on a process of co-localisation, where features from different dimensions are registered in a common spatial map (probably represented in posterior parietal cortex; see Humphreys et al., 2000; cf. Ungerleider & Mishkin, 1982). Parietal (and pulvinar) damage disrupts this co-registration process. As a consequence, features from different dimensions may be misaligned, with the colour (say) of one stimulus assigned to the location where the shape of a different item is registered. Some support for this comes from the data on the effects of pulvinar damage reported by Ward et al. (2002). They found that illusory conjunctions arose on occasions where their patient mislocalised the shape of a stimulus.

Of course, these different accounts of the binding process are not necessarily mutually exclusive. For example, focusing attention on one object may be a means of enhancing temporal synchrony in neuronal firing, and an accurate map of stimulus locations is also a prerequisite for focused spatial attention (see Robertson, 2003). However, whilst FIT emphasises that binding operates only at one attended location at a time, the temporal synchrony and co-localisation accounts are not constrained in this way; for these accounts binding may operate simultaneously on at least a small number of objects and features (see Luck & Vogel, 1997; Wheeler & Treisman, 2002). FIT also suggests that attention to an object is necessary for binding. As we discuss below (see also *Feature coding, binding, and awareness*), there is evidence against this.

Other recent neuropsychological data indicate that we may need to distinguish between different forms of binding in vision. Most relevant here are findings from the phenomenon of extinction, classically associated with parietal damage (see Karnath, 1988). Extinction occurs when a patient can report a single object on the contralesional side of space, but fails to report the same item when another (competing) stimulus appears simultaneously on the ipsilesional side. Extinction can be attributed to the parietal lesion unbalancing any competition for attention between contra- and ipsilesional stimuli (see Duncan, Humphreys, & Ward, 1997). Although patients fail to report the presence of the contralesional stimulus when extinction occurs, the degree of extinction can be affected by the relations between the stimuli in the two fields. In particular, if (1) the stimuli group by some of the basic Gestalt relations between features in the same dimension (e.g., collinearity, common shape, common contrast polarity, common enclosure and connectedness—see Palmer & Rock, 1994), and (2) if they are common components of a known shape, then extinction can be reduced (there is "recovery from extinction"; see Gilchrist, Humphreys, & Riddoch, 1996; Humphreys, 1998, 2001; Humphreys et al., 2000; Mattingley, Davis, & Driver, 1997; Ward, Goodrich, & Driver, 1994, for evidence of effects of Gestalt grouping; see Kumada & Humphreys, 2001; Ward et al., 1994, for effects of stored knowledge). Extinction can also be affected by the relative "goodness" of two shapes, where goodness is determined by the strength of grouping between the elements making up the shape (e.g., the degree of collinearity between line elements; Humphreys & Riddoch, 2003a; Humphreys, Romani, Olson, Riddoch, & Duncan, 1994; Ward & Goodrich, 1996). Interestingly, patients who show effects of grouping within one dimension (e.g., form) can still manifest abnormal numbers of illusory conjunctions when required to report stimulus properties across two or more feature dimensions (e.g., form and colour; Humphreys, 2001; Humphreys et al., 2000). Based on this contrast, we (Humphreys et al., 2000) have proposed a distinction between grouping within a feature dimension and binding across dimensions. The patients showing extinction typically have damage to the posterior parietal cortex, but do not necessarily have damage to primary occipital cortex or to ventral (occipito-temporal) pathways. Hence the grouping effects apparent in these patients may be sustained through these spared cortical regions (see Rees et al., 2000, and Vuilleumier et al., 2001, for some imaging and evoked-potential data on activation in primary occipital cortex under conditions of extinction), where neuronal interactions are sensitive to the statistical regularities between visual elements that underlie Gestalt grouping phenomena (see Gilbert, Ito, Kapadia, & Westheimer, 2000, for physiological data on this). In contrast, the cross-dimension illusory conjunctions found in parietal (and pulvinar) patients typically involve arbitrary pairings of feature values in stimuli (e.g., letters that can take on any colour). The posterior parietal cortex (and pulvinar) may be important for forming these arbitrary bindings, since the features present may only be associated through their common location, and information about spatial co-registration of different feature dimensions is provided by the posterior parietal lobe. On this view, the initial grouping of elements within ventral visual areas can take place even in a patient with impaired spatial attention (due to posterior parietal or pulvinar damage), contrary to the idea from FIT that attention is necessary for features to be linked. The initial grouping processes may also constrain spatial co-registration across feature domains, modulated by the parietal lobe, so that surface properties of stimuli within one feature domain fall within bounded spatial regions. An explicit account of how such a process may occur has been offered by Grossberg and colleagues (e.g., Grossberg & Mingolla, 1985). A framework expressing this idea is illustrated in Figure 3.

Humphreys and Riddoch (2003a) have also used the phenomenon of group-based extinction to examine the relations between grouping and spatial attention. They examined performance when there was extinction of a "poor" object by a "good" object, irrespective of the spatial locations of the stimuli in the field. They found that detection of a letter presented after the objects was better when it fell in the location previously occupied by the good rather than the poor object., suggesting that spatial attention tended to be directed to the location of the good stimulus. This is consistent with the framework outlined in Figure 3, where grouping within the ventral system can feed back to influence the direction of spatial attention modulated by posterior parietal cortex.

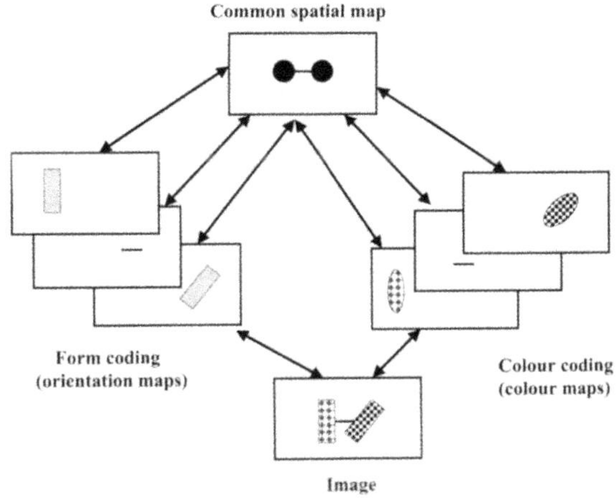

Figure 3. *Example framework indicating the separate coding of form and colour information, along with a proposal for how the two types of information may be bound by co-activating a common spatial map (e.g., in parietal cortex). Units within the map do not themselves code the properties of the stimuli (their form or colour) but feed-back to produce stable (bound) activation at earlier stages of the system.*

In addition to grouping based on the spatial relations between visual elements, there is evidence that grouping is influenced by temporal synchrony between stimuli. Data here come from the apparent opposite finding to extinction, which has been termed anti-extinction by Goodrich and Ward (1997). Anti-extinction occurs when a patient cannot detect a single stimulus presented in their contralesional field, but can report the same item if another stimulus appears on the ipsilesional side. Humphreys, Riddoch, Nys, and Heinke (2002) showed that anti-extinction was sensitive to the two stimuli appearing as onsets, and that it was stronger when the stimuli appeared together relative to when they were staggered in time. They suggested that anti-extinction could reflect temporal grouping between visual onsets. The anti-extinction effect in this case occurred in a patient who also showed illusory conjunctions between arbitrary shapes and colours under prolonged viewing conditions. Thus both spatial grouping (based on Gestalt properties), and temporal grouping (based on temporal synchrony) may take place even when there is impaired binding across feature dimensions. Temporal binding, based on coincident onsets, may be useful for providing a "quick and dirty" registration of visual stimuli. However, it may be detrimental when arbitrary across-domain relations must be formed for each of the many objects that may be present (e.g., when there are multiple letters, each in a different colour), since all the features present would be linked by time. To identify such arbitrary relations, the different features may need to be co-registered on the basis of their locations and any temporal firing patterns adjusted so that only co-located stimuli are synchronised. We speculate that this last process is based on reiterative interactions between the ventral (occipito-temporal) and posterior parietal cortex, and this is disrupted by posterior parietal (and also pulvinar) damage. Hochstein and Ahissar (2002) have recently presented similar ideas when discussing the role of reiterative processes in feature binding.

Feature coding, binding, and awareness. The evidence on recovery from extinction indicates that visual elements can enter into grouping relationships without awareness on the part of the patients. Other arguments for unconscious processing of visual information come from the phenomenon known as "blindsight"—where patients show a capacity to discriminate stimuli of which they are phenomenally unaware (e.g., with discrimination shown when patients are asked to respond by guessing; Weiskrantz, 1986). Blindsight has been found in patients with damage to primary occipital cortex, and it has been demonstrated that such patients can discriminate the locations, sizes, speed, and direction of motion of stimuli of which they are unaware (e.g., Barbur, Ruddock, & Waterfield, 1980; Weiskrantz, Warrington, Sanders, & Marshall, 1974). Weiskrantz (1986) reported that there was little discrimination of the shape of stimuli, though there is evidence of limited coding of colour (see Barbur, Sahraie, Simmons, Weiskrantz, & Williams, 1998; Cowey & Stoerig, 2001). One account of blindsight is that discrimination is based on responses in subcortical structures such as the superior colliculus, which is sensitive to location, movement, and some basic aspects of form (e.g., stimulus size; Barbur et al., 1980). On this view, patients with damage to primary occipital cortex are not aware of stimuli that are processed subcortically, but they can make forced-choice responses on the basis of information processed at that level. Other investigators, though, have reported that higher-level information can be activated by stimuli in the "blind" field of such patients, so that such stimuli influence the computation of illusory contours with stimuli in sighted areas of field (Weiskrantz, 1986) and that both colour and semantic priming may arise from stimuli in the blind field (Cowey, Stoerig, & Hodinott-Hill, 2003; Marcel, 1998). It is possible that such higher-order effects arise from direct inputs from subcortical structures into intermediate visual areas (e.g., V3) that bypass primary visual cortex. It is also the case that more recent studies have suggested that, on at least some occasions, patients can be aware of some aspects of stimuli (particularly their movement) when not experiencing other properties (e.g., their colour; see Zeki & ffytche, 1998; see also G. Riddoch, 1917, for a first description of this). This awareness is linked to activation in regions of association cortex (e.g., area MT), not primary visual cortex (Zeki & ffytche, 1998). Zeki and ffytche have argued that blindsight phenomena, along with conscious residual vision of some properties of stimuli, are caused by V1 lesions uncoupling the relations between discrimination and awareness, so that sometimes there is unawareness of any visual processing (blindsight) and sometimes awareness that something has occurred without linked visual content. Azzopardi and Cowey (1998) also point out that some aspects of blindsight may reflect criterion shifts (particularly with moving stimuli), but this is not always the case (particularly with discrimination of static patterns; Azzopardi & Cowey, 1997). In the latter case, unawareness seems different from normal, near-threshold vision, and seems to reflect a qualitatively distinct process of unconscious perception.

There is also evidence of unconscious colour processing in patients with cerebral achromatopsia following damage to extra-striate cortex (V4). For example, achromatopsic patients can show relatively normal colour-evoked potentials to stimuli (Humphreys et al., 1992) and they can be above chance at making same–different decisions to isoluminant stimuli (see Heywood & Cowey, 1999). It is controversial as to whether tasks such as same–different matching are based on colour-boundary cues that are detected within a magnocellular system insensitive to colour values (Heywood & Cowey, 1999), or whether some aspects of unconscious colour perception are carried by a residual colour (parvocellular) processing channel (Troscianko et al., 1996). Whichever the case, the data indicate that a variety of basic visual features can be computed without awareness, including basic aspects of shape, motion, location, and colour. The evidence on the effects of grouping on extinction also indicates that these basic features can enter into within-domain grouping relations prior to awareness taking place.

Other evidence, again originating from studies of patients with Balint's syndrome, indicates that there can be unconscious registration of cross-dimension binding relations too. Wojciulik and Kanwisher (1998) and Robertson, Treisman, Friedman-Hill, and Grabowecky (1997) have both shown that, even when their Balint's patient was at chance at explicitly discriminating the spatial relations between stimuli (e.g., colour and word identity), he was nevertheless affected by the spatial relations present (e.g., generating Stroop interference if a hue was carried by a word corresponding to a different colour name, but not if the hue was carried by a neutral word). This raises the intriguing question of how information from different visual dimensions can be bound implicitly but not explicitly (e.g., when patients have to report both the colour and the identity of a word). One possibility is that there is initial binding between features from different dimensions, based on the firing of cells in early cortical areas sensitive to a common location. However, higher cortical areas have larger receptive fields and may be prone to binding errors due to pooling inputs from different locations (see Treisman, 1998). The initial, early cortical activation may also be unstable, and affected by small eye movements, etc. To prevent binding errors and to generate a more stable response requires re-entrant feedback so that only a small number of objects are represented along with their features (Hochstein & Ahissar, 2002, see Figure 3). In Balint's syndrome, the early binding process may be preserved but the re-entrant stabilisation process is disrupted by parietal damage. The consequence is that binding can be shown on implicit measures (sensitive to features being initially co-located) even when explicit discrimination responses (made on the basis of the re-entrant binding operation) are disrupted. On this view, implicit feature binding and explicit awareness of feature binding are determined by the same binding processes, which simply become more stable (and available for conscious report) through re-entrant feedback. Consistent with this, Cinel and Humphreys (in press) report that measures of implicit and explicit binding in their Balint's syndrome patient were affected by common variables (e.g., the spatial separation between stimuli). This account then holds that awareness of feature values and their relations is linked to the re-entrant process, which brings together activity from both ventral and more dorsal areas of visual cortex for conscious (explicit) report (see Crick & Koch, 1990, for a similar view, and Pascual-Leone & Walsh, 2001; Silvanto, Cowey, Lavie, & Walsh, 2005, for some evidence).

Finally, we note that this contrast between implicit effects and explicit responses to cross-dimension feature binding reiterates the contrast between cross-dimension binding and within-dimension grouping. When within-dimension grouping takes place in patients showing extinction, it typically makes the stimulus available for conscious report. In contrast, when cross-dimension illusory conjunctions occur, any implicit binding has broken down. Here the patient is conscious of information formed through within-dimension grouping, but then binds the different cross-dimension properties incorrectly. This accords with the proposal that within-dimension grouping arises earlier than, and is independent of, stable cross-dimension grouping based on re-entrant feedback (Figure 3).

Object processing

Shape integration. Studies of recovery from extinction show that, in early stages of vision, there can be grouping of elementary perceptual features—for example, collinear line segments may group to form a single contour. However, to form identifiable shapes, these elementary, grouped features need to be integrated, to form representations of whole objects and their parts. We term this shape integration. Detailed studies of patients with disordered object recognition reveal that this process of shape integration can be selectively disrupted even when elementary grouping processes operate relatively normally. Furthermore, stored knowledge about objects can be preserved in patients with poor shape integration. These case studies demonstrate that the *apperception* process, as identified by Lissauer (1890), can fractionate; a disorder of shape integration can be distinguished from the initial processes involved in grouping elements into contours.

The case of HJA illustrates this. HJA suffered bilateral damage affecting the lingual and fusiform gyri, whilst sparing primary visual cortex. Subsequently he presented with a range of neuropsychological disorders, including a marked deficit in identifying visually presented objects (Humphreys & Riddoch, 1987; M.J. Riddoch & Humphreys, 1987a). His stored knowledge of objects was initially intact, so that he could accurately draw objects from memory. He was also able to generate accurate copies of objects he could not identify, though this could be a painstaking process, producing a shape contour by contour. However, visual object identification was abnormally disrupted by shortening stimulus exposures, by the presence of overlapping contours, and by the presence of internal contours segmenting objects into parts. In an early report on HJA we used these characteristics to define a syndrome of "integrative agnosia," in which there was difficulty in encoding part-relations within objects and in assembling representations of whole objects from their parts (M.J. Riddoch & Humphreys, 1987a). The conditions that exacerbated HJA's deficit in object recognition were all ones that stressed the process of parts-based object identification, by limiting the time for parts coding (short exposures), by requiring that parts be grouped and segmented from other, irrelevant contours (overlapping shapes), and by encouraging parts-based decomposition rather than more wholistic object recognition (when internal contours segmented objects into parts). Similar effects have subsequently been demonstrated with other agnosic patients (e.g., Butter & Trobe, 1994; DeRenzi & Lucchelli, 1993).

More recent work with HJA has gone on to show that early grouping processes, linking form elements into contours, are relatively preserved. For example, HJA had a normal threshold for discriminating the presence of a shape formed from collinear line segments, as the number of noncollinear distractor elements was increased (see Figure 4; Giersch, Humphreys, Boucart, & Kovacs, 2000). On the other hand, he was impaired in circumstances where contours had to be integrated into whole shapes and there was some ambiguity in this integration process. For example, he had abnormal difficulties in performing matching tasks with overlapping and occluded shapes. Interestingly, performance was poorest when there was only a small distance between the edges of parts separated by an occluding element. This suggests that the occluded edge was computed but then disrupted the assignment of visible contours to the correct shapes. This was confirmed in studies of simple copying tasks, where, when presented with occluding figures, HJA sometimes drew in an occluded contour or sometimes segmented between parts separated by this contour (Giersch et al., 2000). Here it seems that early processes of generating contours from form elements—including completing edges from occluded (collinear) segments—continued to operate. In contrast, the higher-order process, of integrating contours into whole shapes, was deficient—particularly when there was some ambiguity in this integration process. These results fit with the proposal that initial grouping

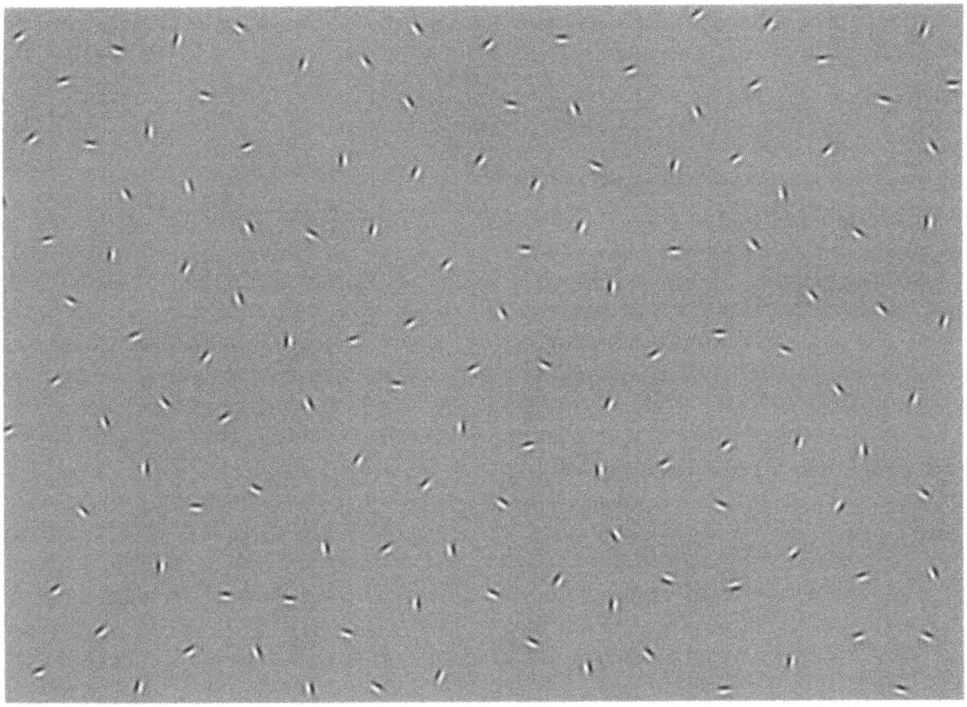

Figure 4. *Example stimuli used in studies assessing the sensitivity of patients to collinear edge contours. The task is to discriminate the location of the circle formed from the local Gabor patches. The number of distractor patches are gradually increased to make discrimination increasingly difficult, so that a threshold for discrimination can be calculated (after Giersch et al., 2000).*

processes operate in early cortical areas spared in HJA (e.g., V1), whilst the higher-order integration process is dependent on subsequent cortical regions (V3, V4 etc.), which integrate and code the relations between the contours grouped in V1 (Gilbert et al., 2000). Similar dissociations between (intact) processes involved in coding form elements into contours, and impaired coding of part-relations between contours, have been reported by Behrmann and Kimchi (2003).

Parallel forms of shape coding: Wholistic and parts-based representations. Although patients such as HJA can be markedly impaired at coding perceptual wholes from their parts, there is also evidence that they can derive wholistic "undifferentiated" representations of objects. M.J. Riddoch and Humphreys (1987a) reported that HJA was better at discriminating silhouettes than line drawings of objects, though only visual information about the overall shape is present in silhouettes (see also Butter & Trobe, 1994). The internal contours of objects, when present, may cue a parts-based analysis that is disrupted for such patients. Furthermore, HJA manifested a "global advantage" when responding to compound letters (Figure 5), indicating that there was relatively rapid coding of the global stimulus in this case (Humphreys, Riddoch, & Quinlan, 1985; cf. Navon, 1977). However, unlike control participants, there were no interference effects from the identity of the irrelevant letter at the other level, either when HJA responded to the global or the local shape. We (Humphreys et al., 1985) proposed that interference was contingent on integrating the undifferentiated representation of the whole with representations of the parts. With such an integrated representation it may be

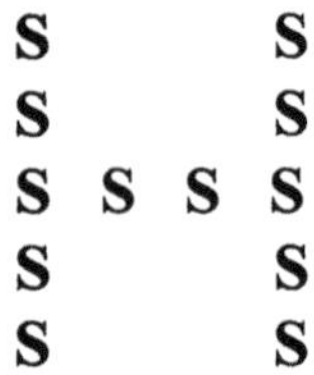

Figure 5. *Example of a compound letter used in studies of hierarchical pattern processing. The task is either to identify the global or the local form.*

difficult to attend to the parts without coding (and suffering interference from) the whole. However, due to his problem in perceptual integration, HJA is relatively spared from interference (see also Behrmann & Kimchi, 2003, patient SM). The implication for normal object processing is that there is parallel coding of undifferentiated wholistic form and of parts-based descriptions of objects. Though normally these processes are integrated to form wholistic representations with articulated parts, this integration process can breakdown following damage to extrastriate, ventral cortex.

Other evidence for the coding of undifferentiated wholistic representations of objects comes from studies of patients with simultanagnosia following bilateral damage to parietal cortex. Shalev and Humphreys (2002) tested one such patient in a task requiring judgements of the spatial relations between the parts of objects or between separate objects. Judgements of part-relations within objects were generally poorer than judgements of the spatial relations between separate objects. However, this pattern was reversed when the patient was encouraged to code the stimulus as a known object. For example, there was relatively poor discrimination when the task was to decide if the small circles fell at the top or the bottom of the oval, in Figure 6, but relatively good discrimination when the task was to decide if the 'eyes' fell at the top or bottom of the 'face' (compared with a baseline where the small circles were outside the oval). Here the patient was unable to derive an explicit representation of the relations between parts and their surrounding whole, though judgements could be made when

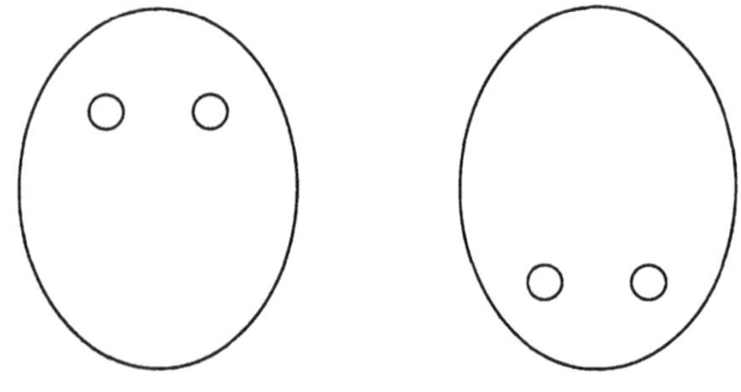

Figure 6. *Example stimuli from Shalev and Humphreys (2002). A patient with Balint's syndrome was able to decide on the relative positions of the small circles within the larger oval when instructed to think of the stimuli as faces, but not when instructed to treat them as shapes.*

the object was recognised, when he essentially responded according to whether the whole stimulus was upright (eye at top) or inverted (eyes at bottom). This dissociation suggests that the familiar form was identified as an undifferentiated whole, without explicit parts-based coding. The preservation of wholistic recognition processes may also explain why some patients can identify at least some objects depicted from standard viewpoints when their ability to group local elements is impaired (see Davidoff & Warrington, 1999; Ricci, Vaishnavi, & Chatterjee, 1999).

This last result has been extended by Riddoch and Humphreys (2004), who examined object recognition in two simultanagnosic patients. Both patients were better at identifying silhouettes than line drawings and at identifying objects that had unique global shapes compared with objects whose parts were important for recognition. Both patients were also somewhat better at identifying living things (e.g., animals) than nonliving things (e.g., tools). In both cases, object recognition seemed dependent on recognising the global form of the stimuli. Living things may be easier to recognise than nonliving things from their undifferentiated global forms, especially if many nonliving things are defined on the basis of the functional relations between their parts (and hence depend on part-decomposition for recognition). Indeed, a parts-based analysis of living things may be inappropriate given that the spatial relations between their parts can change as the stimuli move (see Laeng, Carlesimo, Caltagirone, Capsso, & Miceli, 2002). In such instances, the visual coding problem may contribute to a problem with particular categories of object.

In simultanagnosic patients, with bilateral parietal damage, recognition may be dependent on global shape information because they have difficulty in allocating attention to the parts of objects. Coding the spatial relations between object parts may depend on selective attention, to ensure that individual parts are assigned to the correct relative locations (see Stankiewicz, Hummel, & Cooper, 1998). In contrast, patients with ventral lesions may attend to objects in a relatively normal manner (see Vecera & Behrmann, 1997), but they have a difficulty in integrating parts due to damage to the form processing system. Furthermore, in integrative agnosia, the part information that is derived can cue segmentation of the whole (e.g., generating worse performance with line drawings than silhouettes); the consequence is that such patients may rely on local elements to cue recognition (Humphreys & Riddoch, 1987). When patients attend to local detail, living things, with similar parts, may be more difficult to identify than nonliving things (see Humphreys & Riddoch, 2003b; M.J. Riddoch & Humphreys, 1987a).

Coding orientation and view-invarance. As noted at the start of this article, Warrington and colleagues conducted pioneering work in the 1970s into whether patients were able to recognise and match objects presented across different (often unusual) views. Work following this research attempted to assess the processes that led to view-invariant object recognition, which had broken down in these patients. Farah and Hammond (1988) provided evidence that view-invariant recognition was unlikely to rely purely on mental rotation being used to "normalise" an object back to its familiar orientation. They reported a patient who was poor at mental rotation but who nevertheless could identify misoriented objects, letters, and numbers. A process additional to mental rotation seems required. Humphreys and Riddoch (1984) contrasted the effects of two different kinds of transformation. In a "foreshortened" condition, objects were rotated so that their main axis was foreshortened, but critical distinctive features of the object remained available. In a "minimal feature" condition, objects were rotated to preserve their main axis, but their distinctive features were hidden. Four patients with right parietal damage were found to be impaired at matching in the foreshortened but not the minimal feature condition. In contrast the integrative agnosic, HJA, was relatively poor at minimal feature matching but could match objects whose critical features remained salient. Furthermore, the right parietal patients were facilitated if the foreshortened stimuli were depicted on graph paper giving strong linear perspective cues that coincided with the orientation of the main axis. From these data Humphreys and Riddoch proposed that view-invariant matching could be based on at least two processes. In one process, stimuli could be matched based on their having the same distinctive features, which were extracted irrespective of the overall orientation of the object. In the second, stimuli were matched by extracting the orientation of the main axis and coding the locations of features in relation to the axis. This would yield a view-invariant representation of the whole object. Due to damage to the first (feature-based) process, patients might be reliant on the axis-based process, and thus vulnerable to effects of foreshortening since the main axis was then hard to derive. Due to damage to the second (axis-based) process, patients might be reliant on matching distinctive features and be relatively insensitive to the orientation of the main axis (see also Warrington & James, 1986). It is interesting that the argument for axis-based coding of objects closely matches the proposals made by Marr (1982), who argued that such a process would enable computer vision systems to achieve view-invariant object recognition. However, another interpretation of the evidence is that foreshortening is disruptive because it produces severe changes to the familiar global shapes of objects. This would particularly affect patients who are reliant on global shape for matching and for recognition (cf. Davidoff & Warrington, 1999;

Landis, Regard, Bliestle, & Kleiues, 1988). On this count, depth cues to the main axis of objects may be effective because they prompt a form of problem solving that helps patients code the global shape of objects correctly, but this would be distinct from normal recognition processes.

More recent studies have demonstrated that patients can sometimes recognise objects whilst being poor at judging object orientation—particularly when asked to discriminate between mirror-images of objects (e.g., Davidoff & Warrington, 1999, 2001; Priftis, Rusconi, Umiltà, & Zorzi, 2003; Turnbull, Della Sala, & Beschin, 2002). In each case, the patient had some parietal involvement in their lesion. Warrington and Davidoff (2000) reported that their patient JBA was better at making mirror-image discrimination judgements to objects he could not identify relative to objects he could identify, whilst Davidoff and Warrington's (2001) patient FIM showed improved mirror-image matching of letters compared with objects. It can be argued that the left–right orientation of letters is typically coded in order to discriminate between the stimuli (b vs d, etc.). On the other hand, the left–right orientation (the "handedness") of objects is not typically coded as it is irrelevant to their identity (see Hinton & Parsons, 1981). To judge the handedness of objects requires some extra process to be engaged, probably involving the explicit coding of object parts relative to a view-specific frame. This extra process is modulated by the parietal lobe. In contrast, object recognition may lead to the "throwing away" of any view-specific code that was computed en route to access to stored knowledge being achieved (cf. Marr, 1982). The consequence of this is that patients may be particularly poor at judging the left–right orientation of objects they recognise, since "orientation-stripped" information from the recognition process may be weighted more strongly than any residual ability to carry out explicit orientation coding (see also Cooper & Humphreys, 2000). Other patients with parietal damage show poor representation of object orientation when copying, and in some cases they rotate the object back to its canonical orientation in their drawing (Turnbull, Beschin, & Della Sala, 1997), suggesting that the object recognition process (preserved in such patients) is sensitive to the canonical orientation of objects.

We (M.J. Riddoch et al., 2004) have also reported data on the differential "weighting" of visual codes in a patient with parietal damage. This patient, MH, was selectively poor at detecting a target defined by an orientation difference relative to distractor lines. This problem was most evident when the distractor and target lines grouped, to form an emergent configuration. This can be accounted for if orientation detection in the parietal lobe is disrupted, whilst orientation coding for grouping and pattern recognition, mediated by more ventral visual processes, is spared. If there is a relative increase in the weighting of the pattern recognition process, compared to the process of detecting salient orientation differences, then any emergent pattern will disrupt orientation detection. This result, along with the finding that there can be poor orientation discrimination for objects that are recognised (Warrington & Davidoff, 2000), indicates that visual processing depends on the interplay between dorsal and more ventral visual streams, with the dorsal stream involved in responding to salient orientation differences, in coding objects within an explicit view-based frame and also in directing attention to object parts (see our earlier section on *Parallel forms of shape coding*). Rather than thinking that the dorsal (occipito-parietal) stream is solely concerned with location coding (cf. Ungerleider & Mishkin, 1982), the data suggest that it is intimately involved in visual discrimination and object recognition.

Fractionating the association process: Structural and semantic representations. According to the prevalent view of recognition deficits in the early 1980s, derived from Lissauer's (1890) original framework, the apperceptive stage of processing should be followed by an associative stage in which percepts are mapped onto stored knowledge. However, just as subsequent work has shown that the apperceptive stage can be broken down into several distinct processes (see above), so it has also suggested that associative stages of recognition can fractionate. In 1987 we (M.J. Riddoch & Humphreys, 1987b) reported data from a patient JB with impaired object recognition when tested on semantic matching task with objects (e.g., when asked to decide which two of the following objects would be used together: hammer, chisel, screw). The deficit was modality-specific, since JB found the same task trivial when given the names of the objects. Despite his impaired semantic matching, JB was able to carry out an object decision task at a normal level. This task, modelled on lexical decision tasks in word recognition (e.g., Coltheart, 1978), requires discrimination between real objects and nonobjects. In our study, the nonobjects were constructed by interchanging the parts of real objects so that the nonobjects and objects did not differ in perceptual goodness (see Figure 7). In this case, discrimination may depend on access to stored knowledge about particular objects. The fact that JB performed at a normal level on this test suggests that he could access stored knowledge about the structure of objects, but was then impaired at retrieving semantic knowledge. Since then, a similar pattern of performance has been reported in several patients (Fery & Morais, 2003; Hillis & Caramazza, 1995; Humphreys & Riddoch, 1999; Sheridan & Humphreys, 1993; Stewart, Parkin, & Hunkin, 1992). The dissociation between spared object decision and impaired semantic matching fits with theories that distinguish between stored structural and semantic knowledge about objects (e.g., Humphreys, Riddoch, & Quinlan, 1988; Seymour, 1979). In patients such as JB, there appears to be access to structural knowledge, along with impaired access to semantic knowledge from vision.

However, in other cases, particularly patients with semantic dementia, impairments in semantic

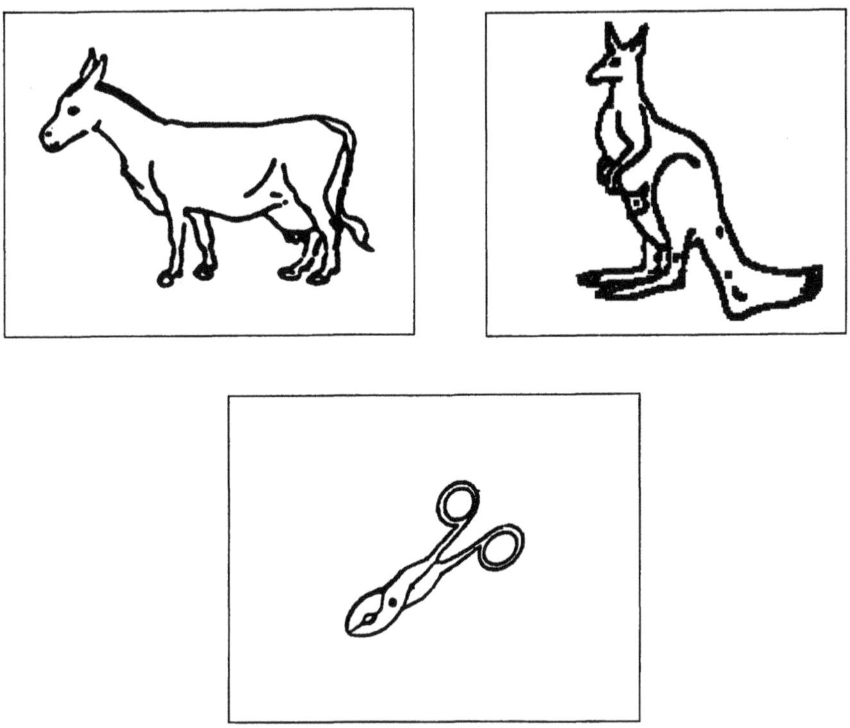

Figure 7. *Example nonobjects from object decision tasks.*

matching are accompanied by deficits in object decision, and in both tasks patients tend to have preserved knowledge about the most typical attributes of stimuli and to lose atypical attributes (e.g., Rogers, Hodges, Lambon Ralph, & Patterson, 2003). Such patterns of co-occurrence can be explained if these patients suffer damage to both structural and semantic knowledge systems, or if these systems interact so that the loss of atypical attributes at a semantic level affects both systems together. In contrast to this, the damage in cases such as JB may produce a form of disconnection in which structural knowledge is functionally isolated from semantic knowledge (Riddoch & Humphreys, 1987b).

Other evidence suggests that stored knowledge about the structure of objects can be distinguished from stored knowledge of object colour. Price and Humphreys (1989) noted that JB (see above) was relatively poor at judging whether drawings of real objects were correctly coloured, despite his spared object decision performance. A similar pattern was reported by Miceli et al. (2001, patient IOC). Interestingly, this patient was able to name colours, indicating that associative links between stored knowledge of objects and their colours may be damaged whilst leaving intact links between colour perception and colour names.

Objects, faces and words. In contrast to the view described above, which holds that there can be forms of associative agnosia without problems in visual perception, Farah (1990) proposed that agnosias were inherently linked to perceptual deficits in patients. Her argument was based in part on evidence from contrasting patterns of deficits across different classes of visual stimuli: objects, faces and words. In a historical review of the literature, Farah noted that researchers had reported "pure" deficits in face recognition (prosopagnosia) and in visual word recognition (alexia) without concomitant impairments with the other classes of stimuli (i.e., spared object and word recognition, in prosopagnosia; spared face and object recognition, in alexia). In contrast, there were no clearly documented cases of "pure" agnosia without associated problems with either faces or words. Farah also pointed out that only a subset of associated deficits had been reported in the literature. For example, patients had been described with prosopagnosia and agnosia (without alexia), with agnosia and alexia (without prosopagnosia), and with all three deficits, whereas the pairing of prosopagnosia and alexia without agnosia was not noted (see Farah, 1990). Farah interpreted these varying patterns of deficit in terms of a two-process account of vision. This account holds that two processing operations take place in parallel: the coding of undifferentiated global forms, and the processing of parts-based representations. Face recognition depends on the global form system, whilst reading depends on the parallel processing of parts (the letters). Objects depend on both types of process, perhaps depending on the objects involved. It follows that selective damage to the processing of undifferentiated wholistic information will disrupt face recognition but not word recognition, whilst lesions to the parts-based process affect visual word recognition but not face recognition. Damage to each process could affect object recognition, but it should not be possible to generate a problem in object recognition without either face or word recognition also being affected.

However, following this parsimonious account, there have been reports of patients with the very patterns of impairment that contravene the proposal. For example, Humphreys and Rumiati (1998) and Rumiati, Humphreys, Riddoch, and Bateman (1994) documented patients with "pure" agnosia, with face and word recognition being spared. Similarly, Buxbaum, Glosser, and Coslett (1996) and De Renzi and di Pellegrino (1998) have reported cases with prosopagnosia and alexia but without marked deficits in object recognition (the converse pattern). These results indicate that there can be relatively independent breakdowns in processing the different classes of object. Where there is a selective deficit in the recognition of one class of object (e.g., objects but not faces or words), basic perceptual processing can be spared, and the deficit can be linked to impaired access to stored perceptual and semantic knowledge (Humphreys & Rumiati, 1998). At the very least, then, there appear to be independent stored representations of knowledge about objects, faces, and words, at the perceptual and perhaps also at a semantic level.

The real object advantage and the re-calibration of visual memories. Prior to 1984, there was only one report of a long-term follow-up of a patient with visual agnosia. This case was documented by Kertesz (1979), who tested the patient across a 10-year interval. He found that object recognition remained disturbed. Where there was an improvement this was found for real objects rather than for drawings. Several more follow-up studies have since been added to the literature, and in each case the patient has been reported as improving on real objects more than on line drawings (M.J. Riddoch, Humphreys, Gannon, Blott, & Jones, 1999; Sparr, Jay, Drislane, & Venna, 1991; Thomas, Forde, Humphreys, & Graham, 2002; Wilson & Davidoff, 1993). In two of these studies, there was also relatively detailed assessment of the patient's long-term visual knowledge. The study of M.J. Riddoch et al. (1999) was carried out with HJA (M.J. Riddoch & Humphreys, 1987a) some 16 years after the stroke that led to his agnosia. When initially tested, HJA had well-preserved long-term visual knowledge about objects he could no longer identify. He could give accurate definitions incorporating the visual characteristics and he could accurately draw the objects from memory. However, when subsequently re-tested, HJA's long-term visual knowledge had deteriorated—for example, his drawings became less precise and his definitions lacked visual properties (see Figure 8). This was not due to some general decrease in his semantic knowledge, as his descriptions of the functional and associative properties of objects if anything increased across the test interval. To account for the results, we (M.J. Riddoch et al., 1999) proposed that long-term visual memory is not simply laid down and maintained in its initial state; rather it must be constantly updated and recalibrated through interaction with on-line perceptual processes. If this recalibration process is disrupted by a perceptual deficit in a patient, then long-term visual memory deteriorates. On this view, perception and memory are interactive processes, with accurate visual perception being needed in order to maintain long-term visual memory. Very similar results were reported by Thomas et al. (2002) in their follow-up study of another agnosic patient.

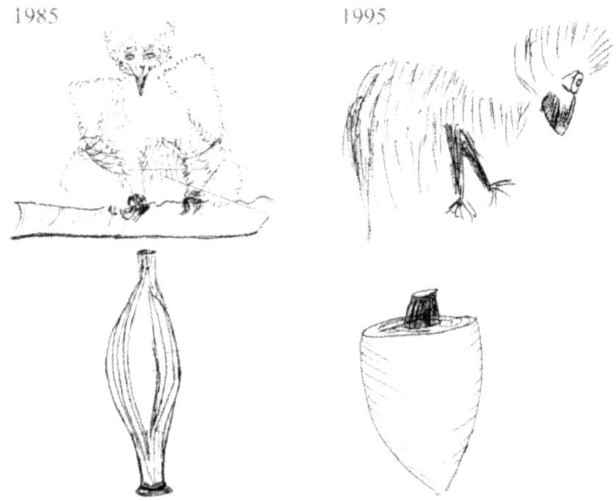

Figure 8. *Examples of HJA's drawing from memory, when tested in 1985 and 1995 (after M.J. Riddoch et al., 1999). The drawings are of an eagle and celery.*

If patients lose the specificity of their visual knowledge about objects, how do they improve at recognising real objects? Chainay and Humphreys (2001) examined the factors that underpin the real-object advantage in agnosia by varying the presence of different depth cues. The usefulness of binocular disparity was manipulated by placing objects at different depths. Motion parallax was varied by either restricting or allowing free head movements. They found an interaction between these factors. There was little effect of eliminating motion parallax when binocular cues could be used (near objects, head restrained), and little effect of reducing binocular disparity when motion parallax was available (far objects, head free). However, recognition was disrupted when both depth cues were restricted (far objects, head restrained). This suggests that real objects are frequently identified better than line drawings because patients can use the depth information available when they view real objects to help organise objects into surfaces at different depths.

Recovery after a lesion may be based at least in part on learning to place increased "weight" on depth cues during the recognition process. Interestingly, in studies of the preserved use of vision for reaching and grasping in agnosia (see the next section: *Action*), performance has been shown to depend strongly on binocular disparity and on motion parallax (Dijkerman, Milner, & Carey, 1999; Mon-Williams, Tresilian, McIntosh, & Milner, 2001). Depth cues from stereopsis and motion parallax can be derived independently of form perception, contributing to both recognition and action when the coding of form information is impaired.

Object perception and imagery. The relative preservation in some agnosics of long-term visual knowledge about objects speaks to one other issue that had been studied relatively little in the neuropsychological literature prior to 1984—the relations between object perception and imagery. As noted above, agnosic patients can have relatively preserved visual knowledge about objects they can no longer identify (see Bartolomeo et al., 1998; Behrmann, Moscovitch, & Winocur, 1994; Behrmann, Winocur, & Moscovitch, 1992; Jankowiak, Kinsbourne, Shalev, & Bachman, 1992; M.J. Riddoch & Humphreys, 1987a; Servos & Goodale, 1995). This good performance can be expressed in a range of tasks typically thought to involve visual imagery (e.g., answering questions such as: Does a lion have a relatively long or short tail? Does a poodle have pointed or floppy ears?; see Kosslyn, 1994). From this it can be argued that imagery operates independently of visual perception, and so imagery can be spared in the face of a perceptual impairment.

This position is not without argument, though. For example, in a study of imagery for faces in patient HJA (Young, Humphreys, Riddoch, Hellawell, & de Haan, 1994), there was intact imagery for individual facial features (hair colour, the presence of glasses or a beard), but poor performance when decisions were made about the relations between features (which two people looked most alike). Similarly, tests of HJA's visuospatial imagery show good recall of individual features of objects, but impaired judgements when features must be spatially related (M.J. Riddoch, Humphreys, Hardy, Blott, & Smith, 2003). Such correlated deficits suggest that perception and imagery rely on a common substrate, revealed when both tasks require the same processes (integrating visual features, in HJA's case). The nature of this substrate is unclear, however. For example, both perception and imagery may rely on some common "visual buffer" that holds information as it is interpreted and/or operated upon (cf. Kosslyn, 1994). Or the same long-term memory representations may be involved when objects are recognised or images retrieved, even if a common buffer is not involved. Whether or not there are common deficits in perception and imagery in a patient may then depend on factors such as the time post-lesion–if there is a deterioration in long-term visual memory over time, then imagery tasks dependent on long-term visual memory too will come to be affected (M.J. Riddoch et al., 1999; Thomas et al., 2002). Alternatively, any associated deficits in patients may stem from separate impairments of perception and imagery (in cases such as HJA). This topic remains in debate (see Bartolomeo, 2002).

Action

Prehensile action in agnosia. In 1984, the cognitive neuropsychology of visual object processing was primarily concerned with object recognition, and there was relatively little work assessing how even simple actions to objects might be effected (e.g., when we reach and grasp a stimulus). Today this perspective has changed, and there has been a shift towards trying to understand how visual information is used to control and access actions to objects, in addition to understanding recognition processes. A good deal of the credit for raising issues concerned with action goes to the case reports of the agnosic patient DF (Milner et al., 1991; see Milner & Goodale, 1995, for a summary). DF became severely agnosic after suffering carbon monoxide poisoning. Unlike many agnosic patients, DF was impaired even at copying the objects she failed to recognise, and she was poor at a wide variety of perceptual judgement tasks (e.g., turning her hand to match the orientation of a bar). This suggests a very early locus to her damage. However, despite her poor performance on perceptual judgement tasks, Milner et al. (1991) noted that DF was able to make correct actions to many stimuli. For example, she could reach and grasp objects appropriately and she could post a letter efficiently through a slot that could appear at different orientations. To account for this, Milner et al. proposed a distinction between the visual information used to support perceptual judgements (impaired in DF and in many agnosics) and the visual information employed in the on-line guidance of action. They argued that the on-line guidance of action was dependent on information computed through the dorsal (occipito-parietal) visual stream, which was relatively spared for DF (see James, Culham, Humphrey, Milner, & Goodale, 2003, for data from brain imaging). This dorsal stream operates independently of the ventral visual stream supporting perceptual judgements and object recognition (damaged in DF's case). Milner and Goodale (1995) also observed that DF's actions were impaired when a short delay was introduced between seeing the object and grasping it. Delayed actions, then, may be dependent on the ventral stream that is lesioned in her case. Here we again see an elaboration on the idea of the functional roles of the dorsal and ventral visual streams in vision (cf. Ungerleider & Mishkin, 1982). The dorsal stream controls on-line, visually guided actions, in addition to coding the locations of objects. The ventral stream controls object recognition and memory-guided (delayed) action.

Another dissociation between perceptual judgements and action has been reported by Priftis et al. (2003), in their case study of a patient with difficulty in discriminating between identical and mirror-image objects after parietal damage. Although impaired at making perceptual judgements about mirror-image reflections, their patient was able to decide which of left- and right-facing objects could be grasped with his right hand. Here there seems a distinction between "dorsal" processes involved in making action judgements to stimuli (spared in this case) and those concerned with generating an explicit representation of the orientation of stimuli relative to the patient's viewpoint (impaired in this case).

Milner and Goodale's (1995) argument, distinguishing the use of vision for on-line action and for recognition, is supported by a double dissociation between DF and patients with "optic ataxia" (e.g., Perenin & Vighetto, 1988). Optic ataxia is associated with damage to the dorsal visual stream. It is characterised by patients being able to recognise objects, whilst showing impaired visually guided reaching and grasping. In one counterintuitive result, Milner and colleagues (Milner, Paulignan, Dijkerman, Michael, & Jeannerod, 1999, 2001, 2003) have demonstrated that reaching and grasping in optic ataxia can improve when patients act on the basis of visual memory (when a delay is introduced between the object being present and the action). This is consistent with optic ataxics having an impaired dorsal stream for immediate action, along with a preserved ventral stream for object recognition and delayed action. In on-line action, optic ataxics appear to be disrupted by the "noise" from their damaged dorsal route. In contrast, in delayed action tasks they can use their preserved "ventral" stream to control action.

Dijkerman, Milner, and Carey (1998) also noted that DF had poor visually guided actions to objects under some circumstances. In particular, she was impaired when she had to use relative position information to guide hand actions (e.g., placing her fingers into the holes in a ball for ten-pin bowling, where the action depends on the relative positions of the holes). Thus ventral information may support actions based on higher-level (relative-position) visual codes. Indeed, it may be that both ventral and dorsal streams modulate the on-line guidance of action, but the ventral stream is relatively slow—so optic ataxics continue to be affected by their impaired dorsal stream even when the ventral stream is relatively preserved (see Rossetti & Pisella, 2002). Other evidence consistent with the ventral stream contributing to on-line action comes from the finding that optic ataxics may grasp familiar objects better than unfamiliar ones (Jeannerod, Decety, & Michel, 1994)—this effect of familiarity presumably being modulated through the ventral visual stream.

Optic aphasia and direct action. In addition to prehensile reaching and grasping actions dissociating from object recognition, neuropsychological evidence also points to patients who seem to be impaired on object recognition tasks, but who are still able to use objects appropriately (not only reaching and grasping them, but making the correct category of action, such as hitting with a hammer or writing with a pen). In the disorder of optic aphasia, patients show relatively preserved gesturing to objects that they fail to name (Freund, 1889; Lhermitte & Beauvois, 1973). Traditionally, the good gestures made by such patients have been taken as evidence that they have a problem solely in name retrieval, with object recognition being preserved. However, this needs to be tested rather than assumed. When detailed tests of semantic knowledge have been conducted, such patients have been impaired (Hillis & Caramazza, 1995; M.J. Riddoch & Humphreys, 1987b). This leaves the question, then, of how can gestures be made to objects, if there is poor access to semantics?

There are several ways to answer this last question. One proposal is that actions to objects can be supported by a "direct" visual route from object structure to action, which is independent of associative/semantic knowledge (M.J. Riddoch & Humphreys, 1987b; Yoon, Heinke, & Humphreys, 2002). Importantly, to explain the often quite specific gestures made in optic aphasia, this account would need to maintain that the direct route is used to generate a specific "class" of action (e.g., drinking with a cup) and not just the prehensile aspects (reaching and grasping)(see Humphreys & Riddoch, 2003c). Alternatively, action may be generated from partially activated semantic information combined with forms of "visual problem solving" in which patients strategically link the visual properties of objects to potential actions (e.g., Hillis & Caramazza, 1995; Hodges, Bozeat, Lambon Ralph, Patterson, & Spatt, 2000).

Results from other disorders are also consistent with the dual-route approach. For example, in "visual apraxia," patients are impaired at making actions to visually presented objects despite often being able to identify the objects and being able to carry out gestures when just given the names of the objects (DeRenzi, Faglioni, & Sorgato, 1982; Pilgrim & Humphreys, 1991; M.J. Riddoch, Humphreys, & Price, 1989). Since the patients can gesture to the name of an object, we may presume that they can generate action from semantic knowledge. They can also access semantics when the object is presented visually. Why is it, then, that visually accessed semantic knowledge is not used to generate action? This pattern can be explained if damage to a direct visual route to action blocks the intact semantic route, so that actions to visually presented objects are selectively affected—this may be analogous to the disruptive effects of on-line vision in optic ataxia (cf. Milner et al., 1999).

In other patients, actions seem to be evoked by the visual properties of objects in an automatic rather than strategic fashion. One example of this is in "utilisation" errors associated with frontal lobe damage (Lhermitte, 1983). These errors involve the patient using an object in an overlearned manner despite task instructions to do otherwise. M.J. Riddoch et al. (M.J. Riddoch, Edwards, Humphreys, West, & Heafield, 1998; M.J. Riddoch, Humphreys, & Edwards, 2000) showed that these errors were sensitive to both the visual properties of objects and to the relations between the objects and the effector used in action—for example, although instructed not to do this, a patient may reach and grasp a cup on the left of their body with their right hand, if the handle faces to the right. Interestingly, the patient studied by Riddoch et al. was unable to make identical cross-body reaches under her own volition, suggesting that the action had to be evoked by the object. In addition, the utilisation errors decreased in frequency if the cup was inverted, though it could still access semantics and be named. Thus the actions were sensitive to having objects in a familiar orientation. This is consistent with actions being activated automatically following the activation of stored representations of familiar objects, with these stored representations being sensitive to object orientation. Crucially, this process operates independently of semantic access to action. Such learned associations between vision and action may exist in addition to processes that link objects to action "on the fly," either using a fast dorsal route to guide the observer's movements or a more deliberate problem-solving strategy.

Summary

The work we have reviewed indicates that views about the neuropsychology of visual object processing have moved a considerable way since 1984, extending our knowledge of visual feature processing, object processing, and how vision is employed for action as well as for object recognition. Studies on feature processing have demonstrated fine-grained problems in processing particular features, as well as selective deficits in grouping features within dimensions, and in the processes that bind visual properties across dimensions. Studies on object processing reveal that apperceptive and associative stages of recognition can be fractionated, as can our stored knowledge about different classes of visual object (objects, faces, and words). In addition, performance can be dependent on the relative "weighting" of independent visual processes (e.g., ventral object recognition and dorsal selection of salient disparities), which can be affected by brain lesions. Other research stresses the parallel use of vision for action and for recognition, with vision being linked in a relatively direct way to both prehensile actions and to learned categories of action (not only reaching and grasping a cup but drinking from it too). In some of these cases, the neuropsychological data can "prise apart" processes that are difficult to separate in studies with normal participants (e.g., differences between stored structural and semantic knowledge), and in other instances neuropsychological descriptions have provided the initial momentum for studies subsequently carried out with normal participants (e.g., on the distinction between vision for recognition and vision for action, where numerous studies with normal participants have subsequently been undertaken examining the effects of visual illusions and stimulus perturbations; see Rossetti & Pisella, 2002, for a review). Such examples illustrate that neuropsychological studies provide an invaluable source of evidence for understanding object processing, indicating which functional processes are *necessary* for recognition and action, and how interactions within complex neural systems can generate emergent behaviour.

PrEview proof published online 2 September 2005

REFERENCES

Ackroyd, K. (2003). *The contribution of neuropsychology to the understanding of depth perception.* Unpublished PhD thesis, University of Birmingham.

Albert, M. L., Reches, D., & Silverberg, R. (1975). Associative visual agnosia without alexia. *Neurology, 25*, 322–326.

Azzopardi, P., & Cowey, A. (1997). Is blindsight like normal, near-threshold vision? *Proceedings of the National Academy of America, 94*, 14190–14194.

Azzopardi, P., & Cowey, A. (1998). Blindsight and visual awareness. *Consciousness and Cognition 7*, 292–311.

Balint, R. (1909). Seelenlahmung des 'Schauens': Optische ataxie, raumliche Störung der Aufmerksamkeit. *Monatschrift für Psychiatrie und Neurologie, 25*, 51–81.

Barbur, J., Ruddock, K. H., & Waterfield, V. A. (1980). Human visual responses in the absence of the geniculo-calcarine projection. *Brain, 103*, 905–928.

Barbur, J. L., Sahraie, A., Simmons, A., Weiskrantz, L., & Williams, S. C. R. (1998). Processing of chromatic signals in the absence of a geniculostriate projection. *Vision Research, 38*, 3447–3453.

Bartolomeo, P. (2002). The relationship between visual perception and visual mental imagery: A reappraisal of the neuropsychological evidence. *Cortex, 38*, 357–378.

Bartolomeo, P., Bachoud-Lévi, A. C., de Gelder, B., Denes, G., Dalla Barba, G., Brugiers, P., & Degos, J. D. (1998). Multiple-domain dissociation between impaired visual perception and preserved mental imagery in a patient with bilateral extrastriate lesions. *Neuropsychologia, 36*, 239–249.

Battelli, L., Casco, C., & Sartori, G. (1997). Dissociation between contour-based and texture-based shape perception: A single case-study. *Visual Cognition, 4*, 275–310.

Battelli, L., Cavanagh, P., & Thornton, I. (2003). Perception of biological motion in parietal patients. *Neuropsychologia, 41*, 1808–1816.

Behrmann, M., & Kimchi, R. (2003). What does visual agnosia tell us about perceptual organization and its relationship to object perception? *Journal of Experimental Psychology: Human Perception and Performance, 29*, 19–42.

Behrmann, M., Moscovitch, M., & Winocur, G. (1994). Intact visual imagery and impaired visual perception in a patient with visual agnosia. *Journal of Experimental Psychology: Human Perception and Performance, 20*, 1068–1087.

Behrmann, M., Winocur, G., & Moscovitch, M. (1992). Dissociation between mental imagery and object recognition in a brain-damaged patient. *Nature, 359*, 636–637.

Bender, M. B., & Feldman, M. (1972). The so-called "visual agnosias." *Brain, 95*, 173–186.

Benson, D. F., & Greenberg, J. P. (1969). Visual form agnosia. *Archives of Neurology, 20*, 82–89.

Butter, C. M., & Trobe, J. D. (1994). Integrative agnosia following progressive multifocal leukoencephalopathy. *Cortex, 30*, 145–158.

Buxbaum, L. J., Glosser, G., & Coslett, H. B. (1996). Relative sparing of object recognition in alexia-prosopagnosia *Brain and Cognition, 32*, 202–205.

Chainay, H., & Humphreys, G. W. (2001). The real object advantage in agnosia: Evidence of a role for shading and depth in object recognition. *Cognitive Neuropsychology, 18*, 175–191.

Cinel, C., & Humphreys, G. W. (in press). On the relations between implicit and explicit spatial binding: Evidence from Balint's syndrome. *Cognitive, Affective and Behavioral Neuroscience.*

Cohen, A., & Rafal, R. D. (1991). Attention and feature integration: Illusory conjunctions in a patient with a parietal lobe lesion. *Psychological Science, 2*, 106–110.

Coltheart, M. (1978). Lexical access in simple reading tasks. In G. Underwood (Ed.), *Strategies of information processing* (pp. 151–216). London: Academic Press.

Cooper, A. C. G., & Humphreys, G. W. (2000). Task-specific effects of orientation information: Neuropsychological evidence. *Neuropsychologia, 38*, 1607–1615.

Cowey, A. (1979). Cortical maps and visual perception: The Grindley memorial lecture. *Quarterly Journal of Experimental Psychology, 31*, 1–17.

Cowey, A. (1985). Aspects of cortical organization related to attention and selective impairments of visual perception. A tutorial review. In M. I. Posner & O. S. M. Marin (Eds.), *Attention and performance XI*. Hillsdale, NJ: Lawrence Erlbaum Associates Inc.

Cowey, A., & Stoerig, P. (2001). Detection and discrimination of chromatic targets in hemianopic macaque monkeys and humans. *European Journal of Neuroscience, 14*, 1320–1330.

Cowey, A., Stoerig, P., & Hodinott-Hill, I. (2003). Chromatic priming in hemianopic visual fields. *Experimental Brain Research, 152*, 95–105.

Cowey, A., & Vaina, L. M. (2000). Blindness to form from motion despite intact static form perception and motion detection. *Neuropsychologia, 38*, 566–578.

Crick, F., & Koch, C. (1990). Towards a neurobiological theory of consciousness. *Seminars in the Neurosciences, 2*, 263–276.

Davidoff, J., & Warrington, E. K. (1999). The bare bones of object recognition: Implications from a case of object recognition impairment. *Neuropsychologia, 37*, 279–292.

Davidoff, J., & Warrington, E. K. (2001). A particular difficulty in discriminating between mirror images. *Neuropsychologia, 39*, 1022–1036.

DeRenzi, E., & di Pellegrino, G. (1998). Prosopagnosia and alexia without object agnosia. *Cortex, 34*, 403–415.

DeRenzi, E., Faglioni, P., & Sorgato, P. (1982). Modality-specific and supramodal mechanisms of apraxia. *Brain, 105*, 301–312.

DeRenzi, E., & Lucchelli, F. (1993). The fuzzy boundaries of apperceptive agnosia. *Cortex, 29*, 187–215.

Dijkerman, H. C., Milner, A. D., & Carey, D. P. (1998). Grasping spatial relationships: Failure to demonstrate allocentric visual coding in a patient with visual form agnosia. *Consciousness and Cognition, 7*, 438–453.

Dijkerman, H. C., Milner, A. D., & Carey, D. P. (1999). Motion parallax enables depth processing for action in a visual form agnostic when binocular vision is unavailable. *Neuropsychologia, 37*, 1505–1510.

Duncan, J., Humphreys, G. W., & Ward, R. (1997). Competitive brain activity in visual attention. *Current Opinion in Neurobiology, 7*, 255–261.

Efron, R. (1968). What is perception? In R. S. Cohen & M. Wartofsky (Eds.), *Boston studies in the philosophy of science, Vol. 4* (pp. 137–173). New York: Humanities Press.

Farah, M. J. (1990). *Visual agnosia.* Cambridge, MA: MIT Press.

Farah, M. J., & Hammond, K. M. (1988). Mental rotation and orientation-invariant object recognition: Dissociable processes. *Cognition, 29*, 29–46.

Fery, P., & Morais, J. (2003). A case study of visual agnosia without perceptual processing or structural description impairment. *Cognitive Neuropsychology, 20*, 595–618.

Freund, D. C. (1889). Lieber Optische Aphasia und Seelenblindheit. *Archive für Psychiatrie und Nervenkrankheiten, 20*, 276–297.

Friedman-Hill, S. R., Robertson, L. C., & Treisman, A. (1995). Parietal contributions to visual feature binding: Evidence from a patient with bilateral lesions. *Science, 269*, 853–855.

Giersch, A., Humphreys, G. W., Boucart, M., & Kovacs, I. (2000). The computation of occluded contours in visual agnosia: Evidence for early computation prior to shape binding and figure-ground coding. *Cognitive Neuropsychology, 17*, 731–759.

Gilbert, C., Ito, M., Kapadia, M., & Westheimer, G. (2000). Interactions between attention, context and learning in primary visual cortex. *Vision Research, 40*, 1217–1226.

Gilchrist, I., Humphreys, G. W., & Riddoch, M. J. (1996). Grouping and extinction: Evidence for low-level modulation of selection. *Cognitive Neuropsychology, 13*, 1223–1256.

Goodrich, S. J., & Ward, R. (1997). Anti-extinction following unilateral parietal damage. *Cognitive Neuropsychology, 14*, 595–612.

Grossberg, S., & Mingolla, E. (1985). Neural dynamics of form perception: Boundary completion, illusory figures and neon color spreading. *Psychological Review, 92*, 173–211.

Heywood, C. A., & Cowey, A. (1999). Cerebral achromatopsia. In G. W. Humphreys (Ed.), *Case studies in the neuropsychology of vision* (pp. 17–40). Hove, UK: Psychology Press.

Heywood, C. A., Wilson, B., & Cowey, A. (1987). A case study of cortical colour "blindness" with relatively intact achromatic discrimination. *Journal of Neurology, Neurosurgery and Psychiatry, 50*, 22–29.

Heywood, C. A., & Zihl, J. (1999). Motion blindness. In G. W. Humphreys (Ed.), *Case studies in the neuropsychology of vision* (pp. 1–17). Hove, UK: Psychology Press.

Hillis, A. E., & Caramazza, A. (1995). Cognitive and neural mechanisms underlying visual and semantic processing: Implications from "optic aphasia." *Journal of Cognitive Neuroscience, 7*, 457–478.

Hinton, G. E., & Parsons, L. M. (1981). Frames of references and mental imagery. In A. D. Baddeley & J. Long (Eds.), *Attention and performance IX.* Hillsdale, NJ: Lawrence Erlbaum Associates Inc.

Hochstein, S., & Ahissar, M. (2002). View from the top: Hierarchies and reverse hierarchies in the visual system. *Neuron, 36*, 791–804.

Hodges, J. R., Bozeat, S., Lambon Ralph, M. A., Patterson, K., & Spatt, J. (2000). The role of conceptual knowledge in object use: Evidence from semantic dementia. *Brain, 123*, 1913–1925.

Holmes, G. (1918). Disturbances of vision by cerebral lesions. *British Journal of Ophthalmology, 2*, 353–384.

Humphreys, G. W. (1998). The neural representation of objects in space: A dual coding account. *Philosophical Transactions of the Royal Society, 353*, 1341–1351.

Humphreys, G. W. (2001). A multi-stage account of binding in vision: Neuropsychological evidence. *Visual Cognition, 8*, 381–410.

Humphreys, G. W., Cinel, C., Wolfe, J., Olson, A., & Klempen, N. (2000). Fractionating the binding process: Neuropsychological evidence distinguishing binding of form from binding of surface features. *Vision Research, 40*, 1569–1596.

Humphreys, G. W., & Riddoch, M. J. (1984). Routes to object constancy: Implications from neurological impairments of object constancy. *Quarterly Journal of Experimental Psychology, 36A*, 385–415.

Humphreys, G. W., & Riddoch, M. J. (1987). *To see but not to see: A case study of visual agnosia.* Hove, UK: Lawrence Erlbaum Associates Ltd.

Humphreys, G. W., & Riddoch, M. J. (1999). Impaired development of semantic memory: Separating semantic from structural knowledge and diagnosing a role for action in establishing stored memories for objects *Neurocase, 5*, 519–532.

Humphreys, G. W., & Riddoch, M. J. (2003a). From *what* to *where*: Neuropsychological evidence for implicit interactions between object- and space-based attention. *Psychological Science, 14*, 487–492.

Humphreys, G. W., & Riddoch, M. J. (2003b). A case series analysis of category-specific deficits of living things: The HIT account. *Cognitive Neuropsychology, 20*, 263–306.

Humphreys, G. W., & Riddoch, M. J. (2003c). From vision to action, and action to vision: A convergent route approach to vision, action and attention. In D. Irwin & B. Ross (Eds.), *The psychology of learning and motivation: Visual cognition. Vol. 42*. New York: Academic Press.

Humphreys, G. W., Riddoch, M. J., Nys, G., & Heinke, D. (2002). Unconscious transient binding by time: Neuropsychological evidence from anti-extinction. *Cognitive Neuropsychology, 19*, 361–380.

Humphreys, G. W., Riddoch, M. J., & Quinlan, P. T. (1985). Interactive processes in perceptual organization: Evidence from visual agnosia. In M. I. Posner & O. S. M. Marin (Eds.), *Attention & performance XI*. Hillsdale, NJ: Lawrence Erlbaum Associates Inc.

Humphreys, G. W., Riddoch, M. J., & Quinlan, P. T. (1988). Cascade processes in picture identification. *Cognitive Neuropsychology, 5*, 67–103.

Humphreys, G. W., Romani, C., Olson, A., Riddoch, M. J., & Duncan, J. (1994). Non-spatial extinction following lesions of the parietal lobe in humans. *Nature, 372*, 357–359.

Humphreys, G. W., & Rumiati, R. I. (1998). Stimulus specificity in visual recognition: Agnosia without prosopagnosia or alexia. *Cognitive Neuropsychology, 15*, 243–278.

Humphreys, G. W., Troscianko, T., Riddoch, M. J. Boucart, M., Donnelly, N., & Harding, G. (1992). Covert processing in different visual recognition systems. In A. D. Milner & M. Rugg (Eds.), *The neuropsychology of consciousness*. London: Academic Press.

James, T. W., Culham, J., Humphrey, G. K., Milner, A. D., & Goodale, M. A. (2003). Ventral occipital lesions impair object recognition but not object-directed grasping: An fMRI study. *Brain, 126*, 2463–2475.

Jankowiak, J., Kinsbourne, M., Shalev, R. S., & Bachman, D. I. (1992). Preserved visual imagery and categorization in a case of associative visual agnosia. *Journal of Cognitive Neuroscience, 4*, 119–131.

Jeannerod, M., Decety, J., & Michel, F. (1994). Impairment of grasping movements following a bilateral posterior parietal lesion. *Neuropsychologia, 32*, 369–380.

Karnath, H. -O. (1988). Deficits in attention in acute and recovered visual hemi-neglect. *Neuropsychologia, 26*, 27–43.

Kertesz, A. (1979). Visual agnosia: A dual deficit of perception and recognition. *Cortex, 15*, 403–419.

Kosslyn, S. M. (1994). *Image and brain: The resolution of the imagery debate*. Cambridge, MA: MIT Press.

Kumada, T., & Humphreys, G. W. (2001). Lexical recovery on extinction: Interactions between visual form and stored knowledge modulate visual selection. *Cognitive Neuropsychology, 18*, 465–478.

Laeng, B., Carlesimo, G. A., Caltagirone, C., Capasso, R., & Miceli, G. (2002). Rigid and nonrigid objects in canonical and noncanonical views: Hemisphere-specific effects on object identification. *Cognitive Neuropsychology, 19*, 697–720.

Landis, T., Regard, M., Bliestle, A., & Kleihues, P. (1988). Prosopagnosia and agnosia for noncanonical views. *Brain, 111*, 1287–1297.

Lhermitte, F. (1983). "Utilisation" behaviour and its relations to lesions of the frontal lobes. *Brain, 106*, 237–255.

Lhermitte, F., & Beauvois, M.-F. (1973). A visual-speech disconnection syndrome. Report of a case with optic aphasia. *Brain, 96*, 695–714.

Lissauer, H. (1890). Ein Fall von Seelenblindheit nebst einem Beitrage zur Theorie derselben. *Archives für Psychiatrie und Nervenkrankheiten, 21*, 222–270.

Luck, S. J., & Vogel, E. K. (1997). The capacity of visual working memory for features and conjunctions. *Nature, 309*, 279–281.

Lueck, C. J., Zeki, S., Friston, K. J., Deiber, M. P., Cope, P., Cunningham, V. J., Lammertsma, A. A., Kennard, C., & Frackowiak, R. S. J. (1989). The colour centre in the cerebral cortex of man. *Nature, 340*(6232), 386–389.

Marcel, A. J. (1998). Blindsight and shape perception: Deficit of visual consciousness or of visual function? *Brain, 121*, 1565–1588.

Marr, D. (1982). *Vision*. San Francisco: W. H. Freeman.

Mattingly, J. B., Davis, G., & Driver, J. (1997). Preattentive filling in of visual surfaces in parietal extinction. *Science, 275*, 671–674.

McLeod, P., Dittrich, W., Driver, J. Perrett, D., & Zihl, J. (1996). Preserved and impaired detection of structure from motion by a "motion-blind" patient. *Visual Cognition, 3*, 363–392.

Miceli, G., Fouch, E., Capasso, R., Shelton, J. R., Tomaiuolo, F., & Caramazza, A. (2001). The dissociation of color from form and function knowledge. *Nature Neuroscience, 4*, 662–667.

Milner, A. D., Dijkerman, H. C., McIntosh, R. D., Rossetti, Y., & Pisella, L. (2003). Delayed reaching and grasping in patients with optic ataxia. *Progress in Brain Research, 142*, 223–240.

Milner, A. D., Dijkerman, H. C., Pisella, L., McIntosh, R. D., Tilikete, C., Vighetto, A., & Rossetti, Y. (2001). Grasping the past: Delay can improve visuomotor performance. *Current Biology, 11*, 1896–1901.

Milner, A. D., & Goodale, M. A. (1995). *The visual brain in action*. New York: Academic Press.

Milner, A. D., Paulignan, Y., Dijkerman, H. C., Michel, F., & Jeannerod, M. (1999). A paradoxical improvement of misreaching in optic ataxia: New evidence for two separate neural systems for visual localization. *Proceedings of the Royal Society, London, B266*, 2225–2229.

Milner, A. D., Perrett, D. I., Johnston, R. S., Benson, P. J., Jordan, T. R., Heeley, D. W., Bettucci, D., Mortara, F., Mutani, R., Terazzi, E., & Davidson, D. L. (1991). Perception and action in visual form agnosia. *Brain, 114*, 405–428.

Mon-Williams, M., Tresilian, J. R., McIntosh, R. D., & Milner, A. D. (2001). Monocular and binocular distance cues: Insights from visual form agnosia I (of III). *Experimental Brain Research, 139*, 127–136.

Moran, J., & Desimone, R. (1985). Selective attention gates visual processing in the extrastriate cortex. *Science, 229*, 782–784.

Navon, D. (1977). Forest before trees: The precedence of global features in visual perception. *Cognitive Psychology, 9*, 353–383.

Palmer, S. E., & Rock, I. (1994). Rethinking perceptual organization: The role of uniform connectedness. *Psychonomic Bulletin and Review, 1*, 29–55.

Pascual-Leone, A., & Walsh, V. (2001). Fast backprojections from the motion to the primary visual area necessary for visual awareness. *Science, 292*, 510–512.

Perenin, M. -T., & Vighetto, A. (1988). Optic ataxia. *Brain, 111*, 643–674.

Pilgrim, E., & Humphreys, G. W. (1991). Impairment of action to visual objects in a case of ideomotor apraxia. *Cognitive Neuropsychology, 8*, 459–473.

Price, C. J., & Humphreys, G. W. (1989). The effects of surface detail on object categorisation and naming. *Quarterly Journal of Experimental Psychology, 41A*, 797–828.

Priftis, K., Rusconi, E., Umiltà, C., & Zorzi, M. (2003). Pure agnosia for mirror stimuli after right inferior parietal lesion. *Brain, 126*, 908–919.

Prinzmetal, W. (1981). Principles of feature integration. *Perception and Psychophysics, 30*, 330–340.

Quaglino, A. (1867). Emiplegia sinistra con amaurosi—Guarigione—Perdita totale della percezione dei colori e della memoria della configurazione degli oggetti. *Giornale d'Oftalmologia Taliano, 10*, 106–112.

Ratcliff, G., & Newcombe, F. (1982). Object recognition: Some deductions from the clinical evidence. In A. W. Ellis (Ed.), *Normality and pathology in cognitive functions*. London: Academic Press.

Rees, G., Wojciulik, E., Clarke, K., Husain, M., Frith, C., & Driver, J. (2000). Unconscious activation of visual cortex in the damaged right hemisphere of a parietal patient with extinction. *Brain, 123*, 1624–1633.

Ricci, R., Vaishnavi, S., & Chatterjee, A. (1999). A deficit of intermediate vision: Experimental observations and theoretical implications. *Neurocase, 5*, 1–12.

Riddoch, G. (1917). Dissociation of visual perceptions due to occipital injuries, with especial reference to appreciation of movement. *Brain, 40*, 15–57.

Riddoch, M. J., Edwards, M. G., Humphreys, G. W., West, R., & Heafield, T. (1998). Visual affordances direct action. Neuropsychological evidence from manual interference. *Cognitive Neuropsychology, 15*, 645–684.

Riddoch, M. J., & Humphreys, G. W. (1987a). A case of integrative visual agnosia. *Brain, 110*, 1431–1462.

Riddoch, M. J., & Humphreys, G. W. (1987b). Visual object processing in optic aphasia: A case of semantic access agnosia. *Cognitive Neuropsychology, 4*, 131–185.

Riddoch, M. J., & Humphreys, G. W. (2004). Object identification in simultanagnosia: When wholes are not the sum of their parts. *Cognitive Neuropsychology, 11*, 423–442.

Riddoch, M. J., Humphreys, G. W., & Edwards, M. (2000). Visual affordances and object selection. In S. Monsell & J. Driver (Eds.), *Attention and performance XVIII*. Cambridge, MA: MIT Press.

Riddoch, M. J., Humphreys, G. W., Gannon, T., Blott, W., & Jones, V. (1999). Memories are made of this: The effects of time on stored visual knowledge in a case of visual agnosia. *Brain, 122*, 537–559.

Riddoch, M. J., Humphreys, G. W., Hardy, E., Blott, W., & Smith, A. (2003). Visual and spatial short-term memory in visual agnosia. *Cognitive Neuropsychology*, *20*, 641–671.

Riddoch, M. J., Humphreys, G. W., Jacobson, S., Pluck, G., Bateman, A., & Edwards, M. G. (2004). Impaired orientation discrimination and localization following parietal damage: On the interplay between dorsal and ventral processes in visual perception. *Cognitive Neuropsychology*, *21*, 597–624.

Riddoch, M. J., Humphreys, G. W., & Price, C. J. (1989). Routes to action: Evidence from apraxia. *Cognitive Neuropsychology*, *6*, 437–454.

Robertson, L. C. (2003). Binding, spatial attention and perceptual awareness. *Nature Reviews Neuroscience*, *4*, 93–102.

Robertson, L. C., Treisman, A., Friedman-Hill, S., & Grabowecky, M. (1997). The interaction of spatial and object pathways: Evidence from Balint's syndrome. *Journal of Cognitive Neuroscience*, *9*, 295–317.

Rogers, T. T., Hodges, J. R., Lambon Ralph, M. A., & Patterson, K. (2003). Object recognition under semantic impairment: The effects of conceptual regularities on perceptual decisions. *Language and Cognitive Processes*, *18*, 625–662.

Rossetti, Y., & Pisella, L. (2002). Several 'vision for action' systems: A guide to dissociating and integrating dorsal and ventral functions. In W. Prinz & B. Hommel (Eds.), *Common mechanisms in perception and action: Attention and performance XIX* (pp. 62–119). Oxford: Oxford University Press.

Rumiati, R. I., Humphreys, G. W., Riddoch, M. J., & Bateman, A. (1994). Visual object agnosia without prosopagnosia or alexia: Evidence for hierarchical theories of visual recognition. *Visual Cognition*, *1*, 181–225.

Servos, P., & Goodale, M. A. (1995). Preserved visual imagery in visual form agnosia. *Neuropsychologia*, *33*, 1383–1394.

Seymour, P. H. K. (1979). *Human visual cognition*. London: Collier MacMillan.

Shalev, L., & Humphreys, G. W. (2002). Implicit location encoding via stored representations of familiar objects: Neuropsychological evidence. *Cognitive Neuropsychology*, *19*, 721–744.

Sheridan, J., & Humphreys, G. W. (1993). A verbal-semantic category-specific recognition deficit. *Cognitive Neuropsychology*, *10*, 143–184.

Silvanto, J., Cowey, A., Lavie, N., & Walsh, V. (2005). Striate cortex (V1) activity gates awareness of motion. *Nature Neuroscience*, *8*, 143–144.

Singer, W., & Gray, C. M. (1995). Visual feature integration and the temporal correlation hypothesis. *Annual Review of Neuroscience*, *18*, 555–586.

Sparr, S. A., Jay, M., Drislane, F. W., & Venna, N. (1991). A historic case of visual agnosia revisited after 40 years. *Brain*, *114*, 789–800

Stankiewicz, B. J., Hummel, J. E., & Cooper, E. E. (1998). The role of attention in priming for left-right reflections of object images: Evidence for a dual representation of object shape. *Journal of Experimental Psychology: Human Perception and Performance*, *24*, 732–744.

Stewart, F., Parkin, A. J., & Hunkin, N. M. (1992). Naming impairments following recovery from herpes simplex encephalitis: Category specific? *Quarterly Journal of Experimental Psychology*, *44A*, 261–284.

Thomas, R. M., Forde, E. M. E., Humphreys, G. W., & Graham, K. S. (2002). The effects of passage of time on a patient with category-specific agnosia. *Neurocase*, *8*, 466–479.

Treisman, A. (1998). Feature binding, perception and object perception. *Philosophical Transaction of the Royal Society*, *B353*, 1295–1306.

Treisman, A., & Schmidt, H. (1982). Illusory conjunction in the perception of objects. *Cognitive Psychology*, *14*, 107–141.

Troscianko, T., Davidoff, J., Humphreys, G. W., Lanids, T., Fahle, M., Greenlee, M., Brugger, P., & Phillips, W. (1996). Human colour discrimination based on a non-parvocellular pathway. *Current Biology*, *16*, 4–14.

Turnbull, O. H., Beschin, N., & Della Sala, S. (1997). Agnosia for object orientation: Implications for theories of object recognition. *Neuropsychologia*, *35*, 153–163.

Turnbull, O. H., Della Sala, S., & Beschin, N. (2002). Agnosia for object orientation: Naming and mental rotation evidence. *Neurocase*, *8*, 296–305.

Ungerleider, L. G., & Mishkin, M. (1982). Two cortical visual systems. In D. J. Ingle, M. A. Goodale, & R. J. W. Mansfield (Eds.), *Analysis of visual behaviour* (pp. 549–585). Cambridge, MA: MIT Press.

Vaina, L. M., & Cowey, A. (1996). Impairment of the perception of second order motion but not first order motion in a patient with unilateral focal brain damage. *Proceedings of the Royal Society, London*, *B263*, 1225–1232.

Vaina, L. M., Cowey, A., & Kennedy, D. (1999). Perception of first and second-order motion: Separable neurological mechanisms? *Human Brain Mapping, 7*, 67–77.

Vaina, L. M., Cowey, A., LeMay, M., Bienfang, D. C., & Kikinis, R. (2002). Visual deficits in a patient with 'kaleidoscopic disintegration of the visual world.' *European Journal of Neurology, 9*, 463–477.

Vaina, L. M., Makris, N., Kennedy, D., & Cowey, A. (1998). The selective impairment of the perception of first-order motion by unilateral cortical brain damage. *Visual Neuroscience, 15*, 333–348.

Vecera, S. P., & Behrmann, M. (1997). Spatial attention does not require preattentive grouping. *Neuropsychology, 11*, 30–43.

Verrey, L. (1888). Hémiachromatopsie droite absolue. *Archives of Ophthalmology (Paris), 8*, 289–301.

Vuilleumier, P., Sagiv, N., Hazeltine, E., Poldrack, R. A., Swick, D., Rafal, R. D., & Gabrieli, J. D. E. (2001). Neural fate of seen and unseen faces in visuospatial neglect: A combined event-related functional MRI and event-related potential study. *Proceedings of the National Academy of Sciences of America, 98*, 3495–3500.

Ward, R., Danziger, S., Owen, V., & Rafal, R. D. (2002). Deficits in spatial coding and feature binding following damage to spatiotopic maps in the human pulvinar. *Nature Neuroscience, 5*, 99–100.

Ward, R., & Goodrich, S. (1996). Differences between objects and non-objects in visual extinction: A competition for attention. *Psychological Science, 7*, 177–180.

Ward, R., Goodrich, S., & Driver, J. (1994). Grouping reduces visual extinction: Neuropsychological evidence for weight-linkage in visual selection. *Visual Cognition, 1*, 101–130.

Warrington, E. K. (1982). Neuropsychological studies of object perception. *Philosophical Transactions of the Royal Society, London, B298*, 15–32.

Warrington, E. K., & Davidoff, J. (2000). Failure at object identification improves mirror image matching. *Neuropsychologia, 38*, 1229–1234.

Warrington, E. K., & James, M. (1967). Disorders of visual perception in patients with localized cerebral lesions. *Neuropsychologia*, 5, 253–266.

Warrington, E. K., & James, M. (1986). Visual object recognition in patients with right hemisphere lesions: Axes or features. *Perception, 15*, 355–366.

Warrington, E. K., & Rabin, P. (1970). Perceptual matching in patients with cerebral lesions. *Neuropsychologia, 8*, 475–487.

Warrington, E. K., & Taylor, A. M. (1973). The contribution of the right parietal lobe to object recognition. *Cortex, 9*, 152–164.

Warrington, E. K., & Taylor, A. M. (1978). Two categorical stages of object recognition. *Perception, 7*, 695–705.

Weiskrantz, L. (1986). *Blindsight.* Oxford: Oxford University Press.

Weiskrantz, L., Warrington, E. K., Sanders, M. D., & Marshall, J. (1974). Visual capacity in the hemianopic field following restricted occipital ablation. *Brain, 97*, 709–728.

Wheeler, M. E., & Treisman, A. (2002). Binding in short-term visual memory. *Journal of Experimental Psychology: General, 131*, 48–64.

Wigan, A. L. (1844). *The duality of the mind: Proved by the structure, functions and diseases of the brain, and by the phenomena of mental derangement, and shown to be essential to moral responsibility.* London: Longman, Brown, Green & Longmans.

Wilson, B. A., & Davidoff, J. (1993). Partial recovery from visual object agnosia: A 10-year follow-up study. *Cortex, 29*, 529–542.

Wojciulik, E., & Kanwisher, N. (1998). Implicit but not explicit feature binding in a Balint's patient. *Visual Cognition, 5*, 157–181.

Yoon, E. Y., Heinke, D., & Humphreys, G. W. (2002). Modelling direct perceptual constraints on action selection: The Naming and Action Model (NAM). *Visual Cognition, 9*, 615–661.

Young, A. W., Humphreys, G. W., Riddoch, M. J., Hellawell, D. J., & de Haan, E. H. F. (1994). Recognition impairments and face imagery. *Neuropsychologia, 32*, 693–702.

Zeki, S. (1978). Functional specialization in the visual cortex of the rhesus monkey. *Nature, 274*, 423–428.

Zeki, S. (1993). *A vision of the brain.* Oxford: Blackwells.

Zeki, S., & ffytche, D. H. (1998). The Riddon syndrome: Insights into the neurobiology of conscious vision. *Brain, 121*, 25–45.

Zeki, S., Watson, J. D. G., Lueck, C. J., Friston, K. J., Kennard, C., & Frackowiak, R. S. J. (1991). A direct demonstration of functional specialization in human visual cortex. *Journal of Neuroscience, 11*, 641–649.

Zihl, J., Von Cramon, D., & Mai, N. (1983). Selective disturbance of movement vision after bilateral brain damage. *Brain, 106*, 313–340.

COGNITIVE NEUROPSYCHOLOGY, 2006, 23 (1), 184–201

On the right (and left) track: Twenty years of progress in studying hemispatial neglect

Laurel J. Buxbaum
Moss Rehabilitation Research Institute and Thomas Jefferson University, Philadelphia, PA, USA

In the last 20 years, several important developments have markedly expanded our understanding of the hemispatial neglect syndrome and, more broadly, of the brain's representation of objects, space, and action. This review follows seven "threads" of scientific development to evaluate some of the changes in our thinking about the mechanisms of neglect, its key characteristics, the spatial frames of reference affected, the psychophysical properties of neglect phenomena, the relationship of neglect in various sensory modalities, the role of deficits in arousal and general attention, and the critical neuroanatomic substrates of the disorder. The progress reviewed illustrates that the cognitive neuropsychology approach to the study of neglect complements insights gleaned from neurophysiological studies in the monkey, functional neuroimaging studies of attention and perception, and other investigative techniques, and thus serves as fertile ground for a convergence approach to cognitive neuroscience.

INTRODUCTION

The dramatic symptoms of the neglect syndrome have captured the interest and imaginations of clinicians and researchers for more than a century. From a clinical perspective, it can be startling to encounter a patient who fails to attend to or orient toward stimuli on the side of space opposite a brain lesion, eat from the contralesional side of her plate, or shave the contralesional half of his face. Neglect of space and the body may be accompanied by a number of other associated phenomena that are fascinating in their own right, including anosognosia (denial of symptoms), anosodiaphoria (indifference to illness or disability), motor impersistence, impairments in directional motor programming (directional akinesia or hypokinesia), visual, auditory, and tactile extinction of contralesional stimuli, and neglect of the contralesional sides of objects (object-based or allocentric neglect). Neglect is also of scientific interest for its potential to inform our understanding of the brain's representation of space, attention, and action. Because neglect-like phenonemona may be observed in the monkey, the neglect syndrome may also speak to the homologies between human and monkey brain, and the evolutionary development of space and action

Correspondence should be addressed to Laurel J. Buxbaum, Moss Rehabilitation Research, Korman Pavilion Suite 213, 1200 W. Tabor Rd., Philadelphia PA 19141, USA (Email: lbuxbaum@einstein.edu).
Supported by NIH-R01-NS36387 and NIDRR H133G030169.

DOI:10.1080/02643290500202698

representation. Finally, the study of neglect serves to complement insights gleaned from functional neuroimaging studies of attention and perception, and thus provides an important example of the benefits of a convergence approach to cognitive neuroscience.

Although first reported in the scientific literature early in the last century (see Benke, Luzzatti, & Vallar, 2004; Riddoch, 1935), neglect was not a topic of substantial scientific interest until the early 1940s, when *Brain* published an important manuscript describing the disorder as an impairment of perceptual space and the body schema (*Brain*, 1941). By the 1980s, hundreds of papers on the topic had established the basic characteristics of the disorder. Over the past 20 years—the length of time that *Cognitive Neuropsychology* has been published—several important developments have greatly expanded our understanding of the disorder and, indeed, of the brain's representation of objects, space, and action.

When we review the literature circa mid-1980s to now, at least seven main threads of ongoing investigation are apparent. The first thread is an exploration of the mechanisms underlying the neglect syndrome, with increasing recognition that no one underlying mechanism can account for the varied and multifarious symptoms that characterise the disorder. The second thread entails assessment of the key characteristics of the neglect syndrome and, along with this, the possibility that there are actually several relatively discrete neglect syndromes or different subtypes of neglect. The third thread, informed by recent developments in studies of normal attention, is an examination of the spatial frames of reference affected by neglect. This includes developments in our knowledge of egocentric frames of reference for spatial coding and action, and object-based mechanisms for allocating attention to objects. The fourth thread is an investigation of the psychophysical properties of neglect phenomena and, related to this, of the degree to which unreported stimuli may be processed. The fifth thread is an assessment of the sensory modalities within which neglect may occur, and the relationship of neglect in these various modalities. The sixth thread is an exploration of the role of deficits in arousal and general attention in the genesis of the neglect syndrome. The seventh thread is an ongoing dialogue about the critical neuroanatomic substrates of the disorder.[1] In the following sections, we follow these threads to provide a sense of changes in contemporary thinking about neglect over the past 20 years, and conclude with some comments about areas of inquiry that appear particularly ripe for future development.

MECHANISMS UNDERLYING THE NEGLECT SYNDROME

For much of the last 20 to 30 years, several competing theories of the mechanisms underlying the neglect syndrome have vied for dominance. Heilman and Valenstein (1972, 1979), for example, proposed that the neglect syndrome is a deficit in the orienting response, i.e., an attention-arousal deficit, caused by disruption of a corticolimbic-reticular loop involved in preparation for action. Kinsbourne (1974, 1987) framed neglect as an imbalance in systems of each hemisphere biasing attention, oculomotor actions, and head turning in a contralateral direction. Bisiach and Luzzatti (1978) noted that patients with left neglect omitted left-sided details when attempting to recall the layout of a familiar location, the Piazza del Duomo in Milan. In a related study, Bisiach and colleagues (Bisiach, Luzzatti, & Perani, 1979) demonstrated that left neglect patients omitted left-sided details from abstract shapes, even when the shapes were exhibited moving from left to right, or right to left, behind a static midline window. Scanning or perceptual deficits could not account for these

[1] There is actually an eighth thread, comprising investigation of treatments of neglect. Several thorough reviews of studies and progress in this area have been reported elsewhere. The interested reader is referred to Bowen, Lincoln, and Dewey (2002), Cicerone et al. (2000), and Pierce and Buxbaum (2002).

data, which instead indicated a specific deficit in the left side of a representation of space. At least one study from this time period identified a single factor accounting for variability among tests of neglect (Halligan, Marshall, & Wade, 1989), reinforcing the idea that a single underlying mechanism might eventually be elucidated.

On the other hand, several investigators writing in the 1980s held that neglect is a multifactorial disorder. DeRenzi (1982) considered several competing theories, and concluded that perceptual, motor, and representational factors may all contribute to the pattern of behaviour. Mesulam (1985), similarly argued that neglect is a "composite phenomenon" that has complex perceptual, motor, and motivational aspects. And although Bisiach (1987) emphasised the representational account, he concluded that sensory, motor, and representational factors are all influential.

Many investigators would agree that the latter view would prove prescient. As the multimodal and complex nature of the neglect syndrome has become increasingly evident (see the next section), there has been growing consensus that no single underlying mechanism can account for the range of demonstrated symptoms. Indeed, most current manuscripts on neglect include an introductory statement about the multifactorial nature of the disorder (e.g., Chatterjee, 1998; Vallar, 2001). To summarise the change in *zeitgeist*, the thrust of many inquiries has shifted from an attempt to determine causality to an exploration of what the observed fractionations and dissociations tell us about the brain's representations of space and action. In this shift the cognitive neuropsychology approach has played a central role (see, e.g., Subbiah & Caramazza, 2000).

CHARACTERISTICS OF THE NEGLECT SYNDROME

The dramatic and varied clinical characteristics of the neglect syndrome were well described even in relatively early accounts (e.g., Critchley, 1953). In the more recent literature, many investigators have noted a great variety of clinical symptoms. Mesulam (1985), for example, observed: "One patient may shave, groom, and dress, only the right side of the body; another may fail to eat food placed on the left side of the tray; another may omit to read the left half of each sentence; still another may fail to copy detail on the left side of a drawing and may show a curious tendency to leave an uncommonly wide margin on the left side of the paper when asked to write" (p. 142).

In addition to the more obvious manifestations of the disorder (many of which are listed in the Introduction), there is evidence that the neglect syndrome may affect primary sensory and motor processing, thus having significant implications for rehabilitation. Patients with neglect have greater motor and sensory impairment than patients without neglect, suggesting that primary sensory and motor deficits may be augmented (or perhaps mimicked) by unilateral attentional and spatial deficits (Barrett, Peterlin, & Heilman, 2003; Buxbaum et al., 2004; Sterzi et al., 1993). For example, detection of unilateral tactile stimuli may be improved by crossing the contralesional limb into the "good" ipsilesional hemispace (Aglioti, Smania, & Peru, 1999; Smania & Aglioti, 1995; Valenza, Seghier, Schwartz, Lazeyras, & Vuilleumier, 2004; Vallar, 1997). Neglect is frequently associated with impaired vision in the contralateral hemispatial field, but in at least some cases the hemianopia may be modified by the direction of gaze, suggesting an interaction with attentional factors (Muller-Oehring et al., 2003; Nadeau & Heilman, 1991).

An intriguing (but not uncontroversial) recent development is the notion that there may actually be a detailed taxonomy underlying the plethora of symptoms, and that the disorder may be fractionated into distinct subtypes, each having a different behavioural profile. The seeds of this idea were sown in earlier studies with monkeys showing that neglect may comprise either a failure to respond with the contralesional limb to stimuli in both contralesional and ipsilesional hemispace (motor neglect), or a deficit in responding to stimuli in contralesional space with either limb (sensory neglect) (Watson, Miller, & Heilman, 1978; see also Heilman & Valenstein,

1972). Heilman and Valenstein (1979) pointed out that neglect patients bisect lines less accurately in left than right hemispace, suggesting a reduced ability to act in contralesional hemispace. Subsequently, Heilman and colleagues (Heilman, Bowers, Coslett, Whelan, & Watson, 1985; Heilman, Bowers, & Watson, 1983; and see Heilman, Valenstein, & Watson 2000, for review) demonstrated that neglect may have a directional component, comprising failure or slowing of movement into and toward the contralesional hemispace with either limb.

Bisiach, Geminiani, Berti, and Rusconi (1990) pointed out that most common tests of neglect confound motor and perceptual demands. Subsequently, a number of methods have been devised to dissociate motor and perceptual components, including pulleys (Bisiach et al., 1990), video monitors (Coslett, Bowers, Fitzpatrick, Haws, & Heilman, 1990), inversion of computer mice (Halligan & Marshall, 1989), mirrors (Tegner & Levander, 1991), overhead projection (Nico, 1996), and computerised response tasks (Buxbaum & Permaul, 2001) (see Figure 1).

There is reason, however, to view at least some of these findings with caution. Harvey, Kramer-McCaffery, Dow, Murphy, and Gilchrist (2002) reported that neglect patients tested on three tasks designed to dissociate motor and perceptual demands (the "overhead task," Nico, 1996, the "pulley test," Bisiach et al., 1990, and the "landmark test," Milner, Harvey, Roberts, & Forster, 1993) performed inconsistently on the three measures. Furthermore, there is no evidence

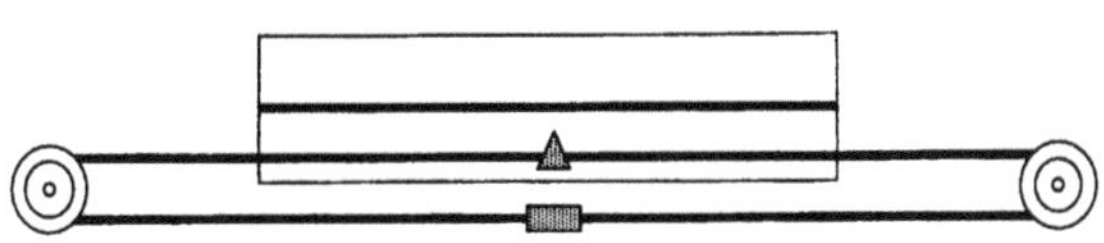

Figure 1. *Schematic drawing of the pulley apparatus used by Bisiach et al. (1990) to investigate dissociations of perceptual and motor neglect. The task is to bisect the depicted line, either by handling a pointer (triangle), or a portion of string (rectangle). In the latter noncongruent condition the direction of hand movement and direction of pointer movement are placed in opposition.*

that the patterns of performance are stable across time.

In addition to motor and perceptual neglect components, the 1980s and 1990s witnessed a flurry of studies drawing distinctions between neglect of the contralesional body (personal neglect: Bisiach, Perani, Vallar, & Berti, 1986; Guariglia & Antonucci, 1992; Halligan & Marshall, 1991; Mennemeier, Wertman, & Heilman, 1992), near space (peripersonal neglect: Halligan & Marshall, 1991; Mennemeier et al., 1992) and far space (extrapersonal neglect: Berti, Smania, & Allport, 2001). Some patients neglect regions in inferior space (Rapcsak, Cimino, & Heilman, 1988). There are patients who exhibit neglect of representational space but not visual space, and vice versa (Coslett, 1997). There are patients whose neglect is multimodal (e.g., visual, auditory, and tactile), and others who exhibit a unimodal deficit. Finally (but clearly not exhaustively), neglect may occur for the left sides of words and objects but the right side of space (Costello & Warrington, 1987).

We (Buxbaum et al., 2004) recently assessed 166 right-hemisphere stroke patients with a battery of tests of percpetual, motor, personal, and peripersonal neglect, and found 12 patterns of co-occurrence of these deficits. The most frequent pattern was peripersonal neglect (defined by performance on traditional tasks such as line bisection and letter cancellation, with (n = 22) or without (n = 19) perceptual neglect (defined by impaired indication of detection of left-sided targets via midline response). Also frequent was peripersonal neglect with associated motor neglect (n = 18), as well as isolated motor neglect (n = 10), defined by impairment in indicating detection of midline targets with a left-sided response. Much more rare was isolated personal neglect, defined by impaired tactile detection of left-sided targets on the body (n = 2). The sheer number of these and other reported patterns and their combinatorial possibilities raise the question of whether all (or even some) of these symptoms reflect distinct subtypes after all, or extreme values in a multidimensional factor space, or even merely "noise" due to poor reliability of the tests employed. We will return to this question later.

NEURAL CODING OF ATTENTION AND SPACE FOR ACTION

Ongoing work on the neurophysiology of spatial and action coding in the monkey has had an important impact on the characterisation of neglect and on the types of research questions that are asked. In the early 1980s, Ungerleider and Mishkin (1982) described the dissociation of ventral and dorsal visual processing streams. The former courses from the occipital lobe through temporal lobe structures, and is specialised largely for the identification of objects (the "what" system), whereas the latter projects from the occipital through the parietal and frontal lobes, and is concerned with the coding of location, independent of identity (the "where" system). On several accounts, knowledge of "which object is where" is derived from the "binding" of information processed by each system (e.g., Coslett & Saffran, 1991; Kahneman, Treisman, & Gibbs, 1992).

Milner and Goodale (1995) have suggested that the dorsal stream is more accurately characterised as the "how" system, in that it represents the locations of objects for the purpose of action. Several investigators have suggested that spatial coding of target location in the dorsal stream is not absolute or fixed with respect to environmental axes, but is a dynamic map that shifts depending upon the locations of the effectors with respect to the target. Furthermore, the spatial locations of objects within the environment are likely to be specified in terms of the movement vector that would be required for a given effector (e.g., eye, head, hand) to "acquire" the target (e.g., Anderson, Snyder, Li, & Stricanne, 1993; Duhamel, Colby, & Goldberg, 1992; Graziano, Yap, & Gross, 1994). The vector, specifying direction and amplitude of movement, is the emergent product of a population of neurons, each programming movement in a slightly different direction. The neuronal population coding may take the form of a "planar gain field" in which the amplitude of the visual response is modulated by effector position (but see Pouget, Deneve, & Duhamel, 2002; Pouget & Sejnowski, 2001, for an alternative account based on a neural architecture that combines basis functions and attractor dynamics). Collectively, effector-centred coding (i.e., centred on the retina, eye, head, shoulder, torso, hand, etc.) can be termed "egocentric" spatial coding, in that it specifies the locations of objects in the environment (and the motor plans required to acquire the objects) with respect to the body and its parts.

These (and other) physiological studies of the dorsal stream in monkeys had a far-reaching impact on investigators' ability to conceptualise and assess the reference frames within which neglect occurs. Without explicitly invoking the notion of frames of reference, Mesulam (1985) discussed the interaction of head- and eye-centred factors in determining the focus of attention. Bisiach, Capitani, and Porta (1985) dissociated trunk and head position of a manipulandum, and found that both body position and line of sight were influential. A number of investigators have attempted to elucidate the frames of reference contributing to neglect by modifying the location of a target in one reference frame while holding it constant in another (e.g., Calvanio, Petrone, & Levine, 1987; Caramazza & Hillis, 1990b; Hillis, Rapp, Benzing, & Caramazza, 1998; Ladavas, 1987). These studies suggest that different reference frames may simultaneously affect attentional allocation.

One of the more recent developments in this area is the understanding that neglect may be defined with respect to the current position of the hand, i.e., via a hand-centred reference frame. In healthy subjects, distractor stimuli in the path of an intended action interfere strongly with response times relative to distractors not in that path (Tipper, Howard, & Jackson, 1997; Tipper, Lortie, & Baylis, 1992). This is consistent with the "premotor theory of attention" (Rizzolatti & Camarda, 1987), which suggests that spatial attention to objects is a function of the intention to program object-directed movements (e.g., of eye or limb). If a left–right attentional asymmetry in hemispatial neglect is modulated by motor intention, then distractors to the left of the hand should compete less strongly for the control of action than stimuli to the right of the hand. We tested this prediction with 8 right hemisphere stroke patients with left neglect and 12 healthy controls, who reached to central targets presented alone and with surrounding distractors from a left start position (distractors right of hand) and right start position (distractors left of hand). With targets, body, head, and eyes in the same positions, patients with neglect, but not controls, showed *interference* from distractors to the right of the hand, and *facilitation* from distractors to the left of the hand. These data are consistent with damage to the portion of a distributed neuronal population that codes movements to relatively left-ward locations. Thus, the intention to program hand movements in a leftward or rightward direction modulates the pattern of attentional allocation to objects in the array (Buxbaum & Permaul, 2001); see also (Corben, Mattingley, & Bradshaw, 2001).

Evidence from studies in the monkey performed in the 1980s and 1990s indicated that representation of "far" space, i.e., the space beyond reach, may be represented by different neuronal populations (e.g., area 8 and LIP) than those involved in coding "near" reaching space (e.g., area 6, 7b, and VIP) (e.g., Colby, Duhamel, & Goldberg, 1995; Rizzolatti et al., 1988). There is also more recent evidence that coding of "far" and "near" space is not simply a function of whether stimuli are within reaching distance. Berti and Frassinetti (2000), for example, reported a patient with near but not far neglect. However, when he bisected far lines with a long stick, neglect appeared and was as severe as near neglect.

SPATIAL FRAMES OF REFERENCE FOR OBJECT RECOGNITION AND OBJECT-BASED NEGLECT

The problem of how attention is allocated to single object representations is also of relevance to the study of neglect, and is an area of inquiry that has also emerged over the last 20 to 30 years. On reflection, "left" and "right" are relative terms, requiring definition of an axis of origin (i.e., a reference frame). In the cognitive psychology literature, the explicit notion of various frames of reference that define the allocation of attention to objects was perhaps most clearly articulated in the writings of David Marr (1980, 1982). Marr provided a detailed theoretical account of the process by which incoming visual information proceeds from what he termed the "primal sketch" to a 3D model explicitly representing the principal axis of familiar objects.

Perhaps the clearest influence of Marr's work can be seen in the literature indicating that neglect may be defined with respect to the left and right of single objects (object-based neglect). It had long been noted that patients copying multi-object scenes would sometimes omit left-sided details from each object, regardless of its location in egocentric space, but some investigators (e.g., Gainotti, D'Erme, Monteleone, & Silveri, 1986) had attributed this to an effect of sequential eye fixations. With the framework of Marr, these results could be explained as neglect that arose with respect to the principal axes of objects.

Bisiach and colleagues (1986) demonstrated that it is possible to observe right neglect on some tasks, along with left neglect for word reading and sometimes object naming (see also Costello & Warrington, 1987). Caramazza and Hillis (1990a, 1990b) unambiguously demonstrated that object-centred neglect could be distinguished from other forms of neglect (see Figure 2). They used mirror-reversed and vertical stimuli to show that a patient with right neglect failed to read the canonical right side of words regardless of where the right side fell in egocentric space. The same pattern of errors was also observed in written spelling, oral spelling, and backward oral spelling. These and other data were suggestive of damage to the word-centred (i.e., object-centred) grapheme representation.

Subsequently, a number of other investigations demonstrated that the left sides of objects may be

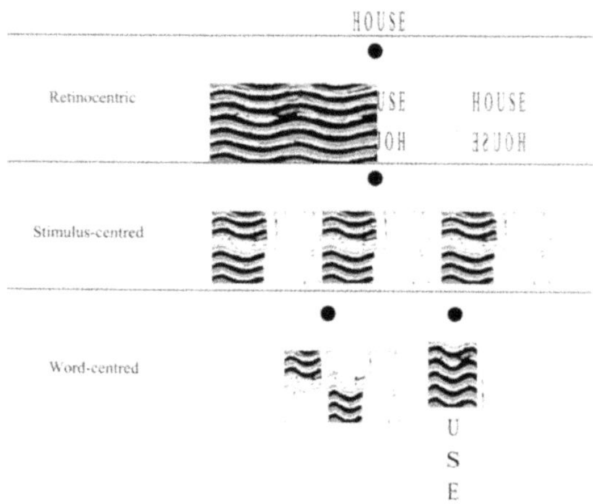

Figure 2. *Figure modified from Subbiah and Caramazza (2000), showing different patterns of neglect that should be observed following damage to retinocentric (coordinate frame centred on the point of fixation), stimulus-centred (coordinate frame centred on the stimulus) and word-centred (coordinate frame centred on the object) levels of representation. The dot corresponds to the fixation point.*

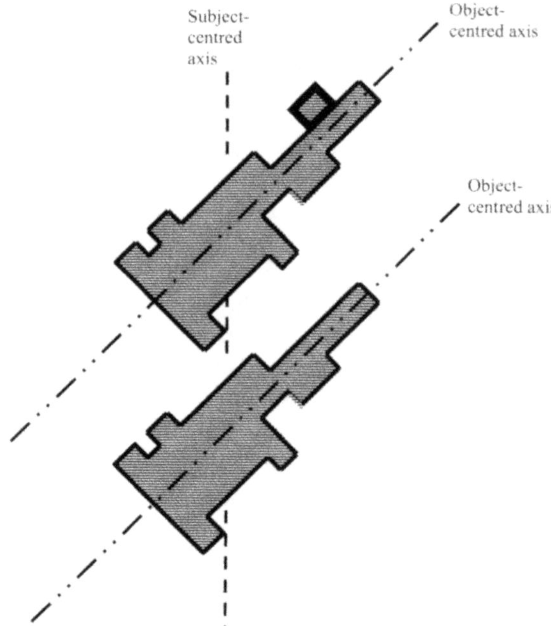

Figure 3. *Drawing of an example item used by Driver and Halligan (1991), with "object-centred" and "body-centred" axes indicated by lines (these lines were not present in the original stimuli). The patient had to indicate whether the shapes were the same or different. In this example, the critical difference between the shapes falls on the right of the subjects' midline, but on the left of the shapes.*

neglected even when the objects are rotated clockwise so that the left sides of the objects fall in right body hemispace (see Figure 3). Behrmann and Tipper (1999) additionally demonstrated that attention may move with a rotating object such that the originally neglected side remains neglected even when it rotates into the "good" side of space.

In the strong sense implied by Marr, object-based neglect is neglect that occurs with respect to the intrinsic principal axes of objects, and thus putatively arises at the level of an internal image of the canonical object, i.e., the structural description. Another possible interpretation of the results with rotated (or mirror-reversed, or vertical) stimuli is that the stimuli are first mentally rotated to their canonical view during the process of object recognition (see Koriat & Norman, 1984) so that the object's left and right, as defined by its vertical axis, are indexed to the viewer's left and right, and subsequently the left is neglected (Buxbaum, Coslett, Montgomery, & Farah, 1996; Cubelli & Speri, 2001; Subbiah & Caramazza, 2000). On this account, although orientation-invariant object-centred representations may be used by the object recognition system to identify the object and locate its vertical axis, spatial attention is nevertheless allocated with respect to an egocentric reference frame. The notion of interactions between object-based and space-based mechanisms of attention is consistent with contributions from both ventral and dorsal stream structures in parsing and attending to space.

PSYCHOPHYSICS AND NEGLECT

It has long been recognised that the magnitude of neglect is influenced by stimulus properties. For example, bisection errors tend to be larger on longer lines (Bisiach, Bulgarelli, Sterzi, & Vallar, 1983). On short lines, patients may even bisect the lines to the left of the true midpoint, a phenomenon referred to as crossover (Halligan &

Marshall, 1988; Marshall & Halligan, 1989). Most theory of neglect have difficulty explaining this effect.

Several recent developments suggest that these phenomena may be described mathematically by power functions. Power functions describe perceived sensory stimuli in terms of an exponential relationship. Chatterjee and colleagues (Chatterjee, 1995; Chatterjee, Mennemeier, & Heliman, 1992; Chatterjee, Ricci, & Calhoun, 2000) have demonstrated that on a variety of tasks, patients with neglect have power functions with exponents that are lower than those of normal patients. This suggests that they do not experience stimuli of increasing magnitude as increasing proportionately.

Another area of developing interest is the temporal modulation of stimulus processing. Temporal order judgments can shed light on whether one stimulus has "prior entry" to conscious perception. Patients with left-sided visual extinction may report that right stimuli precede left unless the latter lead by over 200 ms (Rorden, Mattingley, Karnath, & Driver, 1997). Similarly, patients with auditory extinction judge left-and right-sided acoustic stimuli as being simultaneous when the contralesional sound leads by 270 ms (Karnath, Zimmer, & Lewald, 2002). Patients with tactile extinction are also impaired in determining whether left- or right-sided stimuli have temporal precedence, particularly when left stimuli precede right by 100 to 200 ms (Guerrini, Berlucchi, Bricolo, & Aglioti, 2003). Tactile extinction of left-sided stimuli is greatest when stimuli on the left and right are presented simultaneously (Guerrini et al., 2003) or when right stimuli precede left by 300–900 ms (Cate & Behrmann, 2002), and accuracy improves with longer stimulus onset asynchronies. The similar magnitude of the temporal asynchronies across the three modalities suggests that a temporal delay in awareness for contralesional events may be a ubiquitous aspect of extinction phenomena.

Data relevant to the role of spatial attention in timing also come from investigations of patients with neglect. Basso, Nichelli, Frassinetti, and di Pellegrino (1996) reported a single subject with neglect who consistently underestimated the duration of brief (e.g., 300 and 700 ms) stimuli in the left hemispace. Additionally, Husain, Shapiro, Martin, and Kennard (1997) reported that neglect patients exhibited a significantly longer "attentional blink" for midline visual stimuli than controls; that is, they were impaired at identifying a second target with rapid sequential visual presentation of stimuli (RSVP) unless the two targets were separated by at least 1440 ms.

Snyder and Chatterjee (2004) recently reported a patient with an acute right temporal-parietal stroke in whom they tested the hypothesis that the lesioned region is important in integrating spatial and temporal attention. The patient, like others who have been reported (see above), was biased to judge ipsilesional stimuli as occurring before contralesional stimuli. More importantly, for vertically aligned stimuli, AF judged more accurately the temporal order of successive ipsilesional than contralesional stimuli. Furthermore, his contralesional performance improved with stimuli with larger vertical separation. This suggests that right temporal-parietal damage produces a processing refractory period for stimuli in contralesional space that extends in both space and time.

IMPLICIT PROCESSING OF NEGLECTED INFORMATION

One of the more fascinating developments in the past 20 years is the discovery that neglect patients may implicitly process aspects of stimuli that remain unreported, and that these unreported stimuli may affect performance in systematic ways. One of the earliest relevant observations was that patients with visual extinction are able to make same–different judgments that were more accurate than would be predicted by chance (Volpe, LeDoux, & Gazzaniga, 1979). Another important observation was Marshall and Halligan's (1988) case report of a patient with neglect who, when shown two houses, one with flames shooting from a left-sided window, could not report any differences between the houses, yet systematically preferred the one that was not on fire (see also Bisiach & Rusconi, 1990).

Subsequently, a number of other studies have shown that neglect patients are influenced by unreported stimuli in priming and lexical decision tasks. For example, Berti and Rizzolatti (1992) showed that subjects were able to make "same–different" judgments on visually dissimilar exemplars of the same category, even when one stimulus was neglected. McGlinchey-Berroth, Milberg, Verfaellie, Alexander, and Kilduff (1993) demonstrated that lexical decisions to words at midline were facilitated by unreported semantically related primes in the neglected hemifield. A recent fMRI study demonstrated that extinguished fearful faces activated the left amygdala to the same degree as reported extinguished faces (Vuilleumier et al., 2002). One possible account of the failure to report stimuli that are processed to semantic levels is that conscious awareness requires activated object representations ("what") to be bound to spatial representations ("where") (Coslett & Saffran, 1991; Styles, 1986), and that the latter is deficient in neglect (Berti & Rizzolatti, 1992; Buxbaum & Coslett, 1994; Deouell, 2002).

NEGLECT IN VARIOUS SENSORY MODALITIES/CROSS-MODALITY INTERACTIONS

Even the earliest descriptions of neglect noted that it may affect multiple sensory modalities. Beginning in the 1980s, there was increasing awareness of cross-modal relationships in the neglect syndrome. Bisiach, Cornacchia, Sterzi, and Vallar (1984) noted that neglect patients may be deficient in localising sounds. Rubens (1985) reported amelioration of visual neglect with stimulation of the vestibular system. Subsequently, vestibular stimulation was reported to transiently improve anosognosia (Cappa, Sterzi, Vallar, & Bisiach, 1987) and somatosensory detection (Vallar, Bottini, Rusconi, & Sterzi, 1993). These studies suggest that representation and awareness of space involves input from multiple sensory modalities.

A considerable body of more recent work, only a small portion of which can be reviewed here, confirms these earlier contributions. For example, Dijkerman, Webeling, ter Wal, Groet, and Van Zandvoort (2004) demonstrated in a case study that prism adaptation may have effects on pressure sensitivity and proprioception, consistent with cross-modal benefit of the prisms. Ladavas, Farne, Pavani and colleagues (e.g., Farne, Dematte, & Ladavas, 2003; Farne, Pavani, Meneghello, & Ladavas, 2000; Ladavas, di Pellegrino, Farne, & Zeloni, 1998; Ladavas & Farne, 2004; Ladavas, Pavani, & Farne, 2001; Pavani & Castiello, 2004) have contributed a number of studies exploring the close relationships of auditory, visual, and tactile neglect and extinction (see also Carey, 2000; Kerkhoff, 1999) (see Figure 3). These studies are consistent with multisensory coding of the space closely surrounding the body surface. Subsequent investigations in healthy subjects (e.g., Maravita, Spence, & Driver, 2003), as well as in patients with tactile extinction (Farne, Iriki, & Ladavas, 2005), have indicated that use of tools may change patterns of cross-modal interference between vision and touch (see Figure 4). Thus, peripersonal and extrapersonal coding are dynamic processes related not only to absolute distance coding, but also to the computation of the extension of the body in space.

AROUSAL AND NONLATERALISED ATTENTION

A prominent tradition ascribes many of the characteristic symptoms of neglect after right-hemisphere damage to nonlateralised deficits in attention and arousal. It has long been recognised that right-hemisphere lesioned patients have electroencephalographic slowing and galvanic skin response abnormalities that are more pronounced than those of left-hemisphere damaged patients (Heilman, 1979; Heilman, Schwartz, & Watson, 1978; Heilman & Van Den Abell, 1980). Heilman and colleagues proposed that the right hemisphere is dominant both for arousal and for lateralised attention to the left side of space and

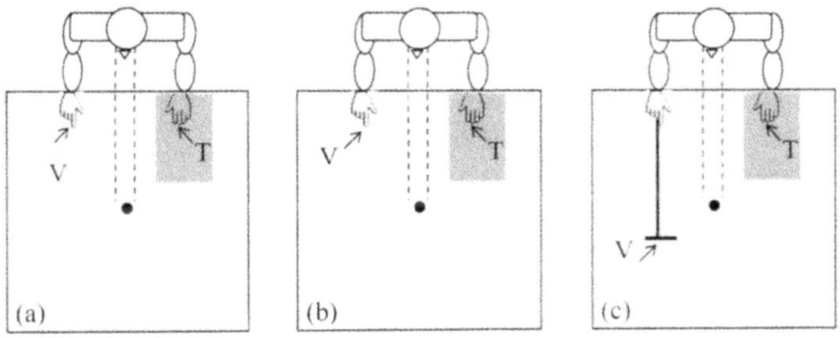

Figure 4. *Schematic illustration of the experimental set-up used by Farne, Iriki, and Ladavas (2005) to assess tactile extinction as a function of cross-modal condition. The visual stimulus (V) could be located near (a) or far (b and c) from the patient's right hand. Tactile (T) stimuli were delivered to the patient's left hand screened from view (grey rectangle). The visual stimulus was presented at the same distant position (60 cm from the hand) in both the far condition (b) without any tool and (c) after passive visual/proprioceptive exposure to a tool 60 cm in length.*

that these deficits may interact to produce many of the disorder's characteristic symptoms.

More recent work has made even more explicit the link between deficits in non-lateralised attention and lateralised spatial disturbances. For example, performance of a non-lateralised secondary task disrupts performance on lateralised neglect tests (Robertson, 1993). The magnitude of abnormality of an "attentional blink" for the second of two sequential stimuli correlates with neglect severity (Husain et al., 1997). Neglect patients also have deficits in detecting stimuli that are transiently presented in either hemifield (Duncan et al., 1999). Finally, they are impaired in sustained attention tasks such as keeping track of tones, and the magnitude of this deficit independently predicts performance on lateralised neglect tests (Robertson et al., 1997). These and other recent data (see also Robertson, 1993, 2001; Robertson, Mattingley, Rorden, & Driver, 1998) suggest that the severity of neglect symptoms is determined in part by the severity of deficits in arousal and attention.

NEUROANATOMIC SUBSTRATES OF NEGLECT

Many reports of the localisation of neglect up to and through the present have depended heavily upon reconstructions of clinical CT or MRI scans of relatively low spatial resolution, sometimes obtained shortly after a stroke when oedema is at a peak. Perhaps not surprisingly, there has been ongoing disagreement about the critical substrates of the disorder. Countless early reports of neglect attributed it to parietal lobe lesions. Subsequent studies documented that neglect may also appear consequent to frontal, subcortical, and white matter lesions as well (Vallar, 2001). A recent study of 1281 acute stroke patients found that 43% of right-hemisphere lesioned patients exhibited neglect, as measured by the NIH stroke scale tactile extinction and visual neglect items. Neglect was associated (in descending order) with lesions of temporal, parietal, frontal, and occipital lobes, basal ganglia, and thalamus, as judged by clinical CT scans obtained 7 days after the stroke (Ringman, Saver, Woolson, Clarke, & Adams, 2004). Similarly, a large study from our laboratory of 166 right-hemisphere stroke patients using categorical modelling (similar to logistic regression) found that, in chronic patients, lesions in the cingulate/orbitofrontal region, inferior parietal, occipital, superior/middle temporal, and inferior/mesial temporal lobe were more likely to be seen in neglect than non-neglect patients (Buxbaum et al., 2004). This is consistent with earlier observations of Heilman, Watson, and colleagues that a distributed neural network mediates spatial attention (Heilman & Valenstein, 1979; Watson, Valenstein, & Heilman, 1981).

Several investigations have focused on whether motor intentional versus perceptual subtypes of neglect may be associated with different lesion loci. Early reports of the distinction suggested that motor intentional neglect may be associated with frontal lesions (e.g., Bisiach et al., 1990; Coslett et al., 1990; Tegner & Levander, 1991). However, more recent reports have emphasised that this subtype may occur with parietal lesions as well (Mattingley, Husain, Rorden, Kennard, & Driver, 1998; Triggs, Gold, Gerstle, Adair, & Heilman, 1994).

One important development affecting localisation of the neuroanatomic substrate of neglect is the emergence of new lesion analysis techniques, some derived from the functional neuroimaging literature. A recent investigation used voxel-based statistical analyses similar to those employed with fMRI and found superior temporal cortex to be the most frequent cortical correlate of neglect. Damage to the putamen, caudate, planum temporale, insula, and pre- and post-central gyrus were also associated with neglect (Karnath, Fruhmann Berger, Kuker, & Rorden, 2004). Another recent investigation used magnetic source imaging to investigate brain activity associated with line bisection and bilateral visual stimulation, and found the greatest number of activity sources in right inferior parietal cortex (Billingsley, Simos, Sarkari, Fletcher, & Papanicolaou, 2004). Converging evidence relevant to the neuroanatomy of spatial attention has also been sought in fMRI studies with healthy subjects. For example, Hopfinger, Buonocore, and Mangun (2000) demonstrated that superior frontal, inferior parietal, and superior temporal cortices were selectively activated by attention-directing cues. Recently, Hillis et al. (2005) demonstrated that allocentric neglect (defined as neglect of the left sides of single objects) in acute patients is associated most strongly with hypoperfusion of right superior temporal gyrus, whereas egocentric neglect is associated with right inferior parietal hypoperfusion. The continued development of sophisticated lesion analysis and statistical techniques as well as continued focus on converging evidence from functional imaging may have a considerable impact on the confidence with which critical lesions may be identified.

CONCLUSIONS: THE PAST 20 YEARS AND THE NEXT 20

We have followed seven threads of inquiry to give a flavour of developments in the study of hemispatial neglect over the past 20 years. One of the clearest developments has been a shift in emphasis from an attempt to elucidate a single underlying mechanism of neglect to an understanding of the complexity and variability of the disorder. Along with this there has been increasing recognition of the potential of the neglect syndrome to inform our knowledge of the complex, multimodal systems underlying perception and action. In both developments, the cognitive neuropsychological approach, entailing detailed analyses of single cases, has been instrumental.

In elucidating the key characteristics of the neglect syndrome, investigators have taken a nonlinear path. For a time, descriptions of neglect as a list of possible symptoms gave way to an attempt to classify discrete subtypes of neglect, and optimistically, an effort to associate each subtype with a neuroanatomic focus. Subsequent work revealed that it was not to be so simple. First, nearly every possible fractionation of the disorder has been reported, raising the possibility that each patient may be as unique as a snowflake. Second, several of the discrete subtypes of neglect (e.g., motor intentional neglect) have been associated with a variety of lesion sites (e.g., frontal, parietal, and subcortical). And third, it has become increasingly clear that neglect performance measures are lacking in the requisite reliability that would permit confident assignment to one group or another. Several new developments appear to be called for here. First, measures used to assess various subtypes of neglect must be cross-validated against one another (see Harvey et al., 2002, for an example). The test–retest reliability of these measures must also be established. In an ongoing investigation in which patients with neglect are assessed on two to five separate occasions, Branch Coslett and I are finding that the pattern of performance on tasks assessing various subtypes is unstable across test sessions in approximately 75% of subjects. Until we have greater confidence that the measures we use to assess neglect subtypes are reliable and valid, it is probably premature to conclude that there is a true taxonomy of neglect subtypes.

One of the most exciting areas of development is the many burgeoning studies indicating that information may be dynamically coded with respect to distinct frames of reference depending upon task demands, the effectors to be used (if any), and the characteristics of the visual information. Here there has been a remarkable synergy between physiologic studies in the monkey and cognitive neuropsychological studies in brain-lesioned patients. In this area there is a great opportunity to profit from additional convergent evidence from the multiple investigative techniques available to cognitive neuroscientists, including functional neuroimaging, event-related recording, and transcranial magnetic stimulation.

In the area of psychophysics, there have been a number of important developments in our knowledge of the spatial and temporal properties of stimuli that gate access to conscious awareness. In the temporal domain, there is increasing evidence that the parietal lobe serves not only to mark the spatial location of visual information, but to integrate spatial and temporal information. These data are consistent with the recent claim of Walsh (2003a, 2003b) that the parietal lobes are crucial for judgments of magnitude in general. Thus, one common denominator in many of the capacities for which the parietal lobe is crucial—including number processing, spatial processing, and time judgment—may be judgment of magnitude. This appears to be an important area for future investigation.

The past 20 years have also seen a great increase in our understanding of the degree to which personal and peripersonal space is represented though integrated multisensory processing. This is another area in which insights from studies of neglect patients have converged remarkably with evidence from neurophysiological studies in monkeys and behavioural and functional neuroimaging methods with healthy humans. In this area, the implications for rehabilitation seem particularly compelling. Indeed, there is evidence from healthy subjects that attention-directing cues in one modality may improve processing in other modalities (Forster, Cavina-Pratesi, Aglioti, & Berlucchi, 2002; Macaluso, Ermer, Frith, & Driver, 2003). Consistent with this, spatial neglect improves with treatments known to affect vestibular and somatosensory processing (see Kerkhoff, 2003, for review). Similarly, wedge-base prisms that shift proprioceptive representations to the left have effects on mental imagery, auditory extinction, and posture, suggesting that the adaptation process may affect multisensory integration and higher, possibly supramodal, representations of space (Rode, Pisella, Rossetti, Farne, & Boisson, 2003). This appears to be an important focus for future studies of treatment effects.

Our knowledge of the role of nonspatial attentional factors in the neglect syndrome has also increased dramatically since the 1980s. Making neglect tasks more resource-demanding in any way (even by increasing verbal demands) increases the spatial bias in responding (Mennemeier, Morris, & Heilman, 2004). Conversely, cueing patients to improve arousal, or even training them to engage in self-cueing, significantly improves spatial neglect (Robertson et al., 1998). Fluctuations in nonspatial attention may also underlie the extreme day-to-day performance variability observed in "stable" chronic neglect patients. This makes it extremely difficult to establish stable baselines for the purpose of studying treatment effects (Buxbaum, Ferraro, Whyte, Gershkoff, & Coslett, 2005). In the future, it will be important to consider more fully the interaction of general attention and arousal with spatial factors when conducting both basic and applied studies.

In the area of identifying the neuroanatomic substrate of neglect, it might be argued that we are little better off than we were 20 years ago. We knew then that neglect was more frequent after right-than left-hemisphere lesions, particularly in the parietal and temporo-parietal regions. Subsequently, a number of studies using sophisticated imaging techniques and statistical methods (e.g., voxel-based statistical methods, fMRI, categorical modelling, magnetic source imaging) have identified a large, distributed network of frontal, parietal, temporal, and subcortical brain regions implicated in neglect. The precise lesions responsible for various patterns of neglect are still unclear; this problem is in turn likely to be influenced by our difficulty in establishing reliable measures of neglect subtypes, and by the tendency of moderate to severe neglect patients to have large brain lesions. But there is reason also to applaud the progress that has been made in lesion analysis and statistical techniques, and to predict optimistically that additional study will lead to increasing convergence.

This review has highlighted a few of the many contributions that make this an exciting time in neglect research. Insights from cognitive neuropsychological studies of neglect patients have played an important role in elucidating a great variety of neglect syndrome manifestations, which result in turn from damage to different aspects of a complex set of perceptual, attentional, cognitive, and motor mechanisms. Thus, the study of patients with neglect continues to play a key role in our developing understanding of object representations, spatial and nonspatial attention, and spatiomotor processing in the healthy brain. These studies provide important synergy with investigations from other areas of cognitive neuroscience that are informing the experimental and statistical methods that are used, and the kinds of research questions that are asked.

PrEview proof published online 18 October 2005

REFERENCES

Aglioti, S., Smania, N., & Peru, A. (1999). Frames of reference for mapping tactile stimuli in brain-damaged patients. *Journal of Cognitive Neuroscience, 11*, 67–79.

Anderson, R. A., Snyder, L. H., Li, C.-S., & Stricanne, B. (1993). Coordinate transformations in the representation of spatial information. *Current Opinion in Neurobiology, 3*, 171–176.

Barrett, A., M., Peterlin, B., & Heilman, K. (2003). Ipsilateral neglect versus hemianopic compensation. *Neurology, 61*, 120–123.

Basso, G., Nichelli, P., Frassinetti, F., & di Pellegrino, G. (1996). Time perception in a neglected space. *Neuroreport, 7*, 2111–2114.

Behrmann, M., & Tipper, S. P. (1999). Attention accesses multiple reference frames: Evidence from visual neglect. *Journal Experimental Psychology: Human Perception and Performance, 25*, 83–101.

Benke, T., Luzzatti, C., & Vallar, G. (2004). Hermann Zingerle's "impaired perception of the own body due to organic brain disorders". 1913. An introductory comment, and an abridged translation. *Cortex, 40*, 265–274.

Berti, A., & Frassinetti, F. (2000). When far becomes near: Remapping of space by tool use. *Journal of Cognitive Neuroscience, 12*, 415–420.

Berti, A., & Rizzolatti, G. (1992). Visual processing without awareness: Evidence from unilateral neglect. *Journal of Cognitive Neuroscience, 4*, 345–351.

Berti, A., Smania, N., & Allport, A. (2001). Coding of far and near space in neglect patients. *NeuroImage, 14*, S98–102.

Billingsley, R. L., Simos, P. G., Sarkari, S., Fletcher, J. M., & Papanicolaou, A. C. (2004). Spatio-temporal brain activation profiles associated with line bisection judgments and double simultaneous visual stimulation. *Behavioural Brain Research, 152*, 97–107.

Bisiach, E., Bulgarelli, C., Sterzi, R., & Vallar, G. (1983). Line bisection and cognitive plasticity of unilateral neglect of space. *Brain Cognition, 2*, 32–38.

Bisiach, E., Capitani, E., & Porta, E. (1985). Two basic properties of space representation in the brain: Evidence from unilateral neglect. *Journal of Neurology, Neurosurgery, and Psychiatry, 48*, 141–144.

Bisiach, E., Cornacchia, L., Sterzi, R., & Vallar, G. (1984). Disorders of perceived auditory lateralization after lesions of the right hemisphere. *Brain, 107*, 37–52.

Bisiach, E., Geminiani, G., Berti, A., & Rusconi, M. L. (1990). Perceptual and premotor factors of unilateral neglect. *Neurology, 40*, 1278–1281.

Bisiach, E., & Luzzatti, C. (1978). Unilateral neglect of representational space. *Cortex, 14*, 129–133.

Bisiach, E., Luzzatti, C., & Perani, D. (1979). Unilateral neglect, representational schema, and consciousness. *Brain, 102*, 609–618.

Bisiach, E., Perani, D., Vallar, G., & Berti, A. (1986). Unilateral neglect: Personal and extrapersonal. *Neuropsychologia, 24,* 759–767.

Bisiach, E., & Rusconi, M. L. (1990). Break-down of perceptual awareness in unilateral neglect. *Cortex, 26,* 643–649.

Bowen, A., Lincoln, N. B., & Dewey, M. (2002). Cognitive rehabilitation for spatial neglect following stroke. *Cochrane Database System Review(2),* CD003586.

Brain, W. (1941). Visual disorientation with special reference to lesions of the right cerebral hemisphere. *Brain, 64,* 244–272.

Buxbaum, L. J., & Coslett, H. B. (1994). Neglect of chimeric figures: Two halves are better than a whole. *Neuropsychologia, 32,* 275–288.

Buxbaum, L. J., Coslett, H. B., Montgomery, M., & Farah, M. J. (1996). Mental rotation may underlie apparent object-based neglect. *Neuropsychologia, 34,* 113–126.

Buxbaum, L. J., Ferraro, M., Veramonti, T., Farne, A., Whyte, J., Ladavas, E., Frassinetti, F., & Coslett, H. B. (2004). Hemispatial neglect: Subtypes, neuroanatomy, and disability. *Neurology, 62,* 749–756.

Buxbaum, L. J., Ferraro, M., Whyte, J., Gershkoff, A., & Coslett, H. (2005). *Amantadine treatment of hemispatial neglect: A double blind, placebo-controlled study.* Manuscript under review.

Buxbaum, L. J., & Permaul, P. (2001). Hand-centered attentional and motor asymmetries in unilateral neglect. *Neuropsychologia, 39,* 653–664.

Calvanio, R., Petrone, P. N., & Levine, D. N. (1987). Left visual spatial neglect is both enviroment-centered and body-centered. *Neurology, 37,* 1179–1183.

Cappa, S., Sterzi, S., Vallar, G., & Bisiach, E. (1987). Remission of hemineglect and anosognosia during vestibular stimulation. *Neuropsychologia, 25,* 775–782.

Caramazza, A., & Hillis, A. (1990a). Levels of representation, co-ordinate frames, and unilateral neglect. *Neuropsychology, 7,* 391–445.

Caramazza, A., & Hillis, A. E. (1990b). Spatial representation of words in the brain implied by studies of a unilateral neglect patient. *Nature, 346,* 267–269.

Carey, D. P. (2000). Multisensory integration: Attending to seen and felt hands. *Current Biology, 10,* R863–865.

Cate, A., & Behrmann, M. (2002). Spatial and temporal influences on extinction. *Neuropsychologia, 40,* 2206–2225.

Chatterjee, A. (1995). Cross-over, completion and confabulation in unilateral spatial neglect. *Brain,* 118, 455–465.

Chatterjee, A. (1998). Motor minds and mental models in neglect. *Brain Cognition, 37,* 339–349.

Chatterjee, A., Mennemeier, M., & Heilman, K. M. (1992). A stimulus-response relationship in unilateral neglect: The power function. *Neuropsychologia, 30,* 1101–1108.

Chatterjee, A., Ricci, R., & Calhoun, J. (2000). Weighing the evidence for cross over in neglect. *Neuropsychologia, 38,* 1390–1397.

Cicerone, K. D., Dahlberg, C., Kalmar, K., Langenbahn, D. M., Malec, J. F., Bergquist, T. F., Felicetti, T., Giacino, J., Harvey, J., Harrington, D., Herzog, J., Kneipp, S., Laatsch, L., & Morse, P. (2000). Evidence-based cognitive rehabilitation: Recommendations for clinical practice. *Archives of Physical Medicine and Rehabilitation, 81,* 1596–1615.

Colby, N. T., Duhamel, J., & Goldberg, M. E. (1995). Oculocentric spatial representation in parietal cortex. *Cerebral Cortex, 5,* 470–481.

Corben, L. A., Mattingley, J. B., & Bradshaw, J. L. (2001). A kinematic analysis of distractor interference effects during visually guided action in spatial neglect. *Journal of the International Neuropsychological Society, 7,* 334–343.

Coslett, H. B. (1997). Neglect in vision and visual imagery: A double dissociation. *Brain, 120,* 1163–1171.

Coslett, H. B., Bowers, D., Fitzpatrick, E., Haws, B., & Heilman, K. M. (1990). Directional hypokinesia and hemispatial inattention in neglect. *Brain, 113,* 475–486.

Coslett, H. B., & Saffran, E. M. (1991). Simultanagnosia: To see but not two see. *Brain, 114,* 1523–1545.

Costello, A. D., & Warrington, E. K. (1987). The dissociation of visuospatial neglect and neglect dyslexia. *Journal of Neurology, Neurosurgery, and Psychiatry, 50,* 1110–1116.

Critchley, M. (1953). *The parietal lobes.* New York: Hafner.

Cubelli, R., & Speri, V. (2001). Naming rotated pictures and the riddle of object-centred neglect. *Cortex, 37,* 159–174.

Deouell, L. Y. (2002). Pre-requisites for conscious awareness: Clues from electrophysiological and behavioral studies of unilateral neglect patients. *Conscious Cognition, 11,* 546–567.

DeRenzi, E. (1982). *Disorders of space exploration.* New York: John Wiley.

Dijkerman, H. C., Webeling, M., ter Wal, J. M., Groet, E., & Van Zandvoort, M. J. (2004). A long-lasting improvement of somatosensory function after prism adaptation, a case study. *Neuropsychologia, 42,* 1697–1702.

Duhamel, J. R., Colby, C. L., & Goldberg, M. E. (1992). The updating of the representation if visual space in parietal cortex by intended eye movements. *Science, 255,* 90–92.

Duncan, J., Bundesen, C., Olson, A., Humphreys, G., Chavda, S., & Shibuya, J. (1999). Systematic analysis of deficits in visual attention. *Journal of Experimental Psychology: General, 128,* 450–478.

Farne, A., Dematte, M. L., & Ladavas, E. (2003). Beyond the window: Multisensory representation of peripersonal space across a transparent barrier. *International Journal of Psychophysiology, 50,* 51–61.

Farne, A., Iriki, A., & Ladavas, E. (2005). Shaping multisensory action space with tools: Evidence from patients with cross-modal extinction. *Neuropychologia, 43,* 238–248.

Farne, A., Pavani, F., Meneghello, F., & Ladavas, E. (2000). Left tactile extinction following visual stimulation of a rubber hand. *Brain, 123,* 2350–2360.

Forster, B., Cavina-Pratesi, C., Aglioti, S. M., & Berlucchi, G. (2002). Redundant target effect and intersensory facilitation from visual-tactile interactions in simple reaction time. *Experimental Brain Research, 143*(4), 480–487.

Gainotti, G., D'Erme, P., Monteleone, D., & Silveri, M. C. (1986). Mechanisms of unilateral spatial neglect in relation to laterality of cerebral lesions. *Brain, 109,* 599–612.

Graziano, M. S. A., Yap, G. S., & Gross, C. G. (1994). Coding of visual space by premotor neurons. *Science, 266,* 1054–1057.

Guariglia, C., & Antonucci, G. (1992). Personal and extrapersonal space: A case of neglect dissociation. *Neuropsychologia, 30,* 1001–1010.

Guerrini, C., Berlucchi, G., Bricolo, E., & Aglioti, S. M. (2003). Temporal modulation of spatial tactile extinction in right-brain-damaged patients. *Journal of Cognitive Neuroscience, 15,* 523–536.

Halligan, P. W., & Marshall, J. C. (1988). How long is a piece of string? A study of line bisection in a case of visual neglect. *Cortex, 24,* 321–328.

Halligan, P., & Marshall, J. (1989). Perceptual cueing and perceptual-motor compatibility in visuo-spatial neglect: A single case study. *Cognitive Neuropsychology, 6,* 423–435.

Halligan, P., Marshall, J. C. (1991). Left neglect for near but not far space in man. *Nature, 350,* 498–500.

Halligan, P., Marshall, J., & Wade, D. (1989). Visuospatial neglect: Underlying factors and test sensitivity. *Lancet, Oct 14,* 908–911.

Harvey, M., Kramer-McCaffery, T., Dow, L., Murphy, P., & Gilchrist, I. (2002). Categorisation of 'perceptual' and 'premotor' neglect patients across different tasks: Is there strong evidence for a dichotomy? *Neuropsychologia, 40,* 1387–1395.

Heilman, K. M., Bowers, D., Coslett, H. B., Whelan, H., & Watson, R. T. (1985). Directional hypokinesia: Prolonged reaction times for leftward movements in patients with right hemisphere lesions and neglect. *Neurology, 35,* 855–859.

Heilman, K. M., Bowers, D., & Watson, R. T. (1983). Performance on hemispatial pointing task by patients with neglect syndrome. *Neurology, 33,* 661–664.

Heilman, K. M., Schwartz, H. D., & Watson, K. T. (1978). Hypoarousal in patients with the neglect syndrome and emotional indifference. *Neurology, 28,* 229–232.

Heilman, K., & Valenstein, E. (1972). Frontal lobe neglect in man. *Neurology, 22,* 660–664.

Heilman, K., & Valenstein, E. (1979). Mechanisms underlying hemispatial neglect. *Annals of Neurology, 5,* 166–170.

Heilman, K. M., Valenstein, E., & Watson, R. T. (2000). Neglect and related disorders. *Seminars in Neurology, 20,* 463–470.

Heilman, K. M., & Van Den Abell, T. (1979). Right hemispheric dominance for mediating cerebral activation. *Neuropsychologia, 17,* 315–321.

Heilman, K. M., & Van Den Abell, T. (1980). Right hemisphere dominance for attention: The mechanisms underlying hemispheric asymmetries of inattention (neglect). *Neurology, 30,* 327–330.

Hillis, A., Newhart, M., Heidler, J., Barker, P., Herskovits, E., & Degaonkar, M. (2005). *Anatomy of spatial attention: insights from perfusion imaging and hemispatial neglect in acute stroke.* Manuscript under review.

Hillis, A. E., Rapp, B., Benzing, L., & Caramazza, A. (1998). Dissociable coordinate frames of unilateral spatial neglect: "viewer-centered" neglect. *Brain and Cognition, 37,* 491–526.

Hopfinger, J. B., Buonocore, M. H., & Mangun, G. R. (2000). The neural mechanisms of top-down attentional control. *Nature Neuroscience, 3,* 284–291.

Husain, M., Shapiro, K., Martin, J., & Kennard, C. (1997). Abnormal temporal dynamics of visual attention in spatial neglect patients. *Nature*, *385*, 154–156.

Kahneman, D., Treisman, A., & Gibbs, B. J. (1992). The reviewing of object files: Object-specific integration of information. *Cognitive Psychology*, *24*, 175–219.

Karnath, H. O., Fruhmann Berger, M., Kuker, W., & Rorden, C. (2004). The anatomy of spatial neglect based on voxelwise statistical analysis: A study of 140 patients. *Cerebral Cortex*, *14*, 1164–1172.

Karnath, H. O., Zimmer, U., & Lewald, J. (2002). Impaired perception of temporal order in auditory extinction. *Neuropsychologia*, *40*, 1977–1982.

Kerkhoff, G. (1999). Multimodal spatial orientation deficits in left-sided visual neglect. *Neuropsychologia*, *37*, 1387–1405.

Kerkhoff, G. (2003). Modulation and rehabilitation of spatial neglect by sensory stimulation. *Progress in Brain Research*, *142*, 257–271.

Kinsbourne, M. (1974). Mechanisms of hemispheric interaction in man. In M. Kinsbourne & W. L. Smith (Eds.), *Hemispheric disconnection and cerebral function*. Springfield, IL: Charles C. Thomas.

Kinsbourne, M. (1987). Mechanisms of unilateral neglect. In M. Jeannerod (Ed.), *Neurophysiological and neuropsychological aspects of spatial neglect* (pp. 69–86). Amsterdam: Elvesier.

Koriat, A., & Norman, J. (1984). What is rotated in mental rotation? *Journal of Experimental Psychology: Learning, Memory & Cognition*, *10*, 421–434.

Ladavas, E. (1987). Is the hemispatial deficit produced by right parietal lobe damage associated with retinal or gravitational coordinates? *Brain*, *110*, 167–180.

Ladavas, E., di Pellegrino, G., Farne, A., & Zeloni, G. (1998). Neuropsychological evidence of an integrated visuotactile representation of peripersonal space in humans. *Journal of Cognitive Neuroscience*, *10*, 581–589.

Ladavas, E., & Farne, A. (2004). Visuo-tactile representation of near-the-body space. *Journal of Physiology Paris*, *98*, 161–170.

Ladavas, E., Pavani, F., & Farne, A. (2001). Auditory peripersonal space in humans: A case of auditory-tactile extinction. *Neurocase*, *7*, 97–103.

Macaluso, E., Eimer, M., Frith, C. D., & Driver, J. (2003). Preparatory states in crossmodal spatial attention: Spatial specificity and possible control mechanisms. *Experimental Brain Research*, *149*, 62–74.

Maravita, A., Spence, C., & Driver, J. (2003). Multisensory integration and the body schema: Close to hand and within reach. *Current Biology*, *13*, R531–R539.

Marr, D. (1980). Visual information processing: The structure and creation of visual representations. *Philosophical Transactions of the Royal Society of London B Biological Sciences*, *290*, 199–218.

Marr, D. (1982). *Vision*. San Francisco: Freeman.

Marshall, J. C., & Halligan, P. W. (1988). Blindsight and insight in visuo-spatial neglect. *Nature*, *336*, 766–767.

Marshall, J. C., & Halligan, P. W. (1989). When right goes left: An investigation of line bisection in a case of visual neglect. *Cortex*, *25*, 503–515.

Mattingley, J. B., Husain, M., Rorden, C., Kennard, C., & Driver, J. (1998). Motor role of human inferior parietal lobe revealed in unilateral neglect patients. *Nature*, *392*, 179–182.

McGlinchey-Berroth, R., Milberg, W. P., Verfaellie, M., Alexander, M., & Kilduff, P. T. (1993). Semantic processing in the neglected visual field: Evidence from a lexical decision task. Cognitive *Neuropsychology*, *10*, 79–108.

Mennemeier, M., Morris, M., & Heilman, K. (2004). Just thinking about targets can aggravate neglect on cancellation tasks. *Neurocase*, *10*, 29–38.

Mennemeier, M., Wertman, E., & Heilman, K. M. (1992). Neglect of near peripersonal space. *Brain*, *115*, 37–50.

Mesulam, M. M. (1985). *Principles of behavioral neurology*. Philadelphia: David Press.

Milner, A. D., & Goodale, M. A. (1995). *The visual brain in action*. Oxford: Oxford University Press.

Milner, A. D., Harvey, M., Roberts, R. C., & Forster, S. V. (1993). Line bisection errors in visual neglect: Misguided action or size distortion? *Neuropsychologia*, *31*, 39–49.

Muller-Oehring, E., Kasten, E., Poggel, D., Schulte, T., Strasburger, H., & Sabel, B. (2003). Neglect and hemianopia superimposed. *Journal of Clinical and Experimental Neuropsychology*, *25*, 1154–1168.

Nadeau, S. E., & Heilman, K. M. (1991). Gaze-dependent hemianopia without hemispatial neglect. *Neurology*, *41*, 1244–1250.

Nico, D. (1996). Detecting directional hypokinesia: The epidiascope technique. *Neuropsychologia*, *34*, 471–474.

Pavani, F., & Castiello, U. (2004). Binding personal and extrapersonal space through body shadows. *Nature Neuroscience*, *1*, 14–16.

Pouget, A., Deneve, S., & Duhamel, J. R. (2002). A computational perspective on the neural basis of multisensory spatial representations. *Nature Reviews: Neuroscience*, *3*, 741–747.

Pouget, A., & Sejnowski, T. J. (2001). Simulating a lesion in a basis function model of spatial representations: Comparison with hemineglect. *Psychological Review*, *108*, 653–673.

Rapcsak, S. Z., Cimino, C. R., & Heilman, K. M. (1988). Altitudinal neglect. *Neurology*, *38*, 277–281.

Riddoch, G. (1935). Visual disorientation in homonymous half-fields. *Brain*, *58*, 376–382.

Ringman, J. M., Saver, J. L., Woolson, R. F., Clarke, W. R., & Adams, H. P. (2004). Frequency, risk factors, anatomy, and course of unilateral neglect in an acute stroke cohort. *Neurology*, *63*, 468–474.

Rizzolatti, G., & Camarda, R. (1987). Neural circuits for spatial attention and unilateral neglect. In M. Jeannerod (Ed.), *Neurophysiological and neuropsychological aspects of spatial neglect* (pp. 289–313). Amsterdam: Elsevier.

Rizzolatti, G., Camarda, R., Fogasi, L., Gentilucci, M., Luppino, G., & Matelli, M. (1988). Functional organization of inferior area 6 in the macaque monkey. *Experimental Brain Research*, *71*, 491–507.

Robertson, I. H. (1993). The relationship between lateralised and non-lateralised attentional deficits in unilateral neglect. In I. H. Robertson & J. C. Marshall (Eds.), *Unilateral neglect: Clinical and experimental studies* (pp. 257–278). Hove, UK: Lawrence Erlbaum Associates Ltd.

Robertson, I. H. (2001). Do we need the "lateral" in unilateral neglect? Spatially nonselective deficits in unilateral neglect and their implications for rehabilitation. *NeuroImage*, *14*, S85–S90.

Robertson, I. H., & Frasca, R. (1992). Attentional load and visual neglect. *International Journal of Neuroscience*, *62*, 45–56.

Robertson, I. H., Manly, T., Beschin, N., Daini, R., Haeske-Dewick, H., Homberg, V., Jehkonen, M., Pizzamiglio, G., Shiel, A., & Weber, E. (1997). Auditory sustained attention is a marker of unilateral spatial neglect. *Neuropsychologia*, *35*, 1527–1532.

Robertson, I. H., Mattingley, J. B., Rorden, C., & Driver, J. (1998). Phasic alerting of neglect patients overcomes their spatial deficit in visual awareness. *Nature*, *395*, 169–172.

Rode, G., Pisella, L., Rossetti, Y., Farne, A., & Boisson, D. (2003). Bottom-up transfer of sensory-motor plasticity to recovery of spatial cognition: Visuomotor adaptation and spatial neglect. *Progress in Brain Research*, *142*, 273–287.

Rorden, C., Mattingley, J. B., Karnath, H. O., & Driver, J. (1997). Visual extinction and prior entry: Impaired perception of temporal order with intact motion perception after unilateral parietal damage. *Neuropsychologia*, *35*, 421–433.

Rubens, A. (1985). Caloric stimulation and unilateral visual neglect. *Neurology*, *35*, 1019–1024.

Smania, N., & Aglioti, S. (1995). Sensory and spatial components of somaesthetic deficits following right brain damage. *Neurology*, *45*, 1725–1730.

Snyder, J. J., & Chatterjee, A. (2004). Spatial-temporal anisometries following right parietal damage. *Neuropsychologia*, *42*, 1703–1708.

Sterzi, R., Bottini, G., Celani, M. G., Righetti, E., Lamassa, M., Ricci, R., & Vallar, G. (1993). Hemianopia, hemiananaesthesia, and hemiplegia after right and left hemisphere damage. A hemispheric difference. *Journal of Neurology, Neurosurgery, and Psychiatry*, *56*, 308–310.

Styles, E. A. (1986). Perceptual integration of identity, location and colour. *Psychological Research*, *48*, 189–200.

Subbiah, I., & Caramazza, A. (2000). Stimulus-centered neglect in reading and object recognition. *Neurocase*, *6*, 13–31.

Tegner, R., & Levander, M. (1991). Through a looking glass: A new technique to demonstrate directional hypokinesia in unilateral neglect. *Brain*, *114*, 1943–1951.

Tipper, S. P., Howard, L. A., & Jackson, S. R. (1997). Selective reaching to grasp: Evidence for distractor interference effects. *Visual Cognition*, *4*, 1–38.

Tipper, S. P., Lortie, C., & Baylis, G. C. (1992). Selective reaching: Evidence for action-centered attention. *Journal of Experimental Psychology: Human Perception and Performance*, *18*, 891–905.

Triggs, W. J., Gold, M., Gerstle, G., Adair, J., & Heilman, K. M. (1994). Motor neglect associated with a discrete parietal lesion. *Neurology*, *44*, 1164–1166.

Ungerleider, L. G., & Mishkin, M. (1982). Two cortical visual systems. In D. J. Ingle, M. A. Goodale, & R. J. W. Manfield (Eds.), *Analysis of visual behavior*. Cambridge, MA: MIT Press.

Valenza, N., Seghier, M. L., Schwartz, S., Lazeyras, F., & Vuilleumier, P. (2004). Tactile awareness and limb position in neglect: Functional magnetic resonance imaging. *Annals of Neurology*, *55*, 139–143.

Vallar, G. (1997). Spatial frames of reference and somatosensory processing: A neuropsychological perspective. *Philosophical Transactions of the Royal Society of London B Biological Science*, *352*, 1401–1409.

Vallar, G. (2001). Extrapersonal visual unilateral spatial neglect and its neuroanatomy. *NeuroImage*, *14*, S52–S58.

Vallar, G., Bottini, G., Rusconi, M. L., & Sterzi, R. (1993). Exploring somatosensory hemineglect by vestibular stimulation. *Brain*, *116*, 71–86.

Volpe, B. T., LeDoux, J. E., & Gazzaniga, M. S. (1979). Spatially oriented movements in the absence of proprioception. *Neurology*, *29*, 139–1313.

Vuilleumier, P., Armony, J. L., Clarke, K., Husain, M., Driver, J., & Dolan, R. J. (2002). Neural response to emotional faces with and without awareness: Event-related fmri in a parietal patient with visual extinction and spatial neglect. *Neuropsychologia*, *40*, 2156–2166.

Walsh, V. (2003a). A theory of magnitude: Common cortical metrics of time, space and quantity. *Trends in Cognitive Science*, *7*, 483–488.

Walsh, V. (2003b). Time: The back-door of perception. *Trends in Cognitive Science*, *7*, 335–338.

Watson, R. T., Miller, B. D., & Heilman, K. M. (1978). Nonsensory neglect. *Annals of Neurology*, *3*, 505–508.

Watson, R. T., Valenstein, E., & Heilman, K. (1981). Thalamic neglect. *Archives of Neurology*, *38*, 501–506.

COGNITIVE NEUROPSYCHOLOGY, 2006, 23 (1), 202–221

The cognitive neuropsychology of everyday action and planning

Myrna F. Schwartz
Moss Rehabilitation Research Institute, Philadelphia, PA, USA

Today's cognitive neuropsychology of everyday action had its inception in studies of ADS—action disorganisation syndrome—that were inspired by the Norman-Shallice theory of controlled and automatic action selection. It is now known that errors in everyday action and planning are commonplace with many types of brain damage, and that effects associated with gradations in severity, and with the presence of distractor objects, are surprisingly uniform across clinically diverse patient samples. The Norman-Shallice model of automatic action selection, having been implemented for two everyday tasks, has had some success in explaining these facts. This paper reviews the patient and modelling studies and discusses some unanswered questions and methodological challenges that confront future research in this area.

THE PAST

Around the time of Cognitive Neuropsychology's founding, developments were taking place in a number of related fields of study that would ultimately make the long-neglected routines of everyday life a legitimate focus for action and planning research. The artificial intelligence community had decades earlier given rise to the computational metaphor of mind and the view that formulating and following a plan has much in common with writing and executing a digital program. In the 1980s, however, computational neuroscientists were beginning to think in terms of parallel processing, and AI engineers were pondering the failure to program robots to do even simple concrete activities. Some within the AI community were writing that whereas the programming metaphor might work for playing chess and other traditional planning tasks, it had little application to concrete, everyday activities. These activities have special characteristics, they rightly concluded, among which are that they are planned *online*, with continuous updating from perception, using well-learned routines and subroutines (Arbib, 1989; Chapman, 1991; Chapman & Agre, 1987; Maes, 1989).

Correspondence should be addressed to Myrna Schwartz, PhD, Moss Rehabilitation Research Institute, 1200 West Tabor Road, Philadelphia, PA 19141-3099, USA (Email: mschwart@einstein.edu).

Preparation of this paper was supported by a grant from the NIDRR Research Field Initiated Program CFDA 84.133G (Laurel Buxbaum, Principal Investigator).

DOI:10.1080/02643290500202623

Cognitive psychologists, meanwhile, were busy fleshing out the distinction between controlled and automatic processing (Posner & Snyder, 1975; Shiffrin & Schneider, 1977) and the related distinction between willed and automatic action selection (Norman & Shallice, 1980, 1986; Shallice, 1972). If anyone needed confirmation that routine action is planned online and often on "automatic pilot," they had only to read James Reason's reports on the action slips and lapses recorded in the diaries kept by his Manchester University undergraduates (Reason, 1979, 1984; Reason & Mycielska, 1982). As anticipated by William James nearly 100 years earlier, the diary entries demonstrated that in states of absent-mindedness or distraction, behaviour is vulnerable to "capture" by routines that are appropriate to the context, though not necessarily the current intention (Ex. 1; see also James, 1950).

I intended to change from jeans to a dress. I undressed completely as if I were going to bed. (1)

The substituted routine—dressing for bed—was initiated and executed without conscious direction. By implication, so is much of our appropriate behaviour. How are we to understand this? The answer would come from the interactive activation paradigm, as it was being applied to skilled sequential action (MacKay, 1985, 1987; Rumelhart & Norman, 1982) and the classification of slips (MacKay, 1985; Norman, 1981; Reason & Mycielska, 1982; Roy, 1982, 1983).

Common to all interactive activation models of routine action planning is the concept of distributed processors acting locally and in parallel. For example, Norman's (1981) Activation-Trigger-Schema (ATS) framework proposed *schemas*, representing procedural knowledge at differing levels of an abstraction hierarchy. High-level schemas represent goals; low-level schemas, actions. Schemas have conditions, specifying when they are to be "triggered" into action. Planning happens when a triggered high-level schema passes activation to its components, representing individual actions. The component actions are then triggered, in turn, by the match between their conditions and the momentary state of the world (including the body). According to the ATS account, capture errors arise because when it comes to strong habits, even partial matches may be enough to trigger the relevant schema(s). Other slips result from a schema being fired at the wrong time (2) or not at all (3). (Examples are from Norman, 1981.)

One day as I was running on my morning trek, I saw a woman ahead. I was counting steps, but as I neared the woman I decided to say "Good morning." When I got to the woman, she smiled and said "Good morning," and I responded "Thirty-three." (2)

I make an error typing a line on the computer, think of typing the special character that deletes the line (@), and then continue typing, only to find that the computer responds with an error message. The @ sign was only thought, not actually typed. (3)

Don Norman and Tim Shallice's influential theory of attention to action began circulating as early as 1980 and was officially published in 1986 (Norman & Shallice, 1980, 1986). The theory unified the aforementioned themes, including the automatic–controlled distinction and the ATS framework applied to action planning and its derailments. It also added some important new wrinkles. First, the theory introduced "contention scheduling" as a mechanism for controlling conflict among incompatible goals or actions activated in parallel. "Contention scheduling" initially was used interchangeably with "competitive activation"—referring to the mutual excitation of compatible schemas and mutual inhibition of incompatible ones. Eventually "contention scheduling" came to refer to the entire system—schema architecture plus processing mechanisms—as it operates in automatic action selection. Henceforth, I'll use the term in this, its broader sense.

The major innovation in the Norman-Shallice theory was, of course, the level of control overlaid on contention scheduling—supervisory attention or, in current parlance, the supervisory system. The theory posed and answered some key questions: *What is supervisory attention?* Answer: The physical embodiment of willed, deliberate planning. *When does it come into play?* When the required action is ill-learned or novel, when avoidance of error is critical, when a strong habit must be resisted, and/or when conscious planning decisions have to be made. *How does it operate?* By modulating the activation values of existing schemas or, if no relevant schemas exist, creating temporary new ones. *Where is it localised in the brain?* In the frontal lobes.

Building on Luria's theory of the frontal lobes (Luria, 1966), Norman-Shallice's conception of an anterior supervisory system propelled neuropsychological research on the frontal lobes throughout the 1980s and 1990s (e.g., Grafman, 1989; Levine, Dawson, Boutet, Schwartz, & Stuss, 2000; Schwartz, Reed, Montgomery, Palmer, & Mayer, 1991; Shallice, 1982; Shallice & Burgess, 1991a, 1991b; Shallice, Burgess, Schon, & Baxter, 1989). The theory continues to influence contemporary research on supervisory/executive functions of prefrontal cortex (PFC) (e.g., Burgess, Veitch, de Lacy Costello, & Shallice, 2000; Levine et al., 2000) and to evolve in response to patient and brain studies (Stuss, Shallice, Alexander, & Picton, 1995).

With respect to the interaction of the supervisory system (SS) and contention scheduling (CS), an implication of the slips research is that without the timely intervention of SS, CS is prone to error. In a similar vein, Shallice and colleagues proposed that some well-known consequences of frontal lobe damage could be understood as resulting from absent or faulty CS modulation, secondary to an impaired SS. The list of frontal lobe symptoms that might be explained in this way included perseveration, poor set switching, and utilisation behaviour (Lhermitte, 1983; Shallice, 1982, 1988; Shallice et al., 1989).

Could the action disorganisation syndrome (ADS) perhaps be explained in this way, as well? The ADS studies began at MossRehab Hospital in Philadelphia in the late 1980s with investigations of patients with frontal strokes or closed head injury (CHI), who evinced disorganised action in the context of eating, grooming, and dressing.

While eating from the hospital breakfast tray, a patient with damage to and beyond the frontal lobes (Pt. H.H.) spooned a pat of butter into his coffee; on another occasion he poured coffee into his bowl of oatmeal (Schwartz et al., 1991). (4)

CHI patient JK started to spread shaving cream onto his toothbrush. When stopped by the therapist, and without specific instruction from her, he reached into his grooming kit, extracted the toothpaste tube, and proceeded to brush his teeth in the appropriate manner (Schwartz et al., 1995; also Schwartz, Mayer, Fitzpatrick-DeSalme, & Montgomery, 1993). (5)

Historically, the phenomenon of patients making frequent and egregious errors of action in the performance of everyday activities was either dismissed as a manifestation of confusion or dementia, or assimilated into the much-debated classifications of "frontal apraxia" and "ideational apraxia." (See Cooper, 2002, and Schwartz & Buxbaum, 1997, for reviews.) The early ADS studies, invoking the Norman-Shallice framework, hypothesised that the ADS might result from dysregulation of CS by a damaged SS or, alternatively, malfunction within CS itself. The cognitive neuropsychology of everyday action grew from this theoretical base.

THE PRESENT

Present understanding of everyday action disorders in brain-damaged populations is informed by the study of patient groups (Buxbaum, Schwartz, & Montgomery, 1998; DeRenzi & Lucchelli, 1988; Giovannetti, Libon, Buxbaum, & Schwartz, 2002; Goldenberg, Daumuller, & Hagmann, 2001; Goldenberg & Hagmann, 1998; Rusted & Sheppard, 2002; Schwartz et al., 1998, 1999; Schwartz, Segal, Veramonti, Ferraro, & Buxbaum, 2002), as well as case studies in the cognitive neuropsychology style (Buxbaum, Schwartz, & Carew, 1997; Buxbaum, Schwartz, Coslett, & Carew, 1995; Buxbaum, Veramonti, & Schwartz, 2000; Forde & Humphreys, 2000, 2002; Humphreys & Forde, 1998; Riddoch, Humphreys, Heslop, & Castermans, 2002; Rumiati, Zanini, Vorano, & Shallice, 2001; Schwartz et al., 1993; Schwartz et al., 1991, 1995).

The hypothesis that the ADS represents faulty CS modulation resulting from an impaired SS has not fared well. For example, Humphreys and Forde (1998) assessed individual patients on a range of everyday tasks. The high rate of errors in two parents (FK and HG) qualified them as having ADS. A third (DS) was as impaired as FK and HG on frontal executive tests but made far fewer errors on the action tests. On the standard assumption that poor performance on frontal executive tests equates to SS impairment, patient DS has impaired SS without ADS; this indicates that SS impairment does not necessarily impede naturalistic action production, through CS modulation or otherwise.

My colleagues and I tested a group of CHI patients on the everyday tasks of the Multi-level Action Test (MLAT; described below) and showed that errors were commonplace in this group. Against the SS-impairment hypothesis was the fact that patients made far more errors than controls even on the simplest conditions of the MLAT, in which the SS was not expected to play much of a role. Moreover, while performance on executive functions tests contributed to a model predicting error rate, a neuroanatomical variable indicating presence/absence of frontal lesions on CT or MRI was not a significant predictor (Schwartz et al., 1998). In subsequent studies, we discovered that groups with other aetiologies of brain damage showed strong and surprising similarities with the CHI group, with respect to their naturalistic action performance.

The alternative hypothesis—that brain damage disrupts everyday action planning by impeding the normal operation of CS—is undergoing close scrutiny in present-day research. For one thing, the publication of an implemented version of CS (Cooper, Schwartz, Yule, & Shallice, in press; Cooper & Shallice, 2000) has improved the specification of the model to the point where specific claims can be evaluated. Additionally, the MLAT studies just alluded to have provided a set of empirical phenomena against which the CS and other models can be evaluated.

The remainder of this section is organised as follows: First, the relevant findings from the MLAT studies are summarised. Next, the implemented CS model is described, with emphasis on how simulated lesions purport to explain the empirical findings. As is frequently the case with modelling studies, the precise specification of architecture and processing assumptions helps frame old issues in new ways. Here, the issues concern the nature of the representations that enter into routine action selection; the mechanisms that underpin sequential selection, and the contribution of the SS to more and less routine action selection. These issues will be discussed in turn, in relation to patient findings and alternative cognitive models.

The MLAT studies

The MLAT required participants to perform a set of naturalistic tasks, including making toast with butter and jam, gift wrapping a package, and preparing a lunchbox with sandwich, snack, and drink. The tasks were performed under varying conditions—one at a time and in pairs (e.g., toast task alone or together with coffee making); and one at a time with and without distractor objects present. Each distractor was chosen to resemble a target object with respect to its function or means of manipulation (e.g., electric tape for cellophane tape; spatula for knife, hot dogs for lunchmeat.) Performance was untimed and scored according to an error taxonomy that included *omission* of key steps and several categories of *commission* error (e.g., sequence error, spatial error, tool omission). Selection of MLAT tasks, definition of errors, and rules for classifying errors were developed through extensive pilot research. Procedures were identical across the three studies. What follows is a selective summary of data collected from patients with closed head injury (CHI; $n = 30$; Schwartz et al., 1998), left-hemisphere stroke (LCVA: $n = 16$; Buxbaum et al., 1998), and right-hemisphere stroke (RCVA; $n = 30$; Schwartz et al., 1999). All patients were recruited from inpatient rehabilitation services at MossRehab Hospital. Readers are encouraged to consult the original papers for further details. Related evidence from a study of progressive dementia patients is available in Giovannetti et al. (2002). A simplified version of the MLAT, called the *Naturalistic Action Test*, was validated on an additional group of 100

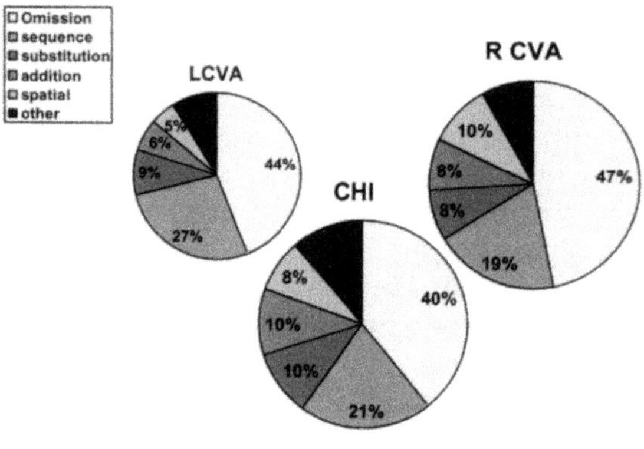

Figure 1. *The distribution of errors by error type is shown for the patient groups that participated in the MLAT studies.*

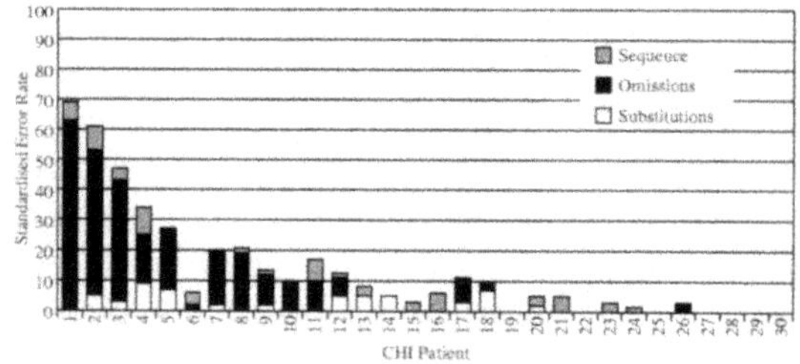

Figure 2. *Standardised scores represent errors per 100 opportunities. The 30 CHI participants in the MLAT study are ordered from left to right in terms of total error score. Only three of the principal error types are shown. Originally presented in Schwartz et al., 1998, the figure was redrawn for Cooper et al. (in press) and is reprinted with kind permission of Psychology Press, www.psypress.co.uk/journals.asp, 2005.*

patients (Schwartz et al., 2002) and is available commercially (Schwartz, Buxbaum, Ferraro, Veramonti, & Segal, 2003).

Error rates and error profiles

In each of the three patient groups, the error rate was at least five times that of age-matched controls.[1] The three groups responded in the same way to manipulations of task complexity. Most surprising was the finding that the different error types occurred in nearly identical proportions across the three groups (Figure 1). Omission errors were most common, followed by sequence errors. Substitution errors and action addition errors occurred in all groups, at lower proportions than omission and sequence errors.

In other analyses, we took account of the fact that different tasks offer different possibilities for error, by converting error frequencies to standardised error scores. A patient's standardised error score for any type of error was defined as the total number of errors of that type produced on a task, divided by the total number of opportunities provided by the task for that type of error, multiplied by 100. Thus, a standardised omission error score of 4 corresponds to a rate of 4 omission errors per 100 opportunities.

Figure 2 shows the standardised scores for omission, sequence, and object substitution errors produced by the patients in the CHI group. The preponderance of omission errors is clear, with sequence errors second most frequent. Graphed data from the other patient groups closely resembled this one.

Errors in relation to severity

Figure 2 illustrates another property common to all the groups we studied, which is that omissions were particularly frequent among the patients who made most errors overall. Low error producers made few omissions, which means that the bulk of their errors were overt, i.e., commission errors. This is numerically expressed by the *commission proportion (CP)*, which is the fraction of commission errors to total errors. Table 1 quantifies this effect using standardised scores. The data in the table are derived from each patient performing two of three tasks (preparing and packing a lunchbox, preparing toast with jam, and wrapping a gift), one with just the items needed for the task, and one with additional distractor items present. Patients who produced no errors are excluded from this analysis. The mean standardised commission proportion for all three patient groups is

[1] The vulnerability to errors of action in CHI and LCVA was expected, based on the literature on ideational and frontal apraxia. The high error rates in RCVA, on the other hand, were completely unexpected. Goldenberg (personal communication, January 2004) has recently completed a study of naturalistic action production in LCVA and RCVA which confirms that the two groups are equally vulnerable to errors of action.

Table 1. *Standardised error rates and standardised commission proportions are shown for three patient groups in the MLAT studies*

Group	*N*	*Mean total standardised errors*	*Mean standardised commission proportion*	
CHI low E	8	5.506	.801	t (13) = 4.2,
CHI high E	7	32.085	.180	$p < .001$
LCVA low E	7	2.365	1.000	t (12) = 17.9,
LCVA high E	7	27.057	.136	$p < .001$
RCVA low E	12	6.555	.808	t (22) = 5.7,
RCVA high E	12	46.160	.186	$p < .001$

Each group was divided into low and high error producers, according to a median split on total error scores. Patients who produced no errors were excluded from the analysis.

significantly higher for low error producers than for high error producers (see Table 1).

Thus, in all patient groups, low error producers tended to produce more commission errors, while high error producers tended to produce more omission errors. The performance of our control subjects continued the pattern, in that controls made fewer errors than the patients and almost all their errors were of the commission type.

The effect of distractor objects

Comparison of standardised error rates in conditions with and without distractor objects showed that in each patient group, the trend was for more omission errors and more object substitutions in the condition with distractors. When the data from the three groups were analysed as one, the increase in omissions was statistically reliable. The object substitution effect remained a nonsignificant trend. (Giovannetti et al., 2002, using similar methods, found that dementia patients produced significantly more substitution errors in the condition with distractors.) While a variety of theoretical positions might predict that the presence of distractor objects would encourage substitution errors, the increase observed in omission errors could not have been predicted.

How is one to understand such uniformity of results across patients with such different aetiologies of damage? The CHI group exhibited the expected frontal lobe symptoms, the LCVA group the expected aphasia and apraxia, the RCVA group the expected attention disturbances; yet, contrary to all expectations, these localised deficits did not cast differential profiles of impairment on the groups' naturalistic action production. Why not? A cardinal tenet of cognitive neuropsychology is that averaging across patients risks obscuring interesting effects or yielding spurious ones. However, this does not begin to explain why the groups looked so similar. One possibility is that the action system, highly complex and with multiple redundant routes and knowledge sources, reacts to any type of damage by reorganising itself in a more or less uniform manner (Schwartz & Buxbaum, 1997). This would mean that the similarities across groups reflect the principles along which the system self-organises after damage but tell us little, if anything, about how the normal system works. An alternative possibility is that in addition to selective deficits to be found across and within the CHI, LCVA, and RCVA groups, there is something they all have in common, which accounts for their uniform action profiles. In several publications, my colleagues and I have hypothesised that this common factor is diminished capacity for a cognitive or attentional resource, which is in limited supply and which is necessary for sustaining goal-directed behaviour over time and in the face of competing goals and inputs (Buxbaum et al., 1998; Schwartz et al., 1999, 2002). The motivation for proposing this was not simply the uniformity of behaviour across patient groups. There was also evidence of a continuum of effects, extending from normals who made occasional action slips, to mild patients who committed normal-looking slips at greater frequencies, and at the extreme end, to ADS-type patients, who committed frequent omissions along with more egregious overt slips. In Schwartz et al. (1999) we summarised the argument with reference to the CHI study. Subsequently, we would extend the argument to the other patient groups, as well:

> To account for the full range of findings, including the continuity of effects between controls, mild, and severe patients, we proposed an account based on limited-capacity resources. According to this account, the neurologically intact controls

responded to the simple task requirements by allocating minimal resources, which made them vulnerable to occasional errors, such as reaching for a distractor when the target was in view, or anticipating an upcoming step in the task ... CHI patients, we proposed, were subject to an additional restriction in resource *capacity*, which made them more vulnerable to errors. As to why omission errors should be especially sensitive to task difficulty and clinical severity, we speculated that severe resource limitations might translate into poverty of effort or, more mechanistically, failure to resolve the competition for schema selection such that none of the candidate action schemas reaches threshold. (Schwartz et al., 1999; p. 52)

Related accounts have been put forward by others. For example, Ian Robertson and colleagues have argued that a critical resource for naturalistic action production is sustained attention, as measured by the Sustained Attention to Response Test (SART), and that reduced capacity for sustained attention explains the vulnerability to action errors in normal individuals and persons with traumatic brain injury (Robertson, Manly, Andrade, Baddeley, & Yiend, 1997). The sustained attention hypothesis is closely related to the original account of ADS, which postulated that a deficit in top-down control over action selection renders behaviour less constrained by goals and more vulnerable to capture by "bottom-up" influences (Schwartz et al., 1991). That hypothesis traces back to Luria's (1996) account of frontal apraxia and reappears in numerous articles on action organisation and disorganisation (e.g., Cooper & Shallice, 2002; MacKay, 1985; Shallice, 1982, 1988). Another related account invokes a role for working memory, and the strength of associations in working memory, as causally implicating the disorganisation of action after frontal lobe damage (Forde & Humphreys, 2000; Humphreys, Forde, & Francis, 2000; Kimberg & Farah, 1993).

It will be evident that these alternative accounts are unlikely to be differentiated, or differentially supported, through behavioural studies of the sort we have been discussing so far. An alternative approach is to develop computationally instantiated models for everyday tasks, so as to see what is required to coax normal-looking behaviour from the model, and how its performance is affected by simulated lesions of different types. It is to such an endeavour that we turn next.

The implemented contention scheduling model

In recent years, Rick Cooper, Tim Shallice, and their colleagues have implemented the CS model for two tasks that have been used with patients: making a cup of instant coffee (Cooper & Shallice, 2000) and preparing and packing a child's lunchbox (Cooper et al., in press). The basic elements of the verbally specified CS model are preserved in the implementations: Action schemas, arranged in a hierarchical network, collect and pass excitatory and inhibitory activation among themselves (competitive activation) and from outside sources. One outside source is the modulatory influence of SS; but as this is meant to be a model of *routine* action selection, the SS is assumed to participate only by providing the initial excitation to the highest level schema.

A second outside source, which is key to the implementation, is triggering (i.e., priming) activation from a network of nodes representing the objects in the immediate environment. In the implementation, compatible schemas and objects prime one another over links that are bidirectional and symmetric. So, for example, a small portable object in view will excite, and be excited by, the low-level *pick up* schema

In the implemented model, all schemas are goal directed, in the sense that schemas specify a set of subgoals, which have subschemas as their components. For our purposes, it is possible to ignore the structural arrangement of goals and schemas and consider schemas and subschemas to interconnect directly. (For further details, see Cooper & Shallice, 2000.) Much of the impact of goals in the model is realised through the preconditions and postconditions associated with individual schemas. Preconditions and postconditions are states of the world that may or may not be true (see Figure 3). When activation in a parent schema reaches the selection threshold, the parent remains selected until the postconditions on all its component schemas are realised, after which it is inhibited, allowing another schema to become selected. This is the essence of goal

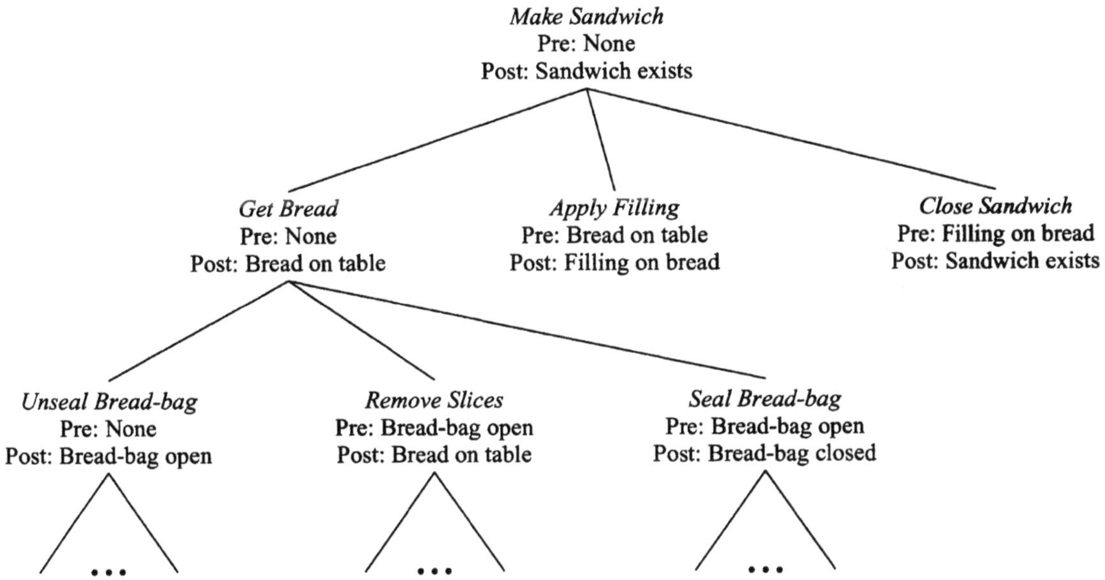

Figure 3. *A portion of the schema hierarchy for the lunchbox task, with preconditions and postconditions specified. [From Cooper, R. (2002). Order and disorder in everyday action: The roles of contention scheduling and supervisory attention. Neurocase, 8, 61–79. With kind permission of Psychology Press, www.psypress.co.uk/journals.asp, 2005.]*

directedness: the organisation of behaviour to achieve specific outcomes.

Preconditions and postconditions constrain the order in which subschemas are selected. A selected parent schema will pass activation to its component schema if and only if the component's precondition is satisfied by the state of world and its postcondition is not. This form of sequencing equips the model with considerable flexibility. When, for example, a schema's postcondition happens to be satisfied at the outset of the task (e.g., because someone has left the milk container opened), then the component schemas will not receive top-down excitation, and action will proceed without attempting it (i.e., without trying to open the already-opened milk container).

The price of this flexibility is that some of the automaticity in CS is lost. The involvement of preconditions and postconditions entails at least rudimentary monitoring of the state of the world and the consequences of action; in the original Norman-Shallice theory, such monitoring activities fell to SS. The lunchbox model also implements a rudimentary error correction mechanism. If, upon schema deselection, the monitoring mechanism finds that its postcondition is not satisfied, the failed schema is reactivated. This has the beneficial consequence of keeping the model moving through the task after an error has occurred.

The implemented model also deviates from the original model in its use of object representations. Actions corresponding to basic level schemas (e.g., *pick up, pour, open*) may involve multiple objects; for example, the act of pouring entails using one object as a source and another as a target. The implemented model therefore assumes one network that represents objects as sources, another that represents objects as targets, and so on. When, for example, the basic level node for *pour* is activated to the point of becoming selected, the model selects the most highly activated *source* and *target* object representations, whereupon it "executes" the action. In effect, this constitutes a form of argument filling for action schemas and a potential locus for object (or better, "argument") substitutions (Schwartz et al., 1991).

The model's behaviour depends on the relative flow of activation within and between the various interactive networks. This behaviour is controlled by 12 parameters, so the model is very complex indeed. Nevertheless, Cooper and colleagues found that within a fairly broad region of the parameter space, the model behaves as it should, meaning that it is accurate and that it produces the expected variability (Cooper & Shallice, 2000; Cooper et al., in press). For example, on some runs of the coffee model, sugar was added before milk, other times it was added after. On some runs, sugar was added from the bowl; on other occasions it was added from packets. This happens because of random noise in the model, which introduces variability in, among other things, which specific schema is selected for a goal (e.g., *sweeten coffee*) when several schemas are equally applicable (*add sugar from bowl*; *add sugar from packet*), and in what order schemas are selected when not subject to ordering constraints. The model has noise operating within and across networks, at a level that is controlled by a parameter. The noise parameter assumes importance in the simulation of patient behaviour, and I will return to it below.

To simulate the empirical findings from the MLAT studies, Cooper et al. (in press) ran the lunchbox model in the basic condition (array contains all and only the relevant objects) and the condition with distractors. The model's performance was scored with the identical error scheme and error algorithm as was used in the patient studies, and error rates were similarly standardised for opportunities. The analyses employed "simulated patients" and a simulated "clinical sample," both of which were biased for severity in the same manner as the actual patient groups, with milder cases over-represented, relative to more severe ones.[2]

Interestingly, most of the simulated lesions failed to reproduce the findings of interest, although in some cases they produced phenomena suggestive of other neurological conditions. For example, a top-down lesion in the schema network was simulated by progressively reducing the amount of activation that passes from the top, down (i.e., from parent to child schemas), while keeping constant the total activation from top-down and bottom-up (object-to-schema) sources. It had been suggested that a functional lesion of this type, which alters the top-down/bottom-up balance within the schema network, might explain the behaviour of some patients with routine action impairment (Schwartz et al., 1991), and such speculation had received some support from the earlier coffee model (Cooper & Shallice, 2000). However, in the lunchbox study, the simulated top-down lesions produced an abundance of omission errors with relatively few commission errors and, in particular, no object substitutions. The presence of distractors resulted in significantly *fewer* omission and commission errors, contrary to the empirical findings. An unexpected, but interesting, result was that with extreme top-down lesions, the model produced repetitive toying behaviour (e.g., perseveratively picking up and then putting down an object). This is a type of action addition that the scoring system does not count, but that is frequently seen in patients with utilisation behaviour (Shallice et al., 1989; see also Schwartz et al., 1991). Here, then, is a possible interpretation of utilisation behaviour that entails the novel prediction that utilisation patients will also exhibit numerous omission errors when tested on routine action tasks.

One interpretation of the uniformity of empirical findings across patient groups is that it reflects a nonspecific disruption to the precision of the action selection system (e.g., on account of diminished resource capacity; see Schwartz et al., 1999). The simulation study tested this hypothesis by increasing N_s, the standard deviation of the noise distribution for the schema network, which has the effect of degrading the effectiveness of selective excitation and inhibition among schemas. The

[2] To create "simulated patients," the model was run many times at settings of a parameter that ranged from small deviations from default (corresponding to mild impairment) to large deviations (corresponding to severe impairment); these runs were then sampled in proportion to the severity distribution of the actual patient sample. The "clinical sample" used the largest possible sample from the simulation data to create another realistically biased sample, which was used to test empirical predictions. Readers are encouraged to consult the original manuscript for further details (Cooper et al., in press).

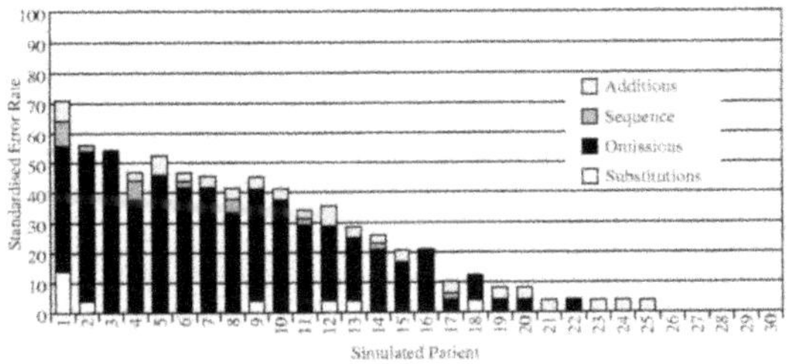

Figure 4. *The error profiles of 30 "simulated patients." Each bar represents one patient performing the Lunchbox task under two conditions—distractor absent and distractor present—with* N_s *(standard deviation of noise in the schema network) ranging from .06 to .17 (default is .01). From Cooper et al. (in press), with kind permission of Psychology Press, www.psypress.co.uk/journals.asp, 2005.*

results appear encouraging. With increasing values of N_s, there was a significant increase in both commission errors and omission errors. Figure 4 presents a bar chart of 30 "simulated patients," for comparison with Figure 2. In the model, as in empirical data, omissions are the most common error, and the proportion of commission errors is substantially greater for the high error producers.[3] Figure 5 shows this clearly. When the "clinical sample" was analysed for the effect of distractor condition, the results were again encouraging, in that standardised error rates were significantly higher in the distractor condition for omissions and object substitutions, but not for sequence or addition additions (Figure 6).

Cooper et al. (in press) also explored the consequences of increasing N_o, the standard deviation of noise in the object network. Given the reciprocal connections between the schema and object networks, it is to be expected that the two simulated lesions would have similar effects, and indeed they did. The major effects reported above for the N_s lesion were largely duplicated for the N_o lesion. However, the N_o lesion produced more object substitutions overall, and statistically comparable error rates of object substitution with and without distractors. The authors explain the effect of an N_o this way:

> The effect of noise in the object representation network is not merely to add noise to schema activations. The transmission of noise to the schema network from the object representation networks is modulated by schema triggering functions. Thus, noise in the object representation networks has the effect of degrading triggering excitation of schemas. It is for this reason that noise in the object representation network cannot be equated with an object recognition deficit. It is also for this reason that noise in the object representation networks does not merely result in increased levels of object substitution errors; such noise affects the activation of schemas—and hence the schemas that are selected—as much as the selection of objects to fill argument roles once a schema has been selected. (Cooper et al., in press)

Thus, the N_o lesion largely reproduced the empirical phenomena, but with some differences from the N_s lesion. This led Cooper et al. to speculate that there might be corresponding differences between anatomically defined patient groups that are too subtle to be discerned by the methods used in the published studies. Specifically, it was hypothesised that the N_s lesion might explain the functional impairment in patients with frontal lesions, while the N_o lesion explains the impairment in patients with left parietotemporal lesions, the site often traditionally implicated in ideational apraxia.

The latter echoes a proposal that Rumiati et al. (2001) made, based on a cognitive assessment of two unusually pure cases of ideational apraxia secondary to left temporoparietal lesions. These patients were shown to recognise by name and by function the objects they misused in action tests. They were also shown to appreciate how scrambled depictions of familiar action sequences should be correctly arranged, thereby demonstrating intact script memories (see below). Having ruled out all the obvious possibilities, Rumiati et al. looked to the implemented coffee model

[3] The quantitative fit is not perfect. For example, relative to the actual patients, the simulated patients have a tendency to make too many omissions. This shows up with particular clarity in the mild severity range, where most actual patients make no omissions but most simulated patients make some. As discussed in Cooper et al. (in press), omissions happen in several ways in the model, one of which is failure to complete a task subsequent to an error. Omissions of this type are commonplace in more severe patients (e.g., Schwartz et al., 1998), but my impression is that they rarely occur in those with mild deficits. Adding further error-correction mechanisms to the model might bring it more in line with the actual rate of omissions throughout the severity range.

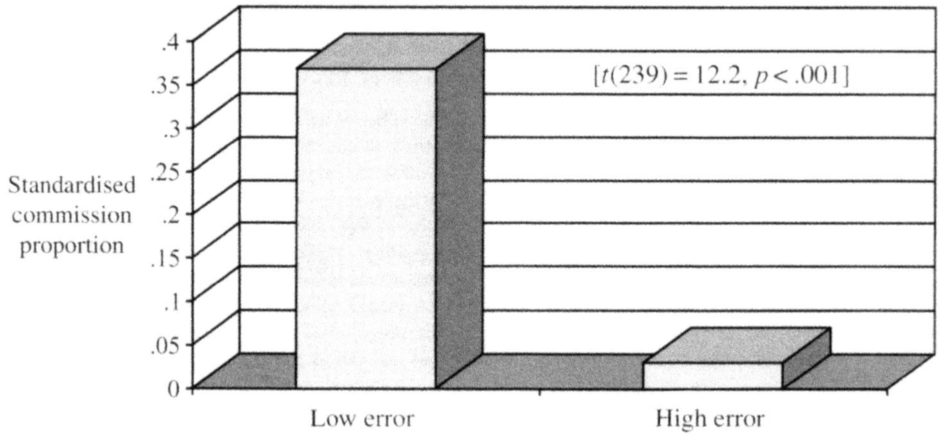

Figure 5. *The standardised commission proportion is the fraction of standardised commission errors to total standardised errors (commissions + omissions). For the lunchbox model with N_s lesions, this was calculated on a realistically biased "clinical sample" of trials with one or more errors (n = 241). This sample was subdivided into low and high error producers, based on total error score (collapsed across distractor condition). Results of a between-group, 2-tailed t-test are shown. Based on data presented in Cooper et al. (in press).*

(Cooper & Shallice, 2000) for an explanation of their patients' ideational apraxia. They speculated that the lesions to left temporoparietal cortex could have disconnected action schemas from their triggering stimulus representations, leading to incorrect schema selection and faulty argument filling (Rumiati et al., 2001). The behaviour of the lunchbox model with the N_o lesion comports with this account.

Issues raised by the implemented CS model

As an account of planning, the implemented CS model is comfortable in its familiarity. It adheres closely to the original formulation and incorporates terms and concepts that have been used to explain planning since the 1960s (e.g., goals, hierarchy, schemas). Its computational armamentarium is also familiar, comprising symbolic preconditions and postconditions inherited from the AI tradition, as well as more recent interactive activation principles. At the same time, the implemented model heralds some important changes in current thinking about planning in the routine action domain, relative to the original Norman-Shallice theory.

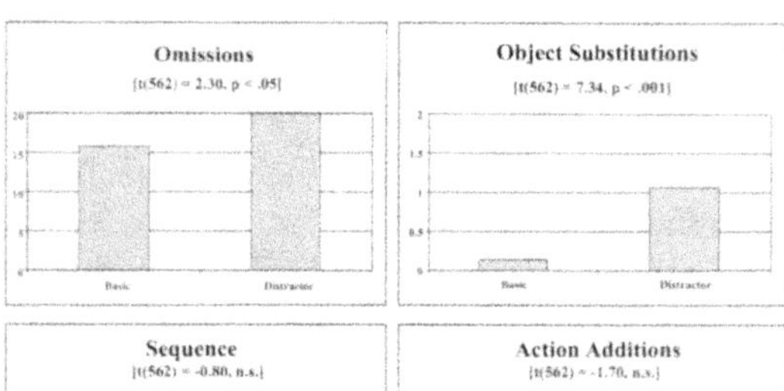

Figure 6. *To examine the effect of distractor condition in the performance of the lunchbox model with N_s lesions, mean standardised error rates were calculated on a "clinical sample" of 564 trials. Results of between-groups, 2 tailed t-tests are shown for four error types. Based on data presented in Cooper et al. (in press).*

Duality or continuum?

First, this work, coupled with the patient findings, demonstrates the futility of trying to draw a sharp line between planning in the contention scheduling vs. supervisory modes. Routine activities of daily life entail constant adjustment and improvisation in relation to the environment and the superordinate plan. We close containers when there is no further need for them and not otherwise; we measure and cut wrapping paper differently when the supply is limited or when there are multiple packages to wrap (Joe, Ferraro, & Schwartz, 2002). Investigators who study eye movements as an index of where and how attention is deployed make a convincing case that attention is continually deployed to the locations and objects that are relevant to task goals; and this holds true for routine activities, no less than for tasks developed for the laboratory (Ballard, Hayhoe, Pook, & Rao, 1997; Hayhoe, 2000; Land & Hayhoe, 2001; Land, Mennie, & Rusted, 1999). The implemented CS models of Cooper and Shallice conform to these realities by having schemas *all the way down the hierarchy* function in a manner that is goal directed and responsive to environmental contingencies. In doing so, the models introduce rudimentary monitoring and error correction—functions previously associated with the supervisory system.

A role for action script memories?

CS, as originally conceived, instantiated *procedural knowledge.* However, equipping the implemented model with goals, preconditions, postconditions, and the mechanisms for checking these against the state of the world, opens the door to a potential role for knowledge of the declarative type. Consistent with this, there is renewed interest in the script-like declarative memory structures of the sort first proposed by Schank (1982; Schank & Abelson, 1977). Recent work has shown that the ability to process action/event script memories (i.e., generate verbal scripts for familiar activities and/or arrange their component steps) is a specialised capacity, which is distinct from the sequential processing of other content domains, and which implicates the PFC (Crozier et al., 1999; Partiot, Grafman, Sadato, Flitman, & Wild, 1996; Rumiati et al., 2001; Sirigu et al., 1995, 1996, 1998; Zanini, Rumiati, & Shallice, 2002). These studies, coupled with evidence of poor script processing in some ADS patients (Buxbaum et al., 1997; Humphreys & Forde, 1998) has led to the hypothesis that ADS is caused by faulty recall of action scripts (Humphreys & Forde, 1998). On the other hand, there is now definitive evidence from Rumiati's lab of a double dissociation between the script and action deficits (Rumiati et al., 2001; Zanini et al., 2002). Therefore, the observed conjunction of the two in some ADS cases is insufficient evidence for a causal link. (For further arguments and discussion, see Joe et al., 2002.)

Varieties of object knowledge

Still another debated issue is how or whether declarative knowledge of object properties features in schema triggering or precondition/postcondition monitoring. It has been shown that the ability of semantic dementia patients to correctly demonstrate the use of single objects correlates with their performance on declarative semantic memory tests for those same objects (Hodges, Bozeat, Lambon Ralph, Patterson, & Spatt, 2000). This suggests that object utilisation draws on the same impaired semantic knowledge base as object naming and comprehension. Yet the literature on multistep action performance documents striking double dissociations between object knowledge as revealed in action performance vs. on standard semantic memory tasks (e.g., Buxbaum et al., 1997; Forde & Humphreys, 2000; Riddoch et al., 2002; Rumiati et al., 2001).

It may be of relevance that the standard semantic memory assessments rarely test the status of *manipulation* knowledge, that is, memory for how objects are grasped and utilised. There is now considerable evidence that manipulation knowledge and function knowledge doubly dissociate, with manipulation knowledge selectively disrupted by frontoparietal lesions in association with apraxia (Buxbaum & Saffran, 1998; Buxbaum & Saffran, 2002; Buxbaum et al., 2000). Recent evidence from neuroimaging confirms the distinctiveness of manipulation knowledge and its frontoparietal localisation (Boronat et al., 2005; Kellenbach, Brett, & Patterson, 2003).

Closely related to stored manipulation knowledge is the capacity to infer from the perceptible structure of an object something about its use. This can take place off-line, as in situations that require choosing among and manipulating novel tools to solve mechanical problems (Goldenberg & Hagmann, 1998). It can also take place online, via the direct pick-up of affordances (Gibson, 1977; Hodges et al., 2000; Riddoch & Humphreys, 1987). The off-line variant (mechanical problem solving) has been shown to be spared in semantic dementia but compromised in corticobasal degeneration, which has its greatest impact on the basal ganglia and parietal lobes (Hodges, Spatt, & Patterson, 1999). In corticobasal patients, as well as in patients with focal left parietal lesions (Goldenberg & Hagmann, 1998), poor mechanical problem solving coincides with ideomotor apraxia. Contrary to the long-held belief that ideomotor apraxia has no real-world consequences, it is now recognised that this condition confers susceptibility to errors on familiar, multistep action tests with objects (Foundas et al., 1995; Goldenberg & Hagman, 1998). A causal connection between errorful multistep performance and impaired manipulation knowledge has yet to be proven; but students of ideational apraxia have long maintained that this connection exists (De Renzi & Lucchelli, 1988; Morlaas, 1928). An intriguing possibility is that the explanation resides in the CS's account of schema triggering by object networks that represent the uses of objects for basic actions.

Action sequencing
Finally, let us consider the nature of action sequencing according to the implemented CS model. In the original specification, sequencing was achieved through the combination of top-down priming and environmental triggering: The priming readied the relevant schemas and the triggering fired them off in sequence. In the implemented model, preconditions and postconditions also play a role, in that executing subschema A and satisfying A's postcondition causes B's precondition to be met, and so on down the line. Triggering activation from the object network supplements this mechanism to yield serially ordered behaviour. Let us now consider two other accounts in play.

Competitive queuing (CQ). Humphreys and Forde (1998; Humphreys et al., 2000) proposed a CQ architecture of the sort that has been used to explain the serial organisation of speech and short-term memory (Hartley & Houghton, 1996; Houghton, 1990). The two ADS patients that Humphreys & Forde (1998) reported were notable for their frequent production of perseverations—in one case, immediate and continuous perseverations (HG); in the other, perseveration of earlier (but not immediate) action steps (FK). As explained by these authors, the CQ architecture offers a natural explanation for the occurrence of these two types of perseveration, and their dissociation in these two patients.

In their comparison of the CQ and CS accounts of action sequencing, Cooper et al. (in press) note that CQ similarly incorporates a selection mechanism that operates on action nodes that interact competitively. The major difference is that the CQ model hypothesises that top-down activation applies an *activation gradient* to schema/action nodes, which instantiates their serial order. It will be interesting to see whether the CQ account of serial ordering in action can be implemented for a schema hierarchy as deep as the one required by the lunchbox task, for example.

A recurrent network model. A model developed by Matt Botvinick and David Plaut (2004) deviates more dramatically from the CS model and from traditional planning models of all types. Theirs is a recurrent network model, which embodies the hypothesis that sequencing actions in the correct order requires continual updating of the mind's representation of current context, and that action errors in normals and patients arise when the representation of context goes awry. To test this hypothesis, Botvinick and Plaut trained the recurrent network model to perform versions of tea making and coffee making and then subjected its representation of context to low levels of noise (simulating normal slips) and high levels of noise (simulating ADS).

Briefly, the model has an input layer that represents features of the object currently in view (interpreted as the target of the action) and the object currently being held (interpreted as the implement). A special input feature represents the task in play. The output layer represents simple actions, including visual search ("locate *x*"). Input is mapped to output over distributed internal representations that are developed though supervised training. Recurrent connections support the maintenance of information about temporal context. One set of connections updates the input based on the last action; a second set updates the hidden units' representation of temporal context.

After 10,000 training epochs, the unlesioned network was able spontaneously and errorlessly to reproduce all the tea sequences or all the coffee sequences it had been trained on, upon being given the initial input from one task or the other. Without hierarchical representations, the internal representations had learned to cope with flexible ordering constraints and with situations in which the same actions (e.g., steps of sugaring) occurred on both the tea and coffee tasks.

At low levels of noise, the model was subject to occasional errors in which actions were repeated or omitted, or else they intruded from the other task. On these occasions, noise caused the internal representation of the current context to resemble a pattern usually associated with a different situation, and the network generated an action appropriate to that situation.

At high levels of noise, the model reproduced types of behaviour familiar from ADS patients, including omission errors (example from the model: *does not add sugar*), anticipations (*pours from an unopened container*), perseverations (*adds cream, then adds sugar, then adds cream*), and object substitutions (*stirs with coffee pack*). Moreover, the model exhibited severity effects like those exhibited by the CHI patients graphed in Figure 2: As the overall error rate increased (with increasing levels of noise) the model's proportion of omissions rose, while the proportion of sequence errors did not.

These and other documented parallels between human performance and performance of the recurrent network model suggest that this is an important and productive avenue for further research. The fact that the model learns through training is a great strength. Its account of action sequencing has commonalities with the intuitively appealing idea of associative chaining; yet its sophisticated representation of temporal context avoids the known pitfalls of traditional chaining models, namely their inflexibility and inability to explain how it can be that a given action may occur in many different contexts.

The success of the model stems from its training history, no less than its architecture. And this leads to the critical question of whether the model can "scale up," that is, cope with the likes of the lunchbox task, with its many steps and deeper hierarchy. To do so may require a training history so extensive and specialised as to forgo any claim to naturalness. Alternatively, it may require greater involvement from higher-level supervisory control processes. But how does a model that explicitly eschews representations of goals and subtask hierarchies link up with supervisory processes, which depend heavily on just such representations? The validity of the recurrent network account of routine action sequencing may ultimately rest on how well it answers this question.

The SS-CS interface

Evidence from numerous sources—human and animal, behavioural and computational—converges on the view that the PFC, playing much the role assigned to it by the Norman-Shallice theory, imposes top-down control through its representation of goals, rules, and internal states. The activity patterns in PFC are thought to configure processing in posterior and subcortical brain areas in accordance with current demands, and *on an as-needed basis* (Miller & Cohen, 2001). As Miller and Cohen put it, "The PFC is critical in situations when the mapping between sensory inputs, thoughts, and actions either are weakly established relative to other existing ones or are rapidly changing" (p. 168). Using its representation of task goals, and input it receives over extensive feedback connections, the PFC responds to instances of conflict, uncertainty, or goal failure by sending excitatory or inhibitory activation to bias the relevant mappings in a manner that achieves the task goals. It is unclear how it could do this on the recurrent network's account of such mappings. In contrast, in the implemented CS model, it is easy to envision how this might work.

First, it must be appreciated that within contemporary PFC theories, the distinction between controlled and automatic processing does not apply to tasks, but rather to task-relevant processes or mappings; furthermore, the distinction is not absolute. Thus, a task-relevant process can proceed independently of PFC control only if the associations between the current internal state, goal, trigger, and response are strong and do not face competition from associations that are stronger. Let us apply this to the CS model's performance of the lunchbox task: The completed sandwich is generally a strong trigger for wrapping; however, in the context of extreme hunger, it might be an even stronger trigger for eating. In that case, the PFC would be called upon to intervene in CS to ensure that the packing schema wins out over the eating schema. Note that the PFC would not be required to plan out all the steps in packing to achieve this end; adding biasing activation to the high-level packing schema would suffice.

The implemented CS model assumes that the SS starts things off by jolting the highest level node, after which the network sustains the necessary level of activation to complete the task through self-activation and activation persistence (along with other processes). The values assigned to the self-activation and persistence parameters ensure that activation is indeed sustained. But consider what would happen in a real-world case where, for example, the person packing the lunchbox has distracting thoughts about an upcoming appointment. In this case, the PFC would presumably provide an additional source of sustaining activation to get the agent through the lunchbox task without error.

The CS model instantiates the principle of competitive activation that is characteristic of all brain processes, and it does so by a balance of self-activation and lateral inhibition—again, controlled by preset parameters. But were the conflict between incompatible schemas to become high enough to trigger conflict monitoring processes in the anterior cingulate, and conflict resolving processes in dorsolateral prefrontal cortex (Botvinick, Nystrom, Fissell, Carter, & Cohen, 1999; Carter et al., 1998, 2000; Gehring & Knight, 2000), the CS model, unlike the recurrent network model, makes it transparent precisely where the biasing activation would be directed. Similarly, if competitive interactions in the object network were to advantage a task-irrelevant object over a task-relevant one, PFC activity could implement selective attention to the relevant object representation, thereby adding to its likelihood of being selected (Desimone & Duncan, 1995).

THE FUTURE

We have seen that today's cognitive neuropsychology of everyday action uses evidence from performance errors to motivate and evaluate models of action and planning. The previous sections enumerated some of the many questions that remain to be answered. In this final section, I would like to address the methodological impediments to this line of work that I hope will be surmounted in the coming years.

As previously discussed, future models of everyday action will be challenged to link up with evolving models of PFC control. However, to make this happen, PFC researchers and modellers need to take greater interest in how the PFC functions in relation to tasks that are hierarchical, multistep, object-based, and on the way to becoming overlearned. We have an impressive array of computational models of classic frontal lobe tasks like the Stroop and Wisconsin Card Sort (e.g., Cohen, Dunbar, & McClelland, 1990; Dehaene & Changeux, 1992; Kimberg & Farah, 1993), which explain how executive functions like conflict monitoring and set switching are realised by frontal networks. It remains to be determined whether and how such functions are called into play by tasks and activities that have the aforementioned properties of everyday activities.

Everyday action is hard to study in the laboratory. There are the obvious practical impediments to simulating real-world conditions, as well as difficulties in establishing the requisite degree of control. If subjects are allowed to perform tasks in their customary manner, they will each do them a bit differently. If we require that they all do the tasks the same way, we must either train subjects beforehand (and risk their becoming bored and noncompliant) or simply instruct them in what steps to follow (which may or may not match a given subject's learning history). Creative energy has gone into development of laboratory procedures that have face validity and that predict real-world performance (Alderman, Burgess, Knight, & Henman, 2003; Manly, Hawkins, Evans, Woldt, & Robertson, 2002; Schwartz et al., 2003). I expect further advances to come from the application of virtual reality technology (Schultheis, Himelstein, & Rizzo, 2002).

There are surprisingly few published techniques for inducing errors in naturalistic tasks. When I began my studies of action errors back in the 1980s, only one technique had been published. Maatson and Baars (1992) extended the "competing plans" technique that had proved successful in eliciting speech errors (Baars, 1976; Baars, Motley, & MacKay, 1975) to the domain of gestures and object-directed action. After viewing pairs of actions, subjects were cued to repeat the actions in either the modelled order or the reverse order, under time pressure. As hypothesised, the induction of competing ordering tendencies caused the subjects to produce frequent errors in which the components of the action were mis-ordered (e.g., put pen in cup, candy in ashtray→put candy in cup, pen in ashtray). Since then, a few other error-elicitation techniques have been tried. In Birmingham, UK, Forde, Humphreys, and their colleagues have experimented with altering working memory demands of routine action tasks in an effort to facilitate or to impede performance by patients (Forde & Humphreys, 2000, 2002) and healthy controls (Humphreys et al., 2000). My colleague, Tania Giovannetti, has been experimenting with the "coffee challenge" task, a timed task that simulates the demands of short-order cooking. When performed in conjunction with demanding secondary task, the coffee-challenge task succeeded in eliciting frequent and varied errors in healthy controls (Giovannetti, Schwartz, Buxbaum, & Holz, 2001). These developments are promising, and it is hoped that we will see more of them in the future.

Conducting research with action-impaired patients poses its own set of challenges. The method of choice for cognitive neuropsychologists is, of course, the single case study. Forde and Humphreys advocate more studies of this type: "We have stressed a detailed analysis of the performance of individual patients at the level of particular tasks. We suggest that this single case approach is particularly useful for understanding the fine-grained constraints that operate on routine behaviour" (Forde & Humphreys, 2002, p. 166). The Birmingham studies speak well for the merits of this approach. However, past work has shown that a given patient's error production can vary dramatically across tasks and task conditions, with respect to both frequency and types of error (Forde & Humphreys, 2002; Schwartz et al., 1991, 1995). Without extensive cross-task testing, single patient research in this performance domain can easily lead to a limited or incorrect understanding. Moreover, since the severity of a deficit can affect the presentation as much or more than the type of deficit, there is more to be learned from studying contrasting cases, differing only in severity or only in type of deficit, than from single cases with ADS.

The availability of virtual reality techniques and new error-elicitation methods would certainly encourage more and better patient research. One technique that is already available and that has been used to great advantage with non-neurological subjects is eye-movement monitoring (Hayhoe, 2000; Land & Hayhoe, 2001; Land et al., 1999). It would be most instructive to know how actions and eye movements are coupled in patients who do and do not make action errors, and how this coupling changes with increased task demands.

I continue to believe that group studies have an important role to play, especially where there are motivated, relevant hypotheses about brain–behaviour or deficit–behaviour correlations. For example, consider the hypothesis that emerged from lesioning the lunchbox model: that the N_s lesion simulates the effect of frontal damage, whereas the N_o lesion simulates a left parietotemporal lesion in its effect on triggering. The modelling results predict that the differential impact of these neurological lesions will be quite subtle, having to do with the relative frequency of object substitutions in conditions with and without distractors. Such a prediction is probably best tested in a contrastive group study that factors out the homogenizing effect of severity. As a second example, consider the growing interest in manipulation-knowledge loss in patients with ideomotor apraxia, and its consequences for naturalistic action production. Perhaps the best way to learn how these behavioural deficits (poor manipulation knowledge, impaired praxis; errors of action) are related, and how they relate to lesion site, is to use regression modelling with a large group of well-tested left-hemisphere patients. My colleague, Laurel Buxbaum, is currently conducting such a study at Moss.

In closing, let us acknowledge that normal studies of complex, everyday action are hard to conduct, and that good patient studies are harder still. Is it worth the effort? If one believes, as I do, that the cognitive demands of complex action do not equate to the sum of its components, and that the disruption of complex action in brain damage is not predictable from the analysis of discrete neurocognitive deficits, then one has no choice but to study complex action directly. Failing to do this, we risk acting like the proverbial drunkard who, having dropped his key on a dark night in the middle of the block, persists in looking for it under the corner street lamp. I hope that in the future, more effort will be expended "up the block," where it may well be darker, but where a bigger payoff awaits.

PrEview proof published online 18 October 2005

REFERENCES

Alderman, N., Burgess, P. W., Knight, C., & Henman, C. (2003). Ecological validity of a simplified version of the multiple errands shopping test. *Journal of the International Neuropsychological Society*, *9*, 31–44.

Arbib, M. A. (1989). *The metaphorical brain. 2: Neural networks and beyond*. New York: John Wiley & Sons.

Baars, B. J. (1976). Spoonerisms as sequencer conflicts: Evidence from artificially elicited errors. *American Journal of Psychology*, *89*, 467–484.

Baars, B. J., Motley, M. T., & MacKay, D. G. (1975). Output editing for lexical status in artificially elicited slips of the tongue. *Journal of Verbal Learning and Verbal Behaviour*, *14*, 382–391.

Ballard, D. H., Hayhoe, M. M., Pook, P. K., & Rao, R. P. N. (1997). Deictic codes for the embodiment of cognition. *Behavioural and Brain Sciences*, *20*, 723–767.

Boronat, C. B., Buxbaum, L. J., Coslett, H. B., Tang, K., Saffran, E. M., Kimberg, D. Y., & Detre, J. A. (2005). Distinctions between manipulation and function knowledge of objects: Evidence from functional magnetic resonance imaging, *Cognitive Brain Research*, *23*, 361–373.

Botvinick, M., Nystrom, L. E., Fissell, K., Carter, C. S., & Cohen, J. D. (1999). Conflict monitoring versus selection-for-action in anterior cingulate cortex. *Nature*, *402*, 179–181.

Botvinick, M., & Plaut, D. (2004). Doing without schema hierarchies: A recurrent connectionist approach to routine sequential action and its pathologies. *Psychological Review*, *111*, 395–429.

Burgess, P. W., Veitch, E., de Lacy Costello, A., & Shallice, T. (2000). The cognitive and neuroanatomical correlates of multitasking. *Neuropychologia*, *38*, 848–863.

Buxbaum, L. J., & Saffran, E. M. (1998). Knowing "how" vs. "what for": A new dissociation. *Brain and Language*, *65*, 73–86.

Buxbaum, L. J., & Saffran, E. M. (2002). Knowledge of object manipulation and object function: Dissociations in apraxic and nonapraxic subjects. *Brain and Language*, *82*, 179–199.

Buxbaum, L. J., Schwartz, M. F., & Carew, T. G. (1997). The role of semantic memory in object use. *Cognitive Neuropsychology*, *14*, 219–254.

Buxbaum, L. J., Schwartz, M. F., Coslett, H., & Carew, T. (1995). Naturalistic action and praxis in callosal apraxia. *Neurocase*, *1*, 3–17.

Buxbaum, L. J., Schwartz, M. F., & Montgomery, M. (1998). Ideational apraxia and naturalistic action. *Cognitive Neuropsychology*, *15*, 617–643.

Buxbaum, L. J., Veramonti, T., & Schwartz, M. (2000). Knowing "what for" but not "how": Function and manipulation artifact-knowledge in apraxia. *Neurocase*, *6*, 83–97.

Carter, C. S., Braver, T. S., Barch, D. M., Botvinick, M., Noll, D., & Cohen, J. D. (1998). Anterior cingulate cortex, error detection, and the online monitoring of performance. *Science*, *280*, 474–479.

Carter, C. S., McDonald, A. M., Botvinick, M. M., Ross, L. L., Stenger, V. A., Noll, D., & Cohen, J. D. (2000). *Parsing executive processes: Strategic vs. evaluative functions of the anterior cingulate cortex*. Paper presented at the Proceedings of the National Academy of Sciences.

Chapman, D. (1991). *Vision, instruction, and action.* Cambridge, MA: MIT Press.
Chapman, D., & Agre, P. E. (1987). Abstract reasoning as emergent from concrete activity. *Proceedings of the 1986 Workshop on Reasoning about Actions and Plans.* Los Altos, CA: Morgan Kaufmann.
Cohen, J. D., Dunbar, K., & McClelland, J. L. (1990). On the control of automatic processes: A parallel distributed processing account of the Stroop effect. *Psychological Review, 97,* 322–361.
Cooper, R. P., (2002). Order and disorder in everyday action: The roles of contention scheduling and supervisory attention. *Neurocase, 8,* 61–79.
Cooper, R. P., Schwartz, M. F., Yule, P., & Shallice, T. (in press). The simulation of action disorganisation in complex activities of daily living. *Cognitive Neuropsychology.*
Cooper, R. P., & Shallice, T. (2000). Contention scheduling and the control of routine activities. *Cognitive Neuropsychology, 17,* 297–338.
Crozier, S., Sirigu, A., Lehéricy, S., Van de Moortele, P., Pillon, B., Grafman, J., Agid, Y., Dubois, B., & LeBihan, D. (1999). Distinct prefrontal activations in processing sequence at the sentence and script level: An fMRI study. *Neuropsychologia, 37,* 1469–1476.
Dehaene, S., & Changeux, J. P. (1992). The Wisconsin Card Sorting test: Theoretical analysis and modeling in a neuronal network. *Cerebral Cortex, 1,* 62–79.
De Renzi, E., & Lucchelli, F. (1988). Ideational apraxia. *Brain, 111,* 1173–1185.
Desimone, R., & Duncan, J. (1995). Neural mechanisms in selective visual attention. *Annual Review of Neuroscience, 18,* 193–222.
Forde, E. M., & Humphreys, G. W. (2000). The role of semantic knowledge and working memory in everyday tasks. *Brain and Cognition, 44,* 214–252.
Forde, E. M. E., & Humphreys, G. W. (2002). Dissociations in routine behaviour across patients and everyday tasks. *Neurocase, 8,* 151–167.
Foundas, A. L., Macauley, B. L., Raymer, A. M., Maher, L. M., Heilman, K. M., & Roth, L. J. G. (1995). Ecological implications of limb apraxia: Evidence from mealtime behaviour. *Journal of the International Neuropsychological Society, 1,* 62–66.
Frith, C. D. (2000). The role of dorsolateral prefrontal cortex in the selection of action, as revealed by functional imaging. In S. Monsell & J. Driver (Eds.), *Attention and performance, XVIII.* Cambridge, MA: MIT Press.
Gehring, W. J., & Knight, R. T. (2000). Prefrontal-cingulate interactions in action monitoring. *Nature Neuroscience, 3,* 516–520.
Gibson, J. J. (1977). The theory of affordances. In R. Shaw & J. Bransford (Eds.), *Perceiving, acting, and knowing: Toward an ecological psychology.* Hillsdale, NJ: Lawrence Erlbaum Associates Inc.
Giovannetti, T., Libon, D. J, Buxbaum, L. J., & Schwartz, M. F. (2002). Naturalistic action impairments in dementia. *Neuropsychologia,* 40, 1220–1232.
Giovannetti, T., Schwartz, M. F., Buxbaum, L. J., & Holz, D. (2001). *Limited resources reduce top-down constraints on object selection in naturalistic action.* Poster presented at the Psychonomics Society Meeting, New Orleans, LA, USA.
Goldenberg, G., Daumuller, M., & Hagmann, S. (2001). Assessment and therapy of complex activities of daily living in apraxia. *Neuropsychological Rehabilitation, 11,* 147–169.
Goldenberg, G., & Hagmann, S. (1998). Tool use and mechanical problem solving in apraxia. *Neuropsychologia, 36,* 581–589.
Grafman, J. (1989). Plans, actions, and mental sets: Managerial knowledge units in the frontal lobe. In E. Perecman (Ed.), *Integrating theory and practice in neuropsychology.* Hillsdale, NJ: Lawrence Erlbaum Associates Inc.
Hartley, T., & Houghton, G. (1996). A linguistically constrained model of short-term memory for non-words. *Journal of Memory and Language, 35,* 1–31.
Hayhoe, M. (2000). Vision using routines: A functional account of vision. *Visual Cognition, 7,* 43–64.
Hodges, J. R., Bozeat, S., Lambon Ralph, M. A., Patterson, K., & Spatt, J. (2000). The role of conceptual knowledge in object use: Evidence from semantic dementia. *Brain, 123,* 1913–1925.
Hodges, J. R., Spatt, J., & Patterson, K. (1999). "What" and "how": Evidence for the dissociation of object knowledge and mechanical problem-solving skills in the human brain. *Proceedings of the National Academy of Science of the United State of America, 96,* 9444–9448.
Houghton, G. (1990). The problem of serial order: A neural network model of sequence learning and recall. In R. Dale, C. Mellish, & M. Zock (Eds.), *Current research in natural language generation* (pp. 287–319). London: Academic Press.
Humphreys, G. W., & Forde, E. M. E. (1998). Disordered action schema and action disorganisation syndrome. *Cognitive Neuropsychology, 15,* 771–811.
Humphreys, G. W., Forde, M. E., & Francis, D. (2000). The organisation of sequential actions. In S. Monsell & J. Driver (Eds.), *Attention and performance* (Vol. 18). Cambridge, MA: MIT Press.

James, W. (1950). *The principles of psychology* (Vol. 1). New York: Dover.

Joe, W., Ferraro, M., & Schwartz, M. F. (2002). Sequencing and interleaving in routine action production. *Neurocase, 8*, 135–150.

Kellenbach, M., Brett, M., & Patterson, K. (2003). Actions speak louder than functions: The importance of manipulability and action in tool representation. *Journal of Cognition Neuroscience, 15*, 30–46.

Kimberg, D. Y., & Farah, M. J. (1993). A unified account of cognitive impairments following frontal lobe damage: The role of working memory in complex, organised behaviour. *Journal of Experimental Psychology: General, 122*, 411–428.

Land, M. F., & Hayhoe, M. (2001). In what ways do eye movements contribute to everyday activities? *Vision Research, 41*, 3559–3565.

Land, M. F., Mennie, N., & Rusted, J. (1999). The roles of vision and eye movements in the control of activities of daily living. *Perception, 28*, 1311–1328.

Levine, B., Dawson, D., Boutet, I., Schwartz, M. L., & Stuss, D. T. (2000). Assessment of strategic self-regulation in traumatic brain injury: Its relationship to injury severity and psychosocial outcome. *Neuropsychology, 14*, 491–500.

Lhermitte, F. (1983). Utilization behaviour and its relation to lesions of the frontal lobes. *Brain, 106*, 237–255.

Luria, A. R. (1966). *Higher cortical functions in man* New York: Basic Books.

MacKay, D. G. (1985). A theory of the representation, organisation and timing of action with implications for sequencing disorders. In E. A. Roy (Ed.), *Neuropsychological studies of apraxia and related disorders* (pp. 111–162). Amsterdam: North-Holland.

MacKay, D. G. (1987). *The organisation of perception and action: A theory for language and other cognitive skills*. New York: Springer.

Maes, P. (1989). How to do the right thing. *Connection Science, 1*, 291–323.

Mattson, M. E., & Baars, B. J. (1992). Laboratory induction of nonspeech action errors. In B. J. Baars (Ed.), *Experimental slips and human error*. New York: Plenum Press.

Manly, T., Hawkins, K., Evans, J., Woldt, K., & Robertson, I. H. (2002). Rehabilitation of executive function: Facilitation of effective goal management on complex tasks using periodic auditory alerts. *Neuropsychologia, 40*, 271–281.

Miller, E. K., & Cohen, J. D. (2001). An integrative theory of prefrontal cortex function. *Annual Review of Neuroscience, 24*, 167–202.

Morlaas, J. (1928). *Contribution a l'etude de l'apraxie* Paris: Legrand.

Norman, D. A. (1981). Categorization of action slips. *Psychological Review, 88*, 1–15.

Norman, D. A., & Shallice, T. (1980). Attention to action: Willed and automatic control of behaviour. *Center for Human Information Processing (CHIP) Technical Report 99*. San Diego: University of California.

Norman, D. A., & Shallice, T. (1986). Attention to action: Willed and automatic control of behaviour. In R. J. Davidson, G. E. Schwartz, & D. Shapiro (Eds.), *Consciousness and self-regulation, Vol. 4* (pp. 1–18). New York: Plenum Press.

Partiot, A., Grafman, J., Sadato, N., Flitman, S., & Wild, K. (1996). Brain activation during script event processing. *NeuroReport*, 7, 761–766.

Posner, M. I., & Snyder, C. R. (1975). Attention and cognitive control. In R. L. Solso (Ed.), *Information processing and cognition* (pp. 55–85). Hillsdale, NJ: Lawrence Erlbaum Associates Inc.

Reason, J. T. (1979). Actions not as planned: The price of automatization. In G. U. R. Stevens (Ed.), *Aspects of consciousness, Vol. 1* (pp. 67–89). London: Academic Press.

Reason, J. T. (1984). Lapses of attention in everyday life. In R. Parasuraman & D. R. Davies (Eds.), *Varieties of attention*. Orlando, FL: Academic Press.

Reason, J. T., & Mycielska, K. (1982). *Absent minded? The psychology of mental lapses and everyday errors*. Englewood Cliffs, NJ: Prentice Hall.

Riddoch, M. J., & Humphreys, G. W. (1987). Visual object processing in a case of optic aphasia. *Cognitive Neuropsychology, 4*, 131–185.

Riddoch, M. J., Humphreys, G. W., Heslop, J., & Castermans, E. (2002). Dissociations between object knowledge and everyday action. *Neurocase, 8*, 100–110.

Robertson, I. H., Manly, T., Andrade, J., Baddeley, B. T., & Yiend, J. (1997). "Oops!"; Performance correlates of everyday attentional failures in traumataic brain injured and normal subjects. *Neuropsychologia, 35*, 747–758.

Roy, E. A. (1982). Action and performance. In A. W. Ellis (Ed.), *Normality and pathology in cognitive functions*. London: Academic Press.

Roy, E. A. (1983). Neuropsychological perspectives on apraxia and related action disorders. In R. A. Magill (Ed.), *Memory and control of action*. Amsterdam: North-Holland.

Rumelhart, D., & Norman, D. A. (1982). Simulating a skilled typist: A study of skilled cognitive-motor performance. *Cognitive Science, 6*, 1–36.

Rumiati, R. I., Zanini, S., Vorano, L., & Shallice, T. (2001). A form of ideational apraxia as a selective deficit of contention scheduling. *Cognitive Neuropsychology, 18*, 617–642.

Rusted, J., & Sheppard, L. (2002). Action-based memory in Alzheimer's disease. *Neurocase, 8*, 111–126.

Schank, R. C. (1982). *Dynamic memory*. Cambridge: Cambridge University Press.

Schank, R. C., & Abelson, R. (1977). *Scripts, plans, and understanding*. Hillsdale, NJ: Lawrence Erlbaum Associates Inc.

Schultheis, M. T., Himelstein, J., & Rizzo, A. A. (2002). Virtual reality and neuropsychology: Upgrading the current tools. *Journal of Head Trauma Rehabilitation, 17*, 378–394.

Schwartz, M. F., & Buxbaum, L. J. (1997). Naturalistic action. In L. J. G. Rothi & K. M. Heilman (Eds.), *Apraxia: The neuropsychology of action* (pp. 269–290). Hove, UK: Psychology Press.

Schwartz, M. F., Buxbaum, L. J., Ferraro, M., Veramonti, T., & Segal, M. (2003). *The Naturalistic Action Test*. Bury St. Edmunds, UK: Thames Valley Test Company.

Schwartz, M. F., Buxbaum, L. J., Montgomery, M. W., Fitzpatrick-DeSalme, E., Hart, T., Ferraro, M., Lee, S. L., & Coslett, H. B. (1999). Naturalistic action production following right hemisphere stroke. *Neuropsychologia, 37*, 51–66.

Schwartz, M. F., Mayer, N. H., Fitzpatrick-DeSalme, E. J., & Montgomery, M. W. (1993). Cognitive theory and the study of everyday action disorders after brain damage. *Journal of Head Trauma Rehabilitation, 8*, 59–72.

Schwartz, M. F., Montgomery, M., Buxbaum, L. J., Lee, S., Carew, T. G., Coslett, H. B., Ferraro, M., Fitzpatrick-DeSalme, E., Hart, T., & Mayer, N. (1998). Naturalistic action impairment in closed head injury. *Neuropsychology, 12*, 13–28.

Schwartz, M. F., Montgomery, M. W., Fitzpatrick-DeSalme, E. J., Ochipa, C., Coslett, H. B., & Mayer, N. H. (1995). Analysis of a disorder of everyday action. *Cognitive Neuropsychology, 12*, 863–892.

Schwartz, M. F., Reed, E. S., Montgomery, M. W., Palmer, C., & Mayer, M. H. (1991). The quantitative description of action disorganisation after brain damage: A case study. *Cognitive Neuropsychology, 8*, 381–414.

Schwartz, M. F., Segal, M. E., Veramonti, T., Ferraro, M., & Buxbaum, L. J. (2002). The Naturalistic Action Test: A standardised assessment for everyday-action impairment. *Neuropsychological Rehabilitation, 12*, 311–339.

Shallice, T. (1972). Dual functions of consciousness. *Psychological Review, 79*, 383–393.

Shallice, T. (1982). Specific impairments of planning. *Philosophical Transactions of the Royal Society of London, B298*, 199–209.

Shallice, T. (1988). *From neuropsychology to mental structure*. Cambridge: Cambridge University Press.

Shallice, T., & Burgess, P. (1991a). Higher-order impairments and frontal lobe lesions in man. In H. S. Levin, H. M. Eisenberg, & A. L. Benton (Eds.), *Frontal lobe function and dysfunction* (pp. 125–138). New York: Oxford University Press.

Shallice, T., & Burgess, P. W. (1991b). Deficits in strategy application following frontal lobe damage in man. *Brain, 114*, 727–741.

Shallice, T., Burgess, P. W., Schon, F., & Baxter, D. M. (1989). The origins of utilization behaviour. *Brain, 112*, 1587–1598.

Shiffrin, R. M., & Schneider, W. (1977). Controlled and automatic human information processing: II. Perceptual learning, automatic attending, and a general theory. *Psychological Review, 84*, 127–190.

Sirigu, A., Cohen, L., Zalla, T., Pradat-Diehl, P., Van Eeckhout, P., Grafman, J., & Agid, Y. (1998). Distinct frontal regions for processing sentence syntax and story grammar. *Cortex, 34*, 771–778.

Sirigu, A., Zalla, T., Pillon, B., Grafman, J., Agid, Y., & Dubois, B. (1995). Selective impairments in managerial knowledge following pre-frontal cortex damage. *Cortex, 31*, 301–316.

Sirigu, A., Zalla, T., Pillon, B., Grafman, J., Agid, Y., & Dubois, B. (1996). Encoding of sequence and boundaries of scripts following prefrontal lesions. *Cortex, 32*, 297–310.

Stuss, D. T., Shallice, T., Alexander, M. P., & Picton, T. W. (1995). A multidisciplinary approach to anterior attentional functions. In J. Grafman, K. J. Holyoak, & F. Boller (Eds.), *Structure and functions of the human prefrontal cortex* (Vol. 769, Annals of the New York Academy of Sciences, pp. 191–211). New York: New York Academy of Sciences.

Zanini, S., Rumiati, R. I., & Shallice, T. (2002). Action sequencing deficit following frontal lobe lesion. *Neurocase, 8*, 88–99.

COGNITIVE NEUROPSYCHOLOGY, 2006, 23 (1), 222–224

SUBJECT INDEX

For Product Safety Concerns and Information please contact our EU
representative GPSR@taylorandfrancis.com
Taylor & Francis Verlag GmbH, Kaufingerstraße 24, 80331 München, Germany

www.ingramcontent.com/pod-product-compliance
Ingram Content Group UK Ltd.
Pitfield, Milton Keynes, MK11 3LW, UK
UKHW052049230926
475702UK00025B/980
* 9 7 8 1 8 4 1 6 9 9 7 2 1 *